VOLUME ONE HUNDRED AND THREE

ADVANCES IN PROTEIN CHEMISTRY AND STRUCTURAL BIOLOGY

Ion Channels as Therapeutic Targets, Part A

VOLUME ONE HUNDRED AND THREE

ADVANCES IN PROTEIN CHEMISTRY AND STRUCTURAL BIOLOGY

Ion Channels as Therapeutic Targets, Part A

Edited by

ROSSEN DONEV

Biomed Consult Ltd
Swansea, United Kingdom

ELSEVIER

AMSTERDAM • BOSTON • HEIDELBERG • LONDON
NEW YORK • OXFORD • PARIS • SAN DIEGO
SAN FRANCISCO • SINGAPORE • SYDNEY • TOKYO
Academic Press is an imprint of Elsevier

Academic Press is an imprint of Elsevier
50 Hampshire Street, 5th Floor, Cambridge, MA 02139, USA
525 B Street, Suite 1800, San Diego, CA 92101-4495, USA
The Boulevard, Langford Lane, Kidlington, Oxford OX5 1GB, UK
125 London Wall, London, EC2Y 5AS, UK

First edition 2016

ISBN: 978-0-12-804794-1
ISSN: 1876-1623

For information on all Academic Press publications
visit our website at http://store.elsevier.com

CONTENTS

CONTRIBUTORS

Helena M. Abelaira
Laboratory of Neurosciences, Graduate Program in Health Sciences, Health Sciences Unit, University of Southern Santa Catarina, Criciuma, Santa Catarina, Brazil

Rashmi K. Ambasta
Molecular Neuroscience and Functional Genomics Laboratory, Delhi Technological University (Formerly DCE), Delhi, India

Adela Banciu
Department of Anatomy, Animal Physiology and Biophysics, Faculty of Biology, University of Bucharest, Bucharest, Romania

Daniel Dumitru Banciu
Department of Anatomy, Animal Physiology and Biophysics, Faculty of Biology, University of Bucharest, Bucharest, Romania

Xu Chen
College of Life Sciences, Shaanxi Normal University, Xi'an, Shaanxi, PR China

Jinke Cheng
Department of Biochemistry and Molecular Cell Biology, Shanghai Jiao Tong University School of Medicine, Shanghai, PR China

Chantelle Fourie
Department of Physiology, Centre for Brain Research, University of Auckland, Auckland, New Zealand

Roman V. Frolov
Division of Biophysics, Department of Physics, University of Oulu, Oulun Yliopisto, Finland

Yan-Lin Fu
Department of Physiology and Biophysics, Case Western Reserve University School of Medicine, Cleveland, Ohio, USA

Lucy Goodman
Department of Physiology, Centre for Brain Research, University of Auckland, Auckland, New Zealand

Zuleide M. Ignácio
Laboratory of Neurosciences, Graduate Program in Health Sciences, Health Sciences Unit, University of Southern Santa Catarina, Criciuma, Santa Catarina, Brazil

Niraj Kumar Jha
Molecular Neuroscience and Functional Genomics Laboratory, Delhi Technological University (Formerly DCE), Delhi, India

Saurabh Kumar Jha
Molecular Neuroscience and Functional Genomics Laboratory, Delhi Technological University (Formerly DCE), Delhi, India

Dhiraj Kumar
Molecular Neuroscience and Functional Genomics Laboratory, Delhi Technological University (Formerly DCE), Delhi, India

Pravir Kumar
Molecular Neuroscience and Functional Genomics Laboratory, Delhi Technological University (Formerly DCE), Delhi, India, and Department of Neurology, Adjunct faculty, Tufts University School of Medicine, Boston, Massachusetts, USA

Kevin Lee
Department of Physiology, Centre for Brain Research, University of Auckland, Auckland, New Zealand

Beulah Leitch
Department of Anatomy, University of Otago, Dunedin, New Zealand

Johanna M. Montgomery
Department of Physiology, Centre for Brain Research, University of Auckland, Auckland, New Zealand

Ting-Wei Mu
Department of Physiology and Biophysics, Case Western Reserve University School of Medicine, Cleveland, Ohio, USA

Yitao Qi
College of Life Sciences, Shaanxi Normal University, Xi'an, Shaanxi, PR China

João Quevedo
Laboratory of Neurosciences, Graduate Program in Health Sciences, Health Sciences Unit, University of Southern Santa Catarina, Criciuma, Santa Catarina, Brazil; Center for Translational Psychiatry; Center of Excellence on Mood Disorders, Department of Psychiatry and Behavioral Sciences, Medical School, and Neuroscience Graduate Program, Graduate School of Biomedical Sciences, The University of Texas Health Science Center at Houston, Houston, Texas, USA

Beatrice Mihaela Radu
Department of Neurological and Movement Sciences, Section of Anatomy and Histology, University of Verona, Verona, Italy, and Department of Anatomy, Animal Physiology and Biophysics, Faculty of Biology, University of Bucharest, Bucharest, Romania

Mihai Radu
Department of Neurological and Movement Sciences, Section of Anatomy and Histology, University of Verona, Verona, Italy, and Department of Life and Environmental Physics, 'Horia Hulubei' National Institute for Physics and Nuclear Engineering, Magurele, Romania

Gislaine Z. Réus
Laboratory of Neurosciences, Graduate Program in Health Sciences, Health Sciences Unit, University of Southern Santa Catarina, Criciuma, Santa Catarina, Brazil

Ana Lúcia S. Rodrigues
Laboratory of Neurobiology of Depression, Department of Biochemistry, Center of Biological Sciences, Federal University of Santa Catarina, Florianópolis, Santa Catarina, Brazil

Susan Schenk
School of Psychology, Victoria University, Wellington, New Zealand

Gerald Seifert
Institute of Cellular Neurosciences, Medical Faculty, University of Bonn, Bonn, Germany

Christian Steinhäuser
Institute of Cellular Neurosciences, Medical Faculty, University of Bonn, Bonn, Germany

Stephanie E. Titus
Center for Translational Psychiatry, Department of Psychiatry and Behavioral Sciences, Medical School, The University of Texas Health Science Center at Houston, Houston, Texas, USA

Talita Tuon
Laboratory of Neurosciences, Graduate Program in Health Sciences, Health Sciences Unit, University of Southern Santa Catarina, Criciuma, Santa Catarina, Brazil

Ya-Juan Wang
Center for Proteomics and Bioinformatics and Department of Epidemiology and Biostatistics, Case Western Reserve University School of Medicine, Cleveland, Ohio, USA

Matti Weckström✠
Division of Biophysics, Department of Physics, University of Oulu, Oulun Yliopisto, Finland

Johannes Weller
Institute of Cellular Neurosciences, Medical Faculty, University of Bonn, Bonn, Germany

Hongmei Wu
College of Life Sciences, Shaanxi Normal University, Xi'an, Shaanxi, PR China

✠ Matti Weckström has died.

PREFACE

Ion channels are pore-forming membrane proteins expressed in almost all cell types. These proteins trigger electrical signaling throughout the body by gating the flow of ions across the cell membrane. Two characteristic features of ion channels distinguish them from other types of ion transporter proteins. First, this is the very high rate of ion transport through the channel compared to other transporter proteins (often 10^6 ions per second or greater) and second, ions pass through channels down their electrochemical gradient without the participation of metabolic energy.

The sequencing of the human genome has identified more than 400 putative ion channels. However, only a fraction of these theoretically identified channels have been cloned and functionally characterized. The widespread tissue distribution of ion channels, along with the multiple physiological consequences of their opening and closing, makes targeting of ion channels very promising targets for development of therapeutics. The potential validation of ion channels as drug targets provides an enormous market opportunity for their reemergence as key targets in drug discovery. However, to realize the great potential of this target class, an understanding of the validation of these targets as well as development of suitable screening technologies that reflect the complexity of ion channel structure and function remains key drivers for exploitation of this opportunity.

In spite of some important drugs targeting ion channels which are today in clinical use, as a class, ion channels remain underexploited in drug discovery. Furthermore, many existing drugs are poorly selective with significant toxicities or suboptimal efficacy. This thematic volume of the *Advances in Protein Chemistry and Structural Biology* is dedicated to ion channels as therapeutic targets and more specifically as promising treatment targets in neurological and psychiatric disorders. Chapter 1 in this volume summarizes current advances about the protein biogenesis process of the Cys-loop receptors. Operating on individual biogenesis steps influences the receptor cell surface level; thus, manipulating the proteostasis network components can regulate the function of the receptors, representing an emerging therapeutic strategy for corresponding channelopathies. Chapter 2 proposes for the first time a novel conceptual framework binding together transient receptor potential (TRP) channels, voltage-gated sodium channels (Nav), and voltage-gated calcium channels (Cav). Authors propose a "flow-excitation model" that takes into account the inputs mediated by TRP and other similar channels, the outputs invariably provided by Cav channels, and the regenerative transmission of signals in the neural networks, for which Nav channels are responsible. This framework is used to examine the function, structure, and pharmacology of these channel classes both at cellular and whole-body physiological level. Building on that basis, the pathologies arising from the direct or indirect malfunction of the channels are discussed. The numerous pharmacological interventions affecting these channels are also described. Part of those are well-established treatments, like treatment of hypertension or some forms of epilepsy, but many others are deeply problematic due to poor drug specificity, ion channel diversity, and widespread expression of the channels in tissues other than those actually targeted.

Chapter 3 reviews the potential role of ion channels in membrane physiology and brain homeostasis where ion channels and their associated factors have been characterized with their functional consequences in neurological diseases. Furthermore, mechanistic role of perturbed ion channels identified in various neurodegenerative disorders is discussed. Finally, ion channel modulators have been investigated for their therapeutic intervention in treating common neurodegenerative disorders. Chapter 4 is dedicated to acid-sensing ion channels (ASICs) which are important pharmacological targets being involved in a variety of pathophysiological processes affecting both the peripheral nervous system (e.g., peripheral pain, diabetic neuropathy) and the central nervous system (e.g., stroke, epilepsy, migraine, anxiety, fear, depression, neurodegenerative diseases). This review discusses the role played by ASICs in different pathologies and the pharmacological agents acting on ASICs that might represent promising drugs. Perspectives and limitations in the use of ASICs antagonists and modulators as pharmaceutical agents are also discussed.

Chapter 5 focuses on the glutamatergic system and its associated receptors that are implicated in the pathophysiology of major depressive disorder. The *N*-methyl-D-aspartate (NMDA), a glutamate receptor, is a binding and/or modulation site for both classical antidepressants and new fast-acting antidepressants. Thus, this review presents evidences describing the effect of antidepressants that modulate NMDA receptors and the mechanisms that contribute to the antidepressant response. Chapter 6 continues on the glutamatergic system. Glutamate is the major neurotransmitter that mediates excitatory synaptic transmission in the brain through activation of alpha-amino-3-hydroxy-5-methyl-4-isoxazole-propionate (AMPA) receptors. These receptors have therefore been identified as a target for the development of therapeutic treatments for neurological disorders including epilepsy, neurodegenerative diseases, autism, and drug addiction. Their therapeutic potential has since declined due to inconsistent results in clinical trials. However, recent advances in basic biomedical research significantly contribute to our knowledge of AMPA receptor structure, binding sites, and interactions with auxiliary proteins. In particular, the large complex of postsynaptic proteins that interact with AMPA receptor subunits has been shown to control AMPA receptor insertion, location, pharmacology, synaptic transmission, and plasticity. Thus, these proteins are now being considered as alternative therapeutic target sites for modulating AMPA receptors in neurological disorders.

Chapter 7 is an experimental example of the role of the intercellular gap junction inwardly rectifying K^+ (Kir) channels and two-pore domain K^+ (K_2P) channels in brain homeostasis maintained by astrocytes. Authors combined functional and molecular analyses to clarify how low pH affects K^+ channel function in astrocytes freshly isolated from the developing mouse hippocampus. No evidence has been found for the presence of ASIC and transient receptor potential vanilloid receptors in hippocampal astrocytes. However, the assembly of astrocytic K^+ channels allows tolerating short, transient acidification, and glial Kir4.1 and K_2P channels can be considered promising new targets in brain diseases accompanied by pH shifts. Chapter 8 in this volume discusses the ion channels modification by small ubiquitin-like modifier (SUMO) proteins and their role in neurological channelopathies, especially the determinants of the channels' regulation. SUMO proteins covalently conjugate lysine residues in a large number of target proteins and modify their functions. SUMO modification (SUMOylation) has emerged as an important regulatory mechanism for protein stability, function, subcellular localization, and protein–protein interactions. It is until recently that the physiological impacts of SUMOylation on the regulation of neuronal K^+ channels have been investigated. It is now clear that this ion channel modification is a key determinant in the function of K^+ channels, and SUMOylation is implicated in a wide range of channelopathies, including epilepsy and sudden death.

Nonetheless, ion channels remain a relatively underexploited family of proteins for therapeutic interventions. A number of recent advances in both technology and biomedical knowledge suggest that these proteins are promising targets for future therapeutic development. Therefore, the aim of this volume is to promote further research in the structure, function, and regulation of different families of ion channels which would result in designing new efficient targeted drugs with significantly fewer adverse effects.

Dr. Rossen Donev
Biomed Consult Ltd
United Kingdom

CHAPTER ONE

Proteostasis Maintenance of Cys-Loop Receptors

Yan-Lin Fu*, Ya-Juan Wang†, Ting-Wei Mu*,[1]
*Department of Physiology and Biophysics, Case Western Reserve University School of Medicine, Cleveland, Ohio, USA
†Center for Proteomics and Bioinformatics and Department of Epidemiology and Biostatistics, Case Western Reserve University School of Medicine, Cleveland, Ohio, USA
[1]Corresponding author: e-mail address: tingwei.mu@case.edu

Contents

Abstract

The Cys-loop receptors play prominent roles in the nervous system. They include γ-aminobutyric acid type A receptors, nicotinic acetylcholine receptors, 5-hydroxytryptamine type-3 receptors, and glycine receptors. Proteostasis represents an optimal state of the cellular proteome in normal physiology. The proteostasis network regulates the folding, assembly, degradation, and trafficking of the Cys-loop receptors, ensuring their efficient functional cell surface expressions. Here, we summarize current advances about the protein biogenesis process of the Cys-loop receptors. Because operating on individual biogenesis steps influences the receptor cell surface level, manipulating the proteostasis network components can regulate the function of the receptors, representing an emerging therapeutic strategy for corresponding channelopathies.

Advances in Protein Chemistry and Structural Biology, Volume 103
ISSN 1876-1623
http://dx.doi.org/10.1016/bs.apcsb.2015.11.002

1. INTRODUCTION

The Cys-loop receptors, belonging to ligand-gated channels family, are activated by neurotransmitters, allowing ion flux through neuronal cell membrane to maintain the neuronal activity of central nervous system (CNS; Lester, Dibas, Dahan, Leite, & Dougherty, 2004). They include γ-aminobutyric acid type A receptors ($GABA_ARs$), nicotinic acetylcholine receptors (nAChRs), 5-hydroxytryptamine type-3 receptors ($5\text{-}HT_3Rs$), and glycine receptors (GlyRs). As the Cys-loop receptors are composed of five homomeric or heteromeric subunits, they are also called pentameric ligand-gated ion channels. The bacterial GLIC and ELIC and the *Caenorhabditis elegans* GluCl are also in this superfamily.

The Cys-loop receptors have prominent roles in the nervous system. As the most studied member, nAChRs are cation channels, permeable to Na^+, K^+, and Ca^{2+} upon activation. They are responsible for synaptic transmission in the CNS, in autonomic ganglias, in the adrenal gland, and at neuromuscular junctions and other peripheral synapses. The receptors are involved in diseases such as Alzheimer's disease (AD), bipolar disease, and myasthenia gravis. nAChRs located at different locations are composed of different sets of subunit subtypes. α1, β1, γ, and δ subunits or α1, β1, δ, and ε subunits form muscle-type nAChRs at a 2:1:1:1 ratio, whereas α2–α10 and β2–β4 subunits compose the most neuronal-type receptors with (α4)3(β2)2, (α4)2(β2)3, or (α7)5 subtypes predominantly found in CNS and α3β4 subtypes in autonomic ganglion and adrenal gland (Gotti et al., 2009; Hogg, Raggenbass, & Bertrand, 2003; Mazzaferro et al., 2014; Palma, Bertrand, Binzoni, & Bertrand, 1996; Wu, Cheng, Jiang, Melcher, & Xu, 2015; Xiao & Kellar, 2004).

$5\text{-}HT_3Rs$, the only inotropic receptor in serotonin receptor family, are also cation channels permeable to Na^+, K^+, and Ca^{2+} upon activation. They are widely located at postsynaptic sites in hippocampus, cortex, substantia nigra, and brain stem. They also exist in the presynaptic GABAergic nerve terminals in the amygdala and CA1 region of the hippocampus, presynaptic glutamatergic synapses, glial cell membranes in the medial nucleus of the solitary tract where they play a major role in regulating the release of neurotransmitters such as GABA, dopamine, glutamate (Connolly, 2008). They are involved in many clinical diseases such as drug addiction, cognitive function, schizophrenia, and satiety control. Its antagonists are used to treat postinfectious irritable bowel syndrome and severe diarrhea-predominant irritable bowel syndrome, chemotherapy-induced vomiting, and radiotherapy-induced and postoperative nausea and vomiting (Wu et al., 2015). The pentameric channels exist either as 5-HT_3A homomeric receptors or as 5-HT_3A/$_3$B heteromeric receptors with a stoichiometry of 3(5-HT_3B):2 (5-HT_3A).

$GABA_A$Rs are chloride channels. They are one of the main targets for anesthesia, epilepsy, anxiety disorders, mood disorders, and schizophrenia (Luscher, Fuchs, & Kilpatrick, 2011). $GABA_A$Rs are expressed postsynaptically, mediating phasic inhibition. They are also expressed at perisynaptic and extrasynaptic sites, mediating the tonic inhibition (Nusser, Hajos, Somogyi, & Mody, 1998). There are abundance interchanges between the receptors locating at postsynaptic and extrasynaptic sites. To date, there are 19 $GABA_A$R subunits belonging to eight classes based on sequence identity. They are α(1–6), β(1–3), γ(1–3), δ, ε, π, θ, and ρ(1–3) (Whiting et al., 1999). There are alternatively spliced variants of several of these subunits. For example, a short form (γ2S) and a long form (γ2L) of γ2 subunits exist, and their difference is that an eight-amino-acid insert exists in the intracellular loop domain (ICD) of the γ2L subunit (Kofuji, Wang, Moss, Huganir, & Burt, 1991; Whiting, McKernan, & Iversen, 1990). The majority of $GABA_A$R subtypes expressed in the brain are composed of α1β2γ2, then α2β3γ2 and α3β3γ2, which form the stoichiometry of 2α:2β:1γ (Vithlani, Terunuma, & Moss, 2011).

Recently, high-resolution structures of the Cys-loop receptors, including nAChR (Unwin, 2005), GluCl (Hibbs & Gouaux, 2011), GLIC (Bocquet et al., 2009), ELIC (Hilf & Dutzler, 2008), 5-HT_3R (Hassaine et al., 2014), $GABA_A$R (Miller & Aricescu, 2014), and GlyR (Du, Lu, Wu, Cheng, & Gouaux, 2015), have been elucidated. The common structural feature of this superfamily is that five subunits form the receptor (Fig. 1A). Each subunit has a large extracellular N-terminal domain, four transmembrane (TM) helices (M1–M4), and a large ICD linking M3 and M4 (Fig. 1B). The signature disulfide bond is formed by two cysteine residues, which are separated by 13 residues. This Cys-loop structure is important in the intersubunit assembly because blocking its formation negatively affects the receptor assembly (Green & Wanamaker, 1997). The N-terminal domains of the five subunits form the ligand-binding domain, which lies in the interfaces of adjacent subunits. The M2 transmembrane helices from five subunits form the channel pore, which allows the flux of specific ions. M1 and M3 helices surround next to M2, and M4 locates in the outermost area of the channel pore. The ICD between

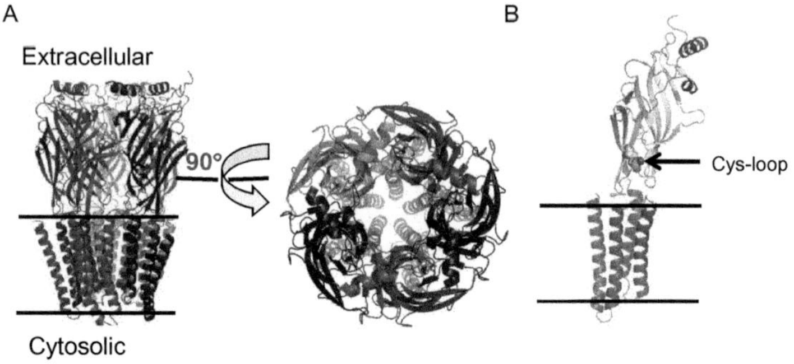

Figure 1 Structural characteristics of the Cys-loop receptors. (A) The Cys-loop receptors are pentameric, forming a central ion pore. (B) Each subunit has a large ER lumen domain, four transmembrane helices, and a large intracellular loop domain (ICD) between TM3 and TM4. The two cysteines that form the signature disulfide bond are shown in sphere model. The cartoons are built from the crystal structures of $GABA_A$ receptors (4COF).

M3 and M4 is important for modulating the trafficking of the receptors and subunit clustering on cell membrane. It also affects the channel conductance by influencing the accessibility of the channel pore to ions (Thompson, Lester, & Lummis, 2010). The TM domains play an important role in channel folding, assembly, and gating.

Proteostasis maintenance of Cys-loop receptors ensures their normal functional (Balch, Morimoto, Dillin, & Kelly, 2008). The proteostasis network regulates their functional cell surface expression levels by operating on their folding, assembly, trafficking, and degradation along protein biogenesis pathways (Fig. 2). To function, individual subunits of Cys-loop receptors need to fold into their native structures and assemble correctly with other subunits in the endoplasmic reticulum (ER). Properly assembled receptors will be able to be transported from the ER through Golgi to cell surface. Unassembled subunits or misfolded subunits will undergo the ER-associated degradation (ERAD) pathway, being retrotranslocated into the cytosol and degraded by the proteasome (Guerriero & Brodsky, 2012; Olzmann, Kopito, & Christianson, 2013; Smith, Ploegh, & Weissman, 2011; Wang, Tayo, et al., 2014). Problems in any step during the biogenesis of the Cys-loop receptors affect the normal surface expression level of the receptors, thus causing diseases. For example, many mutations of human $GABA_A$Rs lead to epilepsy by abolishing the folding, assembly, and trafficking of the mutant receptors (Macdonald, Kang, & Gallagher, 2010). Also, the receptors on the cell surface undergo continuous endocytosis and

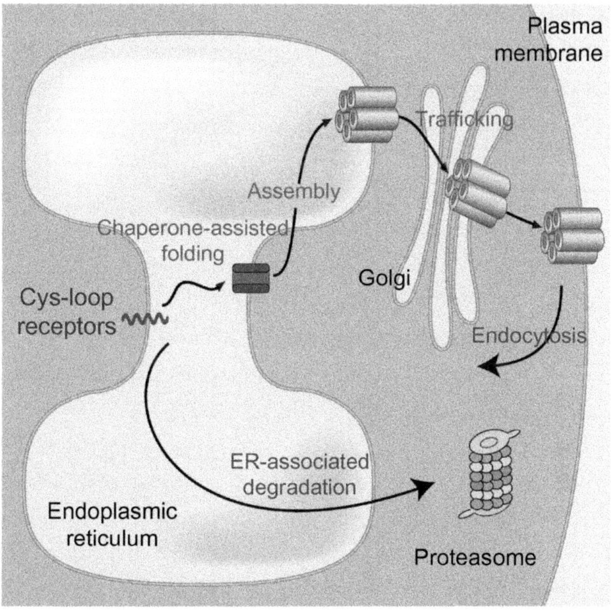

Figure 2 Protein biogenesis pathway of the Cys-loop receptors. The receptor subunit proteins are cotranslationally translocated onto the ER membrane. Molecular chaperones both in the ER and in the cytosol assist their folding. Properly folded subunits assemble into a pentamer, which is then transported from the ER to Golgi and to the plasma membrane. Misfolded proteins and unassembled subunits are degraded by the ER-associated degradation pathway. The receptors on the plasma membrane undergo endocytosis.

membrane insertion. Factors that affect this balance will influence the potency of the receptor-mediated neuron activity.

In this review, we present the proteostasis maintenance of the Cys-loop receptors. We summarize the folding and assembly characteristics of the Cys-loop receptors in the ER and their trafficking from the ER to Golgi. We also discuss the clustering, the endocytosis and recycling of the receptors on the plasma membrane.

2. FOLDING, ASSEMBLY, AND DEGRADATION OF CYS-LOOP RECEPTORS IN THE ER

2.1 Folding and Assembly of Cys-Loop Receptors

The correct synthesis and folding of individual subunits and the subunit assembly at specific forms are required for them to exit the ER for subsequent trafficking to the Golgi and plasma membrane. This is evidenced first by previous study showing that only certain assembly of subunits can form functional surface receptors. Expression of α1, β2, or the long splice variant of γ2 subunits (γ2L) of $GABA_A$Rs alone in the heterologous cells can lead to the formation of homomeric assemblies in the ER, but they fail to exit the ER (Connolly, Krishek, McDonald, Smart, & Moss, 1996). Coexpression of α and β but not α and γ or β and γ can lead to limited functional surface expression of the receptors (Luscher et al., 2011). When α, β, and γ subunits are coexpressed, the formation of 2α and 2β and 1γ subunit is strongly favored against other forms (Luscher et al., 2011). The preference of formation for certain assembly receptor subtypes may be due to the fact that forming the correct assembly structure hides the ER retention signal in the single receptor subunits. The γ2L subunits containing an eight-amino-acid ER retention signal are retained in the ER when expressed alone, whereas the γ2S subunits without this retention signal are able to exit the ER and translocate onto cell surface even when expressed by themselves (Connolly, Uren, et al., 1999). The 5-HT_3B subunits cannot form a homopentamer since this subunit contains the ER retrieval signal, which can only be masked in the presence of the 5-HT_3A subunits (Boyd, Doward, Kirkness, Millar, & Connolly, 2003). Mutation of a motif within a conserved transmembrane domain of nAChR subunits enables them to exit the ER, whereas insertion of this motif to proteins that originally successfully transported to cell surface makes them retained in ER. Assembly of native nAChR subunits into pentameric receptors covers this motif, leading to successful traffick from the ER to cell surface (Wang et al., 2002).

Pathogenic mutations affect the subunit folding or receptor assembly, resulting in loss of functional surface expression of the Cys-loop receptors. For example, the R43Q mutation in the γ2 subunit of $GABA_A$Rs interupts its association with the αβ subunit complex, leading to its retention in the ER (Frugier et al., 2007). $GABA_A$Rs containing only αβ subunits have reduced channel function, leading to childhood absence epilepsy and febrile seizure. The D219N and A322D mutations in the α1 subunit of $GABA_A$Rs are linked to idiopathic generalized epilepsy by affecting the folding and assembly of the subunit, which leads to their enhanced ERAD and impaired surface expression (Gallagher, Ding, Maheshwari, & Macdonald, 2007; Han, Guan, Wang, Hatzoglou, & Mu, 2015). The R177G mutations in the γ2 subunits undermine the subunits folding or assembly and lead to epilepsy phenotype (Todd, Gurba, Botzolakis, Stanic, & Macdonald, 2014). For nAChRs, β4R348C negatively affects the ER exit of nAChRs and leads to reduced agonist-induced currents and amyotrophic lateral sclerosis (Richards et al., 2011). The S143L, C128S, and R147L mutations located at N-terminal extracellular domain of ε subunits for nAChRs influence the subunit assembly and are linked to congenital myasthenic syndromes (Engel, Ohno, & Sine, 1999).

Although it is essential for the Cys-loop receptors to acquire their correct folding and assembly status, these processes are difficult because each receptor, being a pentamer, has a large-molecular weight, which is about 250 kDa, and each subunit has multitransmembrane domains. As a result, the assembly process is generally inefficient and slow. Only 25% of newly synthesized $GABA_ARs$ are assembled into heteromeric receptors, and 30% of the translated α subunits of nAChRs are assembled (Gorrie et al., 1997; Wanamaker, Christianson, & Green, 2003). The half-life of the nAChR assembly is more than 90 min, much longer than 7–10 min, the half-life of influenza hemagglutinin to form homotrimers (Wanamaker et al., 2003). The Green group has determined the assembly models of nAChR by using pulse chase and coimmunoprecipitation assays with subunits sequence-specific antibodies (Wanamaker et al., 2003). However, no folding and assembly models of other Cys-loop receptor are available yet.

The assembly of Cys-loop receptors depends on the N-terminal signal. The N-terminal extension and putative α-helix in the α1, β2, and γ2 subunits of $GABA_ARs$ are required for the intersubunit assembly and thus can affect the cell surface expression level of the receptors (Wong, Tae, & Cromer, 2015). Also, N-terminal extension and α-helix of ρ1 $GABA_C$ receptors, which also belong to Cys-loop receptor family, are also required for the normal assembly, trafficking, and cell surface expression of the receptors (Wong, Tae, & Cromer, 2014). Previous studies determined the specific amino acids located at the N-terminus that are important for the subunit assembly for $GABA_ARs$, nAChRs (Kreienkamp, Maeda, Sine, & Taylor, 1995; Sumikawa, 1992; Sumikawa & Nishizaki, 1994; Tsetlin, Kuzmin, & Kasheverov, 2011), and GlyRs (Kuhse, Laube, Magalei, & Betz, 1993; Tsetlin et al., 2011). However, the assembly of 5-HT_3Rs (Connolly & Wafford, 2004), nAChRs (Avramopoulou, Mamalaki, & Tzartos, 2004), GlyRs (Kuhse et al., 1993), but not $GABA_ARs$ (Buller, Hastings, Kirkness, & Fraser, 1994), depends on N-glycosylation status as all cys-loop channels are glycoproteins. In addition, recent study showed that C-terminal motifs in nAChRs may also be important for subunit assembly (Lo, Botzolakis, Tang, & Macdonald, 2008). A highly conserved aspartate residue at the boundary of the M3–M4 loop and the M4 domain is required for

$GABA_AR$ surface expression. Mutation of this residue interrupts the $GABA_AR$ assembly (Lo et al., 2008).

Many chaperones play a critical role in the folding and assembly process of the Cys-loop receptors. BiP (also known as Grp78), an Hsp70 family protein in the ER, binds the hydrophobic patches of a protein. BiP associates more strongly to misfolded mutant $GABA_A$ receptors harboring an A322D mutation in the α1 subunit compared to the wild-type receptors (Di, Han, Wang, Chance, & Mu, 2013), indicating that BiP acts early in the protein-folding step by binding to the unfolded proteins. Consistently, BiP associates more strongly with unassembled nAChRs subunits (Wanamaker et al., 2003). Calnexin, an ER membrane-bound L-type lectin protein, checks the protein-folding status by recognizing the specific glycan structures on the polypeptide. Increasing the calcium concentration in the ER using L-type calcium channel blockers promotes the trafficking of misfolding-prone mutant α1 subunit harboring the D219N mutation of $GABA_A$ receptors by increasing its interaction with calnexin (Han, Guan, et al., 2015). The binding of a chaperone to the unassembled or unfolded proteins stabilizes the folding intermediates and increases their success rate of proper folding and assembly. ERp57, a protein disulfide isomerase, and calreticulin, an ER soluble homologue of calnexin, associate with nAChRs subunits and may promote the subunit stability (Wanamaker et al., 2003; Wanamaker & Green, 2007). RIC-3 (resistance to inhibitors of acetylcholinesterase 3) is an ER-localized transmembrane protein and serves as a chaperone for $5\text{-}HT_3Rs$. It enhances the folding, assembly, and ER exit of $5\text{-}HT_3R$ (Castillo et al., 2006; Millar, 2008). However, RIC-3's effect on nAChR is relatively unclear yet. Overexpression of RIC-3 enhances the surface expression of α7-nAChRs but reduces that of α4β2-nAChRs by inhibiting the trafficking of the receptors onto cell surface (Castillo et al., 2005).

2.2 ERAD of the Cys-Loop Receptors

The folding and assembly process of the Cys-loop receptors are slow with a high level of failure rate. The subunits that fail to assemble or fold are degraded by ERAD (Olzmann et al., 2013; Smith et al., 2011; Vembar & Brodsky, 2008). Cells utilize this classical pathway to recognize and ubiquitinate unfolded proteins in the ER, extract them to cytosol, and deliver them to protein degradation complex in cytosol called the proteasome. This whole process is accomplished with the synchronized action of a series of both the soluble and membrane ER chaperone proteins and the cytosolic chaperones, which can be collectively called ERAD machinery.

ERAD influences the trafficking and cell surface expression levels of the Cys-loop receptors. PLIC1 negatively regulates $GABA_AR$ degradation by inhibiting ubiquitination (Tsetlin et al., 2011). PLIC1 and its paralog PLIC2 share an ubiquitin-like proteasome-binding domain. The association of this domain with the ICD of $GABA_AR$ subunits slows their ubiquitination and enhances their functional surface expression (Bedford et al., 2001; Luscher et al., 2011; Wu, Wang, Zheleznyak, & Brown, 1999). Ring finger protein 34, an E3 ubiquitin ligase, interacts with the ICD of the γ2 subunits of $GABA_ARs$ and reduces their expression by promoting the degradation of the receptors through both lysosomal and proteasomal degradation pathways (Jin et al., 2014). VCP is a type II member of AAA ATPase. Its prominent function is to extract the ubiquitinated misfolded proteins in the ER to the cytosolic proteasome for degradation. Inhibiting VCP using eeyarestatin I significantly enhances the trafficking of both wild type and mutant α1 subunits harboring the A322D mutation of $GABA_ARs$ (Han, Di, Fu, & Mu, 2015). Furthermore, coapplication of suberanilohydroxamic acid, a proteostasis regulator, with eeyarestatin I additively promotes the forward trafficking of misfolding-prone α1 subunit harboring the A322D mutation of $GABA_ARs$ and enhances their functional cell surface expression (Di et al., 2013; Han, Di, et al., 2015). For nAChRs, blockage of the proteasome function increases their assembly in the ER, leading to their enhanced surface expression in cultured myotubes (Christianson & Green, 2004; Wanamaker et al., 2003). Long-term inhibition of neuronal activity drastically enhances the ubiquitination level of $GABA_ARs$ and decreases their cell surface stability, whereas increasing the level of neuronal activity decreases the ubiquitination of $GABA_ARs$ and promotes their stability on the plasma membrane. Neuron activity itself can regulate the potency of $GABA_AR$-mediated effects through ubiquitination (Saliba, Michels, Jacob, Pangalos, & Moss, 2007). Based on the above evidence, modulating the ERAD rate is a promising way to enhance the surface trafficking of Cys-loop receptors. It will be of great interest to elucidate the ERAD machinery, such as critical E3 ligases and retrotranslocation channels, for the Cys-loop receptors. A tandem mass spectrometry-based proteomics approach identifies potential proteostasis network components for $GABA_A$ receptors, enabling follow-up studies on their ERAD machinery (Wang, Han, Tabib, Yates, & Mu, 2013).

In addition, other factors affect the trafficking of Cys-loop receptors through different mechanisms. For nAChRs, "14-3-3" proteins promote their trafficking through covering the COPI recognition signals and decreasing the ER retention of the receptors (Mrowiec & Schwappach, 2006).

Phosphorylation of α4-nAChR subunits at a protein kinase A (PKA) consensus sequence enhances the interaction of 14-3-3 proteins to the α4 subunits in the ER and promotes the assembly of complete α4β2-nAChRs (Bermudez & Moroni, 2006).

3. TRAFFICKING OF CYS-LOOP RECEPTORS FROM ER TO GOLGI AND TO PLASMA MEMBRANE

Golgi-specific DHHC (Asp-His-His-Cys) zinc finger protein (GODZ), which belongs to DHHC family palmitoyl acyltransferase, specifically palmitoylates the γ2 subunits of $GABA_A$Rs. The palmitoylation is required for targeting the receptors to inhibitory synapses. Knockdown of GODZ causes the loss of $GABA_A$Rs, thus leading to reduced $GABA_A$R-medicated miniature inhibitory synaptic current amplitude and frequency (Fang et al., 2006; Keller et al., 2004; Luscher et al., 2011).

The brefeldin A-inhibited GDP/GTP exchange factor 2 (BIG2) interacts with the ICD of β subunits of $GABA_A$Rs. It enhances the trafficking of β3-containing $GABA_A$Rs by promoting the membrane budding of vesicles from Golgi apparatus (Shin, Morinaga, Noda, & Nakayama, 2004).

The $GABA_A$R-associated protein (GABARAP), which belongs to a ubiquitin-like family protein in mammals and is enriched in Golgi and other somatodendritic membrane compartments, facilitates the trafficking of $GABA_A$Rs in hippocampus neuron onto plasma membrane through connecting the γ subunits with microtubules (Nymann-Andersen et al., 2002; Wang, Bedford, Brandon, Moss, & Olsen, 1999). This GABARAP effect also depends on the interaction of phospholipids to GABARAP (Chen, Chang, Leil, & Olsen, 2007).

Phospholipase C-related catalytically inactive protein (PRIP) is inositol 1,4,5-trisphosphate-binding proteins. It may serve as a bridge protein which connects γ2-containing $GABA_A$Rs with GABARAP and promotes the trafficking of the receptors. Interrupting the interaction of PRIP with γ2 subunits of $GABA_A$Rs decreases the surface expression level of the receptors in both cultured cell lines and neurons (Mizokami et al., 2007).

VILIP-1, a neuronal protein, enhances the surface expression of α4β2-nAChRs in hippocampal neurons by promoting their exit from the trans-Golgi network. This effect is activated by increasing intracellular Ca^{2+}. As a result, it is an important factor that mediates the neuron activity-induced surface expression level change of the receptors (Zhao et al., 2009).

Protein Unc-50, which is found in nematode *C. elegans* but evolutionarily conserved, is needed for the transport of specific types of nAChRs onto the cell surface with unknown mechanism (Eimer et al., 2007).

4. PROTEIN QUALITY CONTROL OF CYS-LOOP RECEPTORS ON THE PLASMA MEMBRANE

4.1 Clustering

Restriction of Cys-loop receptors to designated sites on the postsynaptic plasma membrane is also tightly regulated. This process is important for shaping the postsynaptic sites types and regulating the receptors-mediated inhibitory or excitatory effect.

Gephyrin regulates the clustering of GlyRs and $GABA_A$Rs. Gephyrin is a scaffold protein that mainly accumulates in inhibitory GABAergic and glycinergic synapses in various brain regions. Glycine receptors were the first to be found depending on gephyrin to cluster at postsynaptic sites. Glycerine β loop interacts with E domain of gephyrin. Gephyrin is also involved in the intracellular trafficking and lateral movement of glycine receptors (Fritschy, Harvey, & Schwarz, 2008). Gephyrin-induced clustering of $GABA_A$Rs is subunit-specific. Gephyrin knockout in mice diminishes the number of $\alpha2$, $\alpha3$, $\beta2/3$, and $\gamma2$ subunits-containing synaptic sites, but not the $\alpha1$-, $\alpha5$-containing synaptic sites without affecting the number of total inhibitory synaptic sites (Jacob, Moss, & Jurd, 2008). This could be due to the fact that there are only certain types of $GABA_A$R subunits that can associate with gephyrin. Gephyrin E domain associates with a 10-amino acid hydrophobic motif within the intracellular domain of the $GABA_A$R $\alpha2$, $\alpha3$, and gephyrin also interacts weakly with $\gamma2$, and $\beta3$ subunits (Kneussel et al., 2001; Tretter et al., 2008). Gepyrin is also important in regulating the neuron activity plasticity. Long-term inhibitory potentiation of neurons in visual cortex increases $GABA_A$R-mediated inhibitory postsynaptic currents by inducing the CaMKII phosphorylation of the $GABA_A$R $\beta3^{S383}$ residue and enhances gephyrin clustering of $\beta3$-containing $GABA_A$Rs. Phosphorylation-dependent interaction of Pin, a peptidyl-prolyl isomerase, with gephyrin modulates gephyrin interaction with glycine receptors and thus their clustering (Fritschy et al., 2008). Collybistin, a guanidine exchange factor activating cdc-42, forms a binding complex with gephyrin. Knockout of collybistin in mice does not affect glycinergic synaptic transmission but decreases GABAergic synaptic transmission. Collybistin is not required for gephyrin-mediated GlyR clustering but necessary for gephyrin-mediated clustering of certain $GABA_ARs$ at inhibitory postsynaptic sites (Chiou et al., 2011; Papadopoulos & Soykan, 2011; Saiepour et al., 2010).

$GABA_ARs$ clustering is also mediated by gephyrin-independent pathway. Radixin, which belongs to ERM (ezrin, radixin, moesin) family proteins, is known to mediate the clustering of α5-containing $GABA_ARs$. Depleting of radixin or changing the radixin F-actin-binding motif in neurons disrupts the formation of α5 subunit-containing $GABA_AR$ clustering (Loebrich, Bahring, Katsuno, Tsukita, & Kneussel, 2006).

The clustering of nAChR in neuromuscular junction depends on agrin, a heparan sulfate proteoglycan secreted by the presynaptic motor neuron, and rapsyn, an intracellular scaffolding protein for Wnt signal. Agrin activates the muscle-specific tyrosine kinase MuSK under the assist of rapsyn, resulting in the phosphorylation of the β subunit of nAChRs and the local receptor clustering at the nerve terminus (Lee et al., 2008; Piguet, Schreiter, Segura, Vogel, & Hovius, 2011). 14-3-3 proteins, which, as mentioned above, assists the assembly of α4 subunit-containing nAChRs, could also be involved in the clustering of α3-containing nAChRs at synapses on the surfaces of ganglionic neurons (Rosenberg et al., 2008).

4.2 Endocytosis

Surface receptors undergo consistent recycling between cell surface and intracellular endosomes (Connolly, Kittler, et al., 1999; Connolly, Uren, et al., 1999). The internalized receptors are either recycled back onto cell surface through early and recycling endosomes or degraded through late endosomes in the lysosomes. The regulation of the balance between the internalization and recycling/degradation is also important in regulating the availability of the surface expression of receptors and their mediated neuronal excitatory or inhibitory effect.

For $GABA_ARs$, clathrin adaptor protein AP2 binds to the β and γ subunits, which in turn interact with clathrin, the GTPase dynamin, and other binding partners and form the $GABA_ARs$ containing clathrin-coated pits (Kittler et al., 2000).

Many important factors regulate the endocytosis and recycling process of Cys-loop receptors. For $GABA_ARs$, huntingtin-associated protein 1 (HAP1), which is an adaptor protein for kinesin superfamily motor protein 5 (KIF5) (Twelvetrees et al., 2010), inhibits the degradation of endocytosed β1–3-containing $GABA_ARs$ through the KIF5-dependent trafficking, favors the receptor recycling, and increases their surface expression and receptor-mediated inhibitory effect (Kittler et al., 2004). $GABA_AR$-

interacting factor, GRIF-1, and its paralog TRAK1, also interact with KIF5. They could be involved in the KIF5-dependent trafficking of $GABA_A$Rs (Luscher et al., 2011). BIG2, a guanine exchange factor mentioned earlier, may also involved in the endocytic recycling of GABAARs (Luscher et al., 2011). Inhibiting the lysosomal activity (Arancibia-Carcamo et al., 2009; Kittler et al., 2004), preventing the trafficking of ubiquitinated γ2 subunit-containing $GABA_A$Rs to lysosomes (Arancibia-Carcamo et al., 2009), or disrupting the ubiquitination at lysine residues in the intracellular domain of the γ2 subunit (Arancibia-Carcamo et al., 2009) enhances the accumulation of $GABA_A$Rs at synapses.

Giant ankyrin-G, an extended fibrous polypeptide with 2600 residues, is present in extrasynaptic microdomains on the somatodendritic surfaces of hippocampal and cortical neurons and disrupts $GABA_A$R endocytosis by interacting with the GABARAP (Tseng, Jenkins, Tanaka, Mooney, & Bennett, 2015). This process may be involved in the formation of $GABA_A$R-mediated circuitry in the cerebral cortex. Human mutations in the giant ankyrin exon are linked to autism and severe cognitive dysfunction (Iqbal et al., 2013).

The internalization rate also depends on the extracellular conformation of the $GABA_A$Rs and the presence of $GABA_A$R agonists or antagonists. $GABA_A$Rs that contain the R43Q mutant γ2 subunits have an increased clathrin-mediated and dynamin-dependent endocytosis, which can be reduced by receptor antagonists. Furthermore, receptor agonists enhance the endocytosis of both endogenous and recombinant wild-type $GABA_A$Rs in both cultured neurons and COS-7 cells (Chaumont et al., 2013).

The nAChR agonist, antagonist α-bungarotoxin, and cross-linking anti-nAChR antibodies promote the internalization of nAChRs (Akaaboune, Culican, Turney, & Lichtman, 1999; St John, 2009; St John & Gordon, 2001). This process depends on actin activation, but it still happens without functional clathrin, caveolin, or dynamin (St John, 2009). Neuregulins 1β (NRG1β), which belongs to EGF family, induces the rapid internalization of α7-nAChRs from the surface of these neurons. Its effect relies on tyrosine phosphorylation and activation of actin cytoskeleton.

5. OTHER REGULATIONS OF CYS-LOOP RECEPTORS

5.1 Lipid Involvement in Trafficking and Clustering

Phosphatidylethanolamine is required for the surface expression of $GABA_A$RS in cultured neurons under the assistance of GRBARAP

(Chen & Olsen, 2007). Membrane sphingolipids and other lipids promote the surface expression level of muscle-type nAChRs by affecting the biosynthesis process in ER (Baier & Barrantes, 2007). Decreasing the membrane cholesterol promotes the endocytosis of nAChRs and decreases their cell expression level (Borroni et al., 2007). The underlying mechanism is that membrane lipid serves as lipid rafts, which is required for the trafficking and membrane stabilization of the receptors.

5.2 Phosphorylation Signaling in the Biogenesis of the Receptors

Phosphorylation affects the Cys-loop receptor channel properties (Swope, Moss, Raymond, & Huganir, 1999) and modulates the efficacy of receptor-mediated effect by influencing their trafficking, endocytosis, and recycling process. Neuronal activities that lead to the change in the intracellular calcium signal regulate the activity of kinases and phosphatases, resulting in the altered the biogenesis process and thus the surface expression level of the receptors. For example, enhanced excitatory synaptic activities activate phosphatase calcineurin through Ca^{2+}/calmodulin pathway followed by an increase in intracellular Ca^{2+} concentration. Activated calcineurin dephosphorylates Ser327 in the $GABA_AR$ γ2 subunit, which leads to the enhanced lateral mobility of the receptors, decreased cluster size of $GABA_ARs$, and reduced GABAergic mIPSC (Bannai et al., 2009). Calcineurin is also involved in downregulation of the α2-containing $GABA_AR$ membrane expression level in prolonged seizures activity linked to benzodiazepine pharmacoresistance (Eckel, Szulc, Walker, & Kittler, 2015). PRIP, as mentioned above, modulates the $GABA_AR$ surface expression level by affecting the phosphorylation of the receptors. PRIP inactivates the protein phosphatase 1α (PP1α), which dephosphorylates the $GABA_ARs$ phosphorylated by PKA. As a result, PRIP positively regulates the receptor surface expression and receptor-mediated inhibition effect in hippocampal neuron (Kittler & Moss, 2003; Terunuma et al., 2004; Yoshimura et al., 2001).

Many neurosteroids or neurotrophic factors regulate the surface expression level of receptor by affecting the trafficking, endocytosis, and recycling process. For example, neurosteroids promote the PKC phosphorylation of α4 subunit Ser443 site, which enhances the insertion of the α4 subunit-containing $GABA_ARs$ and leads to increased tonic inhibition (Abramian et al., 2010). However, the same neurosteroid does not have any effect on the α1- and α5-containing $GABA_ARs$, which mediate the phasic inhibition

(Abramian et al., 2014, 2010; Comenencia-Ortiz, Moss, & Davies, 2014). Brain-derived neurotrophic factor induces an initial fast, but short increases in $GABA_A$Rs-induced mIPSC through the phosphorylation of β3 Ser408/409 by PKC and RACK-1 (receptor for activated c-kinase), which leads to decreased endocytosis of the receptors. A following long-lasting down-regulation of $GABA_A$Rs-induced mIPSC is due to increased clathrin-mediated endocytosis of $GABA_A$Rs by dephosphorylating β3 subunits of $GABA_A$Rs (Jovanovic, Thomas, Kittler, Smart, & Moss, 2004).

Phosphorylation also affects the trafficking, endocytosis, and recycling process of nAChRs and 5-HT_3Rs. For example, inhibition of protein tyrosine kinases (PTKs) enhances α7-nAChR-mediated responses to ACh both in oocytes and in hippocampal neurons. The application of a protein tyrosine phosphatase inhibitor leads to the depression of such responses. PTKs promote the exocytosis of α7-containing nAChRs (Cho et al., 2005). Protein tyrosine phosphatases enhance the turnover rate of nAChRs and they are required for proper recycling of nAChRs onto cell surface, whereas activation of the serine/threonine PKA slows the turnover of nAChRs (Bruneau & Akaaboune, 2006; Qu, Moritz, & Huganir, 1990; Sava, Barisone, Di Mauro, Fumagalli, & Sala, 2001; Xu & Salpeter, 1995). PKC enhances the trafficking of the 5-HT_3Rs onto the cell surface and this effect is mediated through an actin-dependent pathway (Sun, Hu, Moradel, Weight, & Zhang, 2003).

6. DISEASE AND THERAPY

Proteostasis deficiency of the Cys-loop receptors causes numerous diseases. For example, deficient trafficking or enhanced internalization of nAChRs is linked to AD, bipolar disease, and myasthenia gravis. Deficiencies in the folding and assembly of $GABA_A$Rs lead to genetic epilepsy. One emerging therapeutic strategy for such diseases is to adapt proteostasis network to restore the function of trafficking-deficient receptors (Balch et al., 2008). Two classes of small molecules are employed: proteostasis regulators and pharmacological chaperones (Mu et al., 2008; Wang, Di, & Mu, 2014). Proteostasis regulators operate on the proteostasis network components to correct the folding and trafficking deficiency. For example, suberanilohydroxamic acid, acting as a proteostasis regulator, enhances the functional cell surface expression of the A322D α1 subunit of $GABA_A$Rs partially by increasing the BiP protein level and the interaction between the calnexin and the mutant α1 subunit in the ER (Di et al., 2013).

Verapamil, an L-type calcium channel blocker, acting as a proteostasis regulator, enhances the function of the D219N α1 subunit of $GABA_AR$s by promoting calnexin-assisted folding (Han, Guan, et al., 2015). Pharmacological chaperones directly bind the receptors, stabilize the assembly intermediates, increase the successful rate of this process, and promote the surface expression level of the receptors. Agonists and antagonists are candidates of pharmacological chaperones for Cys-loop receptors. For example, nicotine and its metabolite cotinine upregulate the surface expression level of nAChRs by serving as pharmacological chaperones, promoting the stabilization of the nAChRs in the ER (Fox, Moonschi, & Richards, 2015; Lester et al., 2009). Similarly, $GABA_AR$ agonists and a competitive antagonist bicuculline enhance the surface expression level of $GABA_AR$s by acting as pharmacological chaperones. The application of brefeldin A, which inhibits the formation of COPI-mediated transport vesicles from ER to Golgi, antagonizes this effect (Eshaq et al., 2010). Combining proteostasis regulators and pharmacological chaperones is expected to achieve better therapeutic effects.

REFERENCES

Abramian, A. M., Comenencia-Ortiz, E., Modgil, A., Vien, T. N., Nakamura, Y., Moore, Y. E., et al. (2014). Neurosteroids promote phosphorylation and membrane insertion of extrasynaptic GABAA receptors. *Proceedings of the National Academy of Sciences of the United States of America*, *111*, 7132–7137.

Abramian, A. M., Comenencia-Ortiz, E., Vithlani, M., Tretter, E. V., Sieghart, W., Davies, P. A., et al. (2010). Protein kinase C phosphorylation regulates membrane insertion of GABAA receptor subtypes that mediate tonic inhibition. *The Journal of Biological Chemistry*, *285*, 41795–41805.

Akaaboune, M., Culican, S. M., Turney, S. G., & Lichtman, J. W. (1999). Rapid and reversible effects of activity on acetylcholine receptor density at the neuromuscular junction in vivo. *Science*, *286*, 503–507.

Arancibia-Carcamo, I. L., Yuen, E. Y., Muir, J., Lumb, M. J., Michels, G., Saliba, R. S., et al. (2009). Ubiquitin-dependent lysosomal targeting of GABA(A) receptors regulates neuronal inhibition. *Proceedings of the National Academy of Sciences of the United States of America*, *106*, 17552–17557.

Avramopoulou, V., Mamalaki, A., & Tzartos, S. J. (2004). Soluble, oligomeric, and ligand-binding extracellular domain of the human alpha7 acetylcholine receptor expressed in yeast: Replacement of the hydrophobic cysteine loop by the hydrophilic loop of the ACh-binding protein enhances protein solubility. *The Journal of Biological Chemistry*, *279*, 38287–38293.

Baier, C. J., & Barrantes, F. J. (2007). Sphingolipids are necessary for nicotinic acetylcholine receptor export in the early secretory pathway. *Journal of Neurochemistry*, *101*, 1072–1084.

Balch, W. E., Morimoto, R. I., Dillin, A., & Kelly, J. W. (2008). Adapting proteostasis for disease intervention. *Science*, *319*, 916–919.

Bannai, H., Levi, S., Schweizer, C., Inoue, T., Launey, T., Racine, V., et al. (2009). Activity-dependent tuning of inhibitory neurotransmission based on GABAAR diffusion dynamics. *Neuron, 62*, 670–682.

Bedford, F. K., Kittler, J. T., Muller, E., Thomas, P., Uren, J. M., Merlo, D., et al. (2001). GABA(A) receptor cell surface number and subunit stability are regulated by the ubiquitin-like protein Plic-1. *Nature Neuroscience, 4*, 908–916.

Bermudez, I., & Moroni, M. (2006). Phosphorylation and function of alpha4beta2 receptor. *Journal of Molecular Neuroscience, 30*, 97–98.

Bocquet, N., Nury, H., Baaden, M., Le Poupon, C., Changeux, J. P., Delarue, M., et al. (2009). X-ray structure of a pentameric ligand-gated ion channel in an apparently open conformation. *Nature, 457*, 111–114.

Borroni, V., Baier, C. J., Lang, T., Bonini, I., White, M. M., Garbus, I., et al. (2007). Cholesterol depletion activates rapid internalization of submicron-sized acetylcholine receptor domains at the cell membrane. *Molecular Membrane Biology, 24*, 1–15.

Boyd, G. W., Doward, A. I., Kirkness, E. F., Millar, N. S., & Connolly, C. N. (2003). Cell surface expression of 5-hydroxytryptamine type 3 receptors is controlled by an endoplasmic reticulum retention signal. *The Journal of Biological Chemistry, 278*, 27681–27687.

Bruneau, E. G., & Akaaboune, M. (2006). The dynamics of recycled acetylcholine receptors at the neuromuscular junction in vivo. *Development, 133*, 4485–4493.

Buller, A. L., Hastings, G. A., Kirkness, E. F., & Fraser, C. M. (1994). Site-directed mutagenesis of N-linked glycosylation sites on the gamma-aminobutyric acid type A receptor alpha 1 subunit. *Molecular Pharmacology, 46*, 858–865.

Castillo, M., Mulet, J., Gutierrez, L. M., Ortiz, J. A., Castelan, F., Gerber, S., et al. (2005). Dual role of the RIC-3 protein in trafficking of serotonin and nicotinic acetylcholine receptors. *The Journal of Biological Chemistry, 280*, 27062–27068.

Castillo, M., Mulet, J., Gutierrez, L. M., Ortiz, J. A., Castelan, F., Gerber, S., et al. (2006). Role of the RIC-3 protein in trafficking of serotonin and nicotinic acetylcholine receptors. *Journal of Molecular Neuroscience, 30*, 153–156.

Chaumont, S., Andre, C., Perrais, D., Boue-Grabot, E., Taly, A., & Garret, M. (2013). Agonist-dependent endocytosis of gamma-aminobutyric acid type A (GABAA) receptors revealed by a gamma2(R43Q) epilepsy mutation. *The Journal of Biological Chemistry, 288*, 28254–28265.

Chen, Z. W., Chang, C. S., Leil, T. A., & Olsen, R. W. (2007). C-terminal modification is required for GABARAP-mediated GABA(A) receptor trafficking. *The Journal of Neuroscience, 27*, 6655–6663.

Chen, Z. W., & Olsen, R. W. (2007). GABAA receptor associated proteins: A key factor regulating GABAA receptor function. *Journal of Neurochemistry, 100*, 279–294.

Chiou, T. T., Bonhomme, B., Jin, H., Miralles, C. P., Xiao, H., Fu, Z., et al. (2011). Differential regulation of the postsynaptic clustering of gamma-aminobutyric acid type A (GABAA) receptors by collybistin isoforms. *The Journal of Biological Chemistry, 286*, 22456–22468.

Cho, C. H., Song, W., Leitzell, K., Teo, E., Meleth, A. D., Quick, M. W., et al. (2005). Rapid upregulation of alpha7 nicotinic acetylcholine receptors by tyrosine dephosphorylation. *The Journal of Neuroscience, 25*, 3712–3723.

Christianson, J. C., & Green, W. N. (2004). Regulation of nicotinic receptor expression by the ubiquitin-proteasome system. *The EMBO Journal, 23*, 4156–4165.

Comenencia-Ortiz, E., Moss, S. J., & Davies, P. A. (2014). Phosphorylation of GABAA receptors influences receptor trafficking and neurosteroid actions. *Psychopharmacology, 231*, 3453–3465.

Connolly, C. N. (2008). Trafficking of 5-HT(3) and GABA(A) receptors (Review). *Molecular Membrane Biology, 25*, 293–301.

Connolly, C. N., Kittler, J. T., Thomas, P., Uren, J. M., Brandon, N. J., Smart, T. G., et al. (1999). Cell surface stability of gamma-aminobutyric acid type A receptors. Dependence on protein kinase C activity and subunit composition. *The Journal of Biological Chemistry*, *274*, 36565–36572.

Connolly, C. N., Krishek, B. J., McDonald, B. J., Smart, T. G., & Moss, S. J. (1996). Assembly and cell surface expression of heteromeric and homomeric gamma-aminobutyric acid type A receptors. *The Journal of Biological Chemistry*, *271*, 89–96.

Connolly, C. N., Uren, J. M., Thomas, P., Gorrie, G. H., Gibson, A., Smart, T. G., et al. (1999). Subcellular localization and endocytosis of homomeric gamma2 subunit splice variants of gamma-aminobutyric acid type A receptors. *Molecular and Cellular Neurosciences*, *13*, 259–271.

Connolly, C. N., & Wafford, K. A. (2004). The Cys-loop superfamily of ligand-gated ion channels: The impact of receptor structure on function. *Biochemical Society Transactions*, *32*, 529–534.

Di, X. J., Han, D. Y., Wang, Y. J., Chance, M. R., & Mu, T. W. (2013). SAHA enhances Proteostasis of epilepsy-associated alpha1(A322D)beta2gamma2 GABA(A) receptors. *Chemistry & Biology*, *20*, 1456–1468.

Du, J., Lu, W., Wu, S., Cheng, Y., & Gouaux, E. (2015). Glycine receptor mechanism elucidated by electron cryo-microscopy. *Nature*, *526*, 224–229.

Eckel, R., Szulc, B., Walker, M. C., & Kittler, J. T. (2015). Activation of calcineurin underlies altered trafficking of alpha2 subunit containing GABAA receptors during prolonged epileptiform activity. *Neuropharmacology*, *88*, 82–90.

Eimer, S., Gottschalk, A., Hengartner, M., Horvitz, H. R., Richmond, J., Schafer, W. R., et al. (2007). Regulation of nicotinic receptor trafficking by the transmembrane Golgi protein UNC-50. *The EMBO Journal*, *26*, 4313–4323.

Engel, A. G., Ohno, K., & Sine, S. M. (1999). Congenital myasthenic syndromes: Recent advances. *Archives of Neurology*, *56*, 163–167.

Eshaq, R. S., Stahl, L. D., Stone, R., 2nd, Smith, S. S., Robinson, L. C., & Leidenheimer, N. J. (2010). GABA acts as a ligand chaperone in the early secretory pathway to promote cell surface expression of GABAA receptors. *Brain Research*, *1346*, 1–13.

Fang, C., Deng, L., Keller, C. A., Fukata, M., Fukata, Y., Chen, G., et al. (2006). GODZ-mediated palmitoylation of GABA(A) receptors is required for normal assembly and function of GABAergic inhibitory synapses. *The Journal of Neuroscience*, *26*, 12758–12768.

Fox, A. M., Moonschi, F. H., & Richards, C. I. (2015). The nicotine metabolite, cotinine, alters the assembly and trafficking of a subset of nicotinic acetylcholine receptors. *The Journal of Biological Chemistry*, *290*, 24403–24412.

Fritschy, J. M., Harvey, R. J., & Schwarz, G. (2008). Gephyrin: Where do we stand, where do we go? *Trends in Neurosciences*, *31*, 257–264.

Frugier, G., Coussen, F., Giraud, M. F., Odessa, M. F., Emerit, M. B., Boue-Grabot, E., et al. (2007). A gamma 2(R43Q) mutation, linked to epilepsy in humans, alters GABAA receptor assembly and modifies subunit composition on the cell surface. *The Journal of Biological Chemistry*, *282*, 3819–3828.

Gallagher, M. J., Ding, L., Maheshwari, A., & Macdonald, R. L. (2007). The GABA(A) receptor alpha 1 subunit epilepsy mutation A322D inhibits transmembrane helix formation and causes proteasomal degradation. *Proceedings of the National Academy of Sciences of the United States of America*, *104*, 12999–13004.

Gorrie, G. H., Vallis, Y., Stephenson, A., Whitfield, J., Browning, B., Smart, T. G., et al. (1997). Assembly of GABAA receptors composed of alpha1 and beta2 subunits in both cultured neurons and fibroblasts. *The Journal of Neuroscience: The Official Journal of the Society for Neuroscience*, *17*, 6587–6596.

Gotti, C., Clementi, F., Fornari, A., Gaimarri, A., Guiducci, S., Manfredi, I., et al. (2009). Structural and functional diversity of native brain neuronal nicotinic receptors. *Biochemical Pharmacology*, *78*, 703–711.

Green, W. N., & Wanamaker, C. P. (1997). The role of the cystine loop in acetylcholine receptor assembly. *The Journal of Biological Chemistry*, *272*, 20945–20953.

Guerriero, C. J., & Brodsky, J. L. (2012). The delicate balance between secreted protein folding and endoplasmic reticulum-associated degradation in human physiology. *Physiological Reviews*, *92*, 537–576.

Han, D. Y., Di, X. J., Fu, Y. L., & Mu, T. W. (2015). Combining valosin-containing protein (VCP) inhibition and suberanilohydroxamic acid (SAHA) treatment additively enhances the folding, trafficking, and function of epilepsy-associated gamma-aminobutyric acid, type A (GABAA) receptors. *The Journal of Biological Chemistry*, *290*, 325–337.

Han, D. Y., Guan, B. J., Wang, Y. J., Hatzoglou, M., & Mu, T. W. (2015). L-type calcium channel blockers enhance trafficking and function of epilepsy-associated alpha1(D219N) subunits of GABA receptors. *ACS Chemical Biology*, *10*, 2135–2148.

Hassaine, G., Deluz, C., Grasso, L., Wyss, R., Tol, M. B., Hovius, R., et al. (2014). X-ray structure of the mouse serotonin 5-HT3 receptor. *Nature*, *512*, 276–281.

Hibbs, R. E., & Gouaux, E. (2011). Principles of activation and permeation in an anion-selective Cys-loop receptor. *Nature*, *474*, 54–60.

Hilf, R. J., & Dutzler, R. (2008). X-ray structure of a prokaryotic pentameric ligand-gated ion channel. *Nature*, *452*, 375–379.

Hogg, R. C., Raggenbass, M., & Bertrand, D. (2003). Nicotinic acetylcholine receptors: From structure to brain function. *Reviews of Physiology, Biochemistry and Pharmacology*, *147*, 1–46.

Iqbal, Z., Vandeweyer, G., van der Voet, M., Waryah, A. M., Zahoor, M. Y., Besseling, J. A., et al. (2013). Homozygous and heterozygous disruptions of ANK3: At the crossroads of neurodevelopmental and psychiatric disorders. *Human Molecular Genetics*, *22*, 1960–1970.

Jacob, T. C., Moss, S. J., & Jurd, R. (2008). GABA(A) receptor trafficking and its role in the dynamic modulation of neuronal inhibition. *Nature Reviews. Neuroscience*, *9*, 331–343.

Jin, H., Chiou, T. T., Serwanski, D. R., Miralles, C. P., Pinal, N., & De Blas, A. L. (2014). Ring finger protein 34 (RNF34) interacts with and promotes gamma-aminobutyric acid type-A receptor degradation via ubiquitination of the gamma2 subunit. *The Journal of Biological Chemistry*, *289*, 29420–29436.

Jovanovic, J. N., Thomas, P., Kittler, J. T., Smart, T. G., & Moss, S. J. (2004). Brain-derived neurotrophic factor modulates fast synaptic inhibition by regulating GABA(A) receptor phosphorylation, activity, and cell-surface stability. *The Journal of Neuroscience*, *24*, 522–530.

Keller, C. A., Yuan, X., Panzanelli, P., Martin, M. L., Alldred, M., Sassoe-Pognetto, M., et al. (2004). The gamma2 subunit of GABA(A) receptors is a substrate for palmitoylation by GODZ. *The Journal of Neuroscience*, *24*, 5881–5891.

Kittler, J. T., Delmas, P., Jovanovic, J. N., Brown, D. A., Smart, T. G., & Moss, S. J. (2000). Constitutive endocytosis of GABAA receptors by an association with the adaptin AP2 complex modulates inhibitory synaptic currents in hippocampal neurons. *The Journal of Neuroscience: The Official Journal of the Society for Neuroscience*, *20*, 7972–7977.

Kittler, J. T., & Moss, S. J. (2003). Modulation of GABAA receptor activity by phosphorylation and receptor trafficking: Implications for the efficacy of synaptic inhibition. *Current Opinion in Neurobiology*, *13*, 341–347.

Kittler, J. T., Thomas, P., Tretter, V., Bogdanov, Y. D., Haucke, V., Smart, T. G., et al. (2004). Huntingtin-associated protein 1 regulates inhibitory synaptic transmission by modulating gamma-aminobutyric acid type A receptor membrane trafficking. *Proceedings of the National Academy of Sciences of the United States of America*, *101*, 12736–12741.

Kneussel, M., Brandstatter, J. H., Gasnier, B., Feng, G., Sanes, J. R., & Betz, H. (2001). Gephyrin-independent clustering of postsynaptic GABA(A) receptor subtypes. *Molecular and Cellular Neurosciences*, *17*, 973–982.

Kofuji, P., Wang, J. B., Moss, S. J., Huganir, R. L., & Burt, D. R. (1991). Generation of two forms of the gamma-aminobutyric acidA receptor gamma 2-subunit in mice by alternative splicing. *Journal of Neurochemistry*, *56*, 713–715.

Kreienkamp, H. J., Maeda, R. K., Sine, S. M., & Taylor, P. (1995). Intersubunit contacts governing assembly of the mammalian nicotinic acetylcholine receptor. *Neuron*, *14*, 635–644.

Kuhse, J., Laube, B., Magalei, D., & Betz, H. (1993). Assembly of the inhibitory glycine receptor: Identification of amino acid sequence motifs governing subunit stoichiometry. *Neuron*, *11*, 1049–1056.

Lee, Y., Rudell, J., Yechikhov, S., Taylor, R., Swope, S., & Ferns, M. (2008). Rapsyn carboxyl terminal domains mediate muscle specific kinase-induced phosphorylation of the muscle acetylcholine receptor. *Neuroscience*, *153*, 997–1007.

Lester, H. A., Dibas, M. I., Dahan, D. S., Leite, J. F., & Dougherty, D. A. (2004). Cys-loop receptors: New twists and turns. *Trends in Neurosciences*, *27*, 329–336.

Lester, H. A., Xiao, C., Srinivasan, R., Son, C. D., Miwa, J., Pantoja, R., et al. (2009). Nicotine is a selective pharmacological chaperone of acetylcholine receptor number and stoichiometry. Implications for drug discovery. *The AAPS Journal*, *11*, 167–177.

Lo, W. Y., Botzolakis, E. J., Tang, X., & Macdonald, R. L. (2008). A conserved Cys-loop receptor aspartate residue in the M3-M4 cytoplasmic loop is required for GABAA receptor assembly. *The Journal of Biological Chemistry*, *283*, 29740–29752.

Loebrich, S., Bahring, R., Katsuno, T., Tsukita, S., & Kneussel, M. (2006). Activated radixin is essential for GABAA receptor alpha5 subunit anchoring at the actin cytoskeleton. *The EMBO Journal*, *25*, 987–999.

Luscher, B., Fuchs, T., & Kilpatrick, C. L. (2011). GABAA receptor trafficking-mediated plasticity of inhibitory synapses. *Neuron*, *70*, 385–409.

Macdonald, R. L., Kang, J. Q., & Gallagher, M. J. (2010). Mutations in GABAA receptor subunits associated with genetic epilepsies. *The Journal of Physiology*, *588*, 1861–1869.

Mazzaferro, S., Gasparri, F., New, K., Alcaino, C., Faundez, M., Iturriaga Vasquez, P., et al. (2014). Non-equivalent ligand selectivity of agonist sites in (alpha4beta2)2alpha4 nicotinic acetylcholine receptors: A key determinant of agonist efficacy. *The Journal of Biological Chemistry*, *289*, 21795–21806.

Millar, N. S. (2008). RIC-3: A nicotinic acetylcholine receptor chaperone. *British Journal of Pharmacology*, *153*(Suppl. 1), S177–S183.

Miller, P. S., & Aricescu, A. R. (2014). Crystal structure of a human GABAA receptor. *Nature*, *512*, 270–275.

Mizokami, A., Kanematsu, T., Ishibashi, H., Yamaguchi, T., Tanida, I., Takenaka, K., et al. (2007). Phospholipase C-related inactive protein is involved in trafficking of gamma2 subunit-containing GABA(A) receptors to the cell surface. *The Journal of Neuroscience*, *27*, 1692–1701.

Mrowiec, T., & Schwappach, B. (2006). 14-3-3 proteins in membrane protein transport. *Biological Chemistry*, *387*, 1227–1236.

Mu, T. W., Ong, D. S., Wang, Y. J., Balch, W. E., Yates, J. R., 3rd, Segatori, L., et al. (2008). Chemical and biological approaches synergize to ameliorate protein-folding diseases. *Cell*, *134*, 769–781.

Nusser, Z., Hajos, N., Somogyi, P., & Mody, I. (1998). Increased number of synaptic GABA (A) receptors underlies potentiation at hippocampal inhibitory synapses. *Nature*, *395*, 172–177.

Nymann-Andersen, J., Wang, H., Chen, L., Kittler, J. T., Moss, S. J., & Olsen, R. W. (2002). Subunit specificity and interaction domain between GABA(A) receptor-associated protein (GABARAP) and GABA(A) receptors. *Journal of Neurochemistry, 80*, 815–823.

Olzmann, J. A., Kopito, R. R., & Christianson, J. C. (2013). The mammalian endoplasmic reticulum-associated degradation system. *Cold Spring Harbor Perspectives in Biology, 5*, a013185.

Palma, E., Bertrand, S., Binzoni, T., & Bertrand, D. (1996). Neuronal nicotinic alpha 7 receptor expressed in Xenopus oocytes presents five putative binding sites for methyllycaconitine. *The Journal of Physiology, 491*(Pt. 1), 151–161.

Papadopoulos, T., & Soykan, T. (2011). The role of collybistin in gephyrin clustering at inhibitory synapses: Facts and open questions. *Frontiers in Cellular Neuroscience, 5*, 11.

Piguet, J., Schreiter, C., Segura, J. M., Vogel, H., & Hovius, R. (2011). Acetylcholine receptor organization in membrane domains in muscle cells: Evidence for rapsyn-independent and rapsyn-dependent mechanisms. *The Journal of Biological Chemistry, 286*, 363–369.

Qu, Z. C., Moritz, E., & Huganir, R. L. (1990). Regulation of tyrosine phosphorylation of the nicotinic acetylcholine receptor at the rat neuromuscular junction. *Neuron, 4*, 367–378.

Richards, C. I., Srinivasan, R., Xiao, C., Mackey, E. D., Miwa, J. M., & Lester, H. A. (2011). Trafficking of alpha4* nicotinic receptors revealed by superecliptic phluorin: Effects of a beta4 amyotrophic lateral sclerosis-associated mutation and chronic exposure to nicotine. *The Journal of Biological Chemistry, 286*, 31241–31249.

Rosenberg, M. M., Yang, F., Giovanni, M., Mohn, J. L., Temburni, M. K., & Jacob, M. H. (2008). Adenomatous polyposis coli plays a key role, in vivo, in coordinating assembly of the neuronal nicotinic postsynaptic complex. *Molecular and Cellular Neurosciences, 38*, 138–152.

Saiepour, L., Fuchs, C., Patrizi, A., Sassoe-Pognetto, M., Harvey, R. J., & Harvey, K. (2010). Complex role of collybistin and gephyrin in GABAA receptor clustering. *The Journal of Biological Chemistry, 285*, 29623–29631.

Saliba, R. S., Michels, G., Jacob, T. C., Pangalos, M. N., & Moss, S. J. (2007). Activity-dependent ubiquitination of GABA(A) receptors regulates their accumulation at synaptic sites. *The Journal of Neuroscience, 27*, 13341–13351.

Sava, A., Barisone, I., Di Mauro, D., Fumagalli, G., & Sala, C. (2001). Modulation of nicotinic acetylcholine receptor turnover by tyrosine phosphorylation in rat myotubes. *Neuroscience Letters, 313*, 37–40.

Shin, H. W., Morinaga, N., Noda, M., & Nakayama, K. (2004). BIG2, a guanine nucleotide exchange factor for ADP-ribosylation factors: Its localization to recycling endosomes and implication in the endosome integrity. *Molecular Biology of the Cell, 15*, 5283–5294.

Smith, M. H., Ploegh, H. L., & Weissman, J. S. (2011). Road to ruin: Targeting proteins for degradation in the endoplasmic reticulum. *Science, 334*, 1086–1090.

St John, P. A. (2009). Cellular trafficking of nicotinic acetylcholine receptors. *Acta Pharmacologica Sinica, 30*, 656–662.

St John, P. A., & Gordon, H. (2001). Agonists cause endocytosis of nicotinic acetylcholine receptors on cultured myotubes. *Journal of Neurobiology, 49*, 212–223.

Sumikawa, K. (1992). Sequences on the N-terminus of ACh receptor subunits regulate their assembly. *Brain Research. Molecular Brain Research, 13*, 349–353.

Sumikawa, K., & Nishizaki, T. (1994). The amino acid residues 1–128 in the alpha subunit of the nicotinic acetylcholine receptor contain assembly signals. *Brain Research. Molecular Brain Research, 25*, 257–264.

Sun, H., Hu, X. Q., Moradel, E. M., Weight, F. F., & Zhang, L. (2003). Modulation of 5-HT3 receptor-mediated response and trafficking by activation of protein kinase C. *The Journal of Biological Chemistry, 278*, 34150–34157.

Swope, S. L., Moss, S. J., Raymond, L. A., & Huganir, R. L. (1999). Regulation of ligand-gated ion channels by protein phosphorylation. *Advances in Second Messenger and Phosphoprotein Research*, *33*, 49–78.

Terunuma, M., Jang, I. S., Ha, S. H., Kittler, J. T., Kanematsu, T., Jovanovic, J. N., et al. (2004). GABAA receptor phospho-dependent modulation is regulated by phospholipase C-related inactive protein type 1, a novel protein phosphatase 1 anchoring protein. *The Journal of Neuroscience*, *24*, 7074–7084.

Thompson, A. J., Lester, H. A., & Lummis, S. C. (2010). The structural basis of function in Cys-loop receptors. *Quarterly Reviews of Biophysics*, *43*, 449–499.

Todd, E., Gurba, K. N., Botzolakis, E. J., Stanic, A. K., & Macdonald, R. L. (2014). GABAA receptor biogenesis is impaired by the gamma2 subunit febrile seizure-associated mutation, GABRG2(R177G). *Neurobiology of Disease*, *69*, 215–224.

Tretter, V., Jacob, T. C., Mukherjee, J., Fritschy, J. M., Pangalos, M. N., & Moss, S. J. (2008). The clustering of GABA(A) receptor subtypes at inhibitory synapses is facilitated via the direct binding of receptor alpha 2 subunits to gephyrin. *The Journal of Neuroscience: The Official Journal of the Society for Neuroscience*, *28*, 1356–1365.

Tseng, W. C., Jenkins, P. M., Tanaka, M., Mooney, R., & Bennett, V. (2015). Giant ankyrin-G stabilizes somatodendritic GABAergic synapses through opposing endocytosis of GABAA receptors. *Proceedings of the National Academy of Sciences of the United States of America*, *112*, 1214–1219.

Tsetlin, V., Kuzmin, D., & Kasheverov, I. (2011). Assembly of nicotinic and other Cys-loop receptors. *Journal of Neurochemistry*, *116*, 734–741.

Twelvetrees, A. E., Yuen, E. Y., Arancibia-Carcamo, I. L., MacAskill, A. F., Rostaing, P., Lumb, M. J., et al. (2010). Delivery of GABAARs to synapses is mediated by HAP1-KIF5 and disrupted by mutant huntingtin. *Neuron*, *65*, 53–65.

Unwin, N. (2005). Refined structure of the nicotinic acetylcholine receptor at 4A resolution. *Journal of Molecular Biology*, *346*, 967–989.

Vembar, S. S., & Brodsky, J. L. (2008). One step at a time: Endoplasmic reticulum-associated degradation. *Nature Reviews. Molecular Cell Biology*, *9*, 944–957.

Vithlani, M., Terunuma, M., & Moss, S. J. (2011). The dynamic modulation of GABA(A) receptor trafficking and its role in regulating the plasticity of inhibitory synapses. *Physiological Reviews*, *91*, 1009–1022.

Wanamaker, C. P., Christianson, J. C., & Green, W. N. (2003). Regulation of nicotinic acetylcholine receptor assembly. *Annals of the New York Academy of Sciences*, *998*, 66–80.

Wanamaker, C. P., & Green, W. N. (2007). Endoplasmic reticulum chaperones stabilize nicotinic receptor subunits and regulate receptor assembly. *The Journal of Biological Chemistry*, *282*, 31113–31123.

Wang, H., Bedford, F. K., Brandon, N. J., Moss, S. J., & Olsen, R. W. (1999). GABA(A)-receptor-associated protein links GABA(A) receptors and the cytoskeleton. *Nature*, *397*, 69–72.

Wang, Y. J., Di, X. J., & Mu, T. W. (2014). Using pharmacological chaperones to restore proteostasis. *Pharmacological Research*, *83*, 3–9.

Wang, Y. J., Han, D. Y., Tabib, T., Yates, J. R., 3rd, & Mu, T. W. (2013). Identification of GABA(C) receptor protein homeostasis network components from three tandem mass spectrometry proteomics approaches. *Journal of Proteome Research*, *12*, 5570–5586.

Wang, Y. J., Tayo, B. O., Bandyopadhyay, A., Wang, H., Feng, T., Franceschini, N., et al. (2014). The association of the vanin-1 N131S variant with blood pressure is mediated by endoplasmic reticulum-associated degradation and loss of function. *PLoS Genetics*, *10*, e1004641.

Wang, J. M., Zhang, L., Yao, Y., Viroonchatapan, N., Rothe, E., & Wang, Z. Z. (2002). A transmembrane motif governs the surface trafficking of nicotinic acetylcholine receptors. *Nature Neuroscience*, *5*, 963–970.

Whiting, P. J., Bonnert, T. P., McKernan, R. M., Farrar, S., Le Bourdelles, B., Heavens, R. P., et al. (1999). Molecular and functional diversity of the expanding GABA-A receptor gene family. *Annals of the New York Academy of Sciences*, *868*, 645–653.

Whiting, P., McKernan, R. M., & Iversen, L. L. (1990). Another mechanism for creating diversity in gamma-aminobutyrate type A receptors: RNA splicing directs expression of two forms of gamma 2 phosphorylation site. *Proceedings of the National Academy of Sciences of the United States of America*, *87*, 9966–9970.

Wong, L. W., Tae, H. S., & Cromer, B. A. (2014). Role of the rho1 GABA(C) receptor N-terminus in assembly, trafficking and function. *ACS Chemical Neuroscience*, *5*, 1266–1277.

Wong, L. W., Tae, H. S., & Cromer, B. A. (2015). Assembly, trafficking and function of alpha1beta2gamma2 GABAA receptors are regulated by N-terminal regions, in a subunit-specific manner. *Journal of Neurochemistry*, *134*, 819–832.

Wu, Z. S., Cheng, H., Jiang, Y., Melcher, K., & Xu, H. E. (2015). Ion channels gated by acetylcholine and serotonin: Structures, biology, and drug discovery. *Acta Pharmacologica Sinica*, *36*, 895–907.

Wu, A. L., Wang, J., Zheleznyak, A., & Brown, E. J. (1999). Ubiquitin-related proteins regulate interaction of vimentin intermediate filaments with the plasma membrane. *Molecular Cell*, *4*, 619–625.

Xiao, Y., & Kellar, K. J. (2004). The comparative pharmacology and up-regulation of rat neuronal nicotinic receptor subtype binding sites stably expressed in transfected mammalian cells. *The Journal of Pharmacology and Experimental Therapeutics*, *310*, 98–107.

Xu, R., & Salpeter, M. M. (1995). Protein kinase A regulates the degradation rate of Rs acetylcholine receptors. *Journal of Cellular Physiology*, *165*, 30–39.

Yoshimura, K., Takeuchi, H., Sato, O., Hidaka, K., Doira, N., Terunuma, M., et al. (2001). Interaction of p130 with, and consequent inhibition of, the catalytic subunit of protein phosphatase 1alpha. *The Journal of Biological Chemistry*, *276*, 17908–17913.

Zhao, C. J., Noack, C., Brackmann, M., Gloveli, T., Maelicke, A., Heinemann, U., et al. (2009). Neuronal Ca2+ sensor VILIP-1 leads to the upregulation of functional alpha4beta2 nicotinic acetylcholine receptors in hippocampal neurons. *Molecular and Cellular Neurosciences*, *40*, 280–292.

CHAPTER TWO

Harnessing the Flow of Excitation: TRP, Voltage-Gated Na^+, and Voltage-Gated Ca^{2+} Channels in Contemporary Medicine

Roman V. Frolov[1], Matti Weckström✠
Division of Biophysics, Department of Physics, University of Oulu, Oulun Yliopisto, Finland
[1]Corresponding author: e-mail address: roman.frolov@oulu.fi

Contents

✠ Matti Weckström has died.

Advances in Protein Chemistry and Structural Biology, Volume 103
ISSN 1876-1623
http://dx.doi.org/10.1016/bs.apcsb.2015.11.001

Abstract

Cellular signaling in both excitable and nonexcitable cells involves several classes of ion channels. Some of them are of minor importance, with very specialized roles in physiology, but here we concentrate on three major channel classes: TRP (transient receptor potential channels), voltage-gated sodium channels (Nav), and voltage-gated calcium channels (Cav). Here, we first propose a conceptual framework binding together all three classes of ion channels, a "flow-of-excitation model" that takes into account the inputs mediated by TRP and other similar channels, the outputs invariably provided by Cav channels, and the regenerative transmission of signals in the neural networks, for which Nav channels are responsible. We use this framework to examine the function, structure, and pharmacology of these channel classes both at cellular and also at whole-body physiological level. Building on that basis we go through the pathologies arising from the direct or indirect malfunction of the channels, utilizing ion channel defects, the channelopathies. The pharmacological interventions affecting these channels are numerous. Part of those are well-established treatments, like treatment of hypertension or some forms of epilepsy, but many other are deeply problematic due to poor drug specificity, ion channel diversity, and widespread expression of the channels in tissues other than those actually targeted.

1. INTRODUCTION

Electrical excitation in the living tissue is enabled by concerted action of ion channels. With rare exceptions, its actuators are cationic channels permeable to calcium and sodium, initiating receptor and action potentials (APs) in response to direct or secondary stimulation of receptors, neurons, and myocytes. This excitation is then countered and terminated by potassium channels, while altered intracellular ionic homeostasis is restored by action of nonconductive ionic pumps and various transporter proteins. Here, we focus on three diverse but molecularly related classes of ion channels that principally enable the electrical responses of excitable cells—voltage-gated Na^+ (Nav), voltage-gated Ca^{2+} (Cav), and predominantly ligand-activated transient receptor potential (TRP) channels. Nav and Cav channels have been extensively studied for several decades and feature prominently in medicine, while TRP channels were discovered relatively recently, and their therapeutic potential is not yet fully understood and remains virtually untapped.

In this chapter, we endeavor to cover the three classes of channels from the integral systems angle. We begin with introduction of a general concept of differential ion channel expression according to the functional place of the excitatory channel in the nervous system ("flow-of-excitation" model). This concept helps to understand (1) channel expression patterns, (2) physiological necessity to have different channels in specific loci of neural circuitries, and (3) the differences in biophysical properties between different excitatory channels. From this basis, we zoom on each channel class, describing their structure, expression, properties, regulation, and involvement in physiological processes. Finally, the role of these channels in medicine is examined: their targeting by drugs and channelopathies. To retain proper focus, we do not attempt to review other channel families serving as excitatory inputs into neural circuits, such as cyclic nucleotide-gated (CNG) channels, acid-sensing ion channels (ASIC), and ligand-gated receptor channels. In addition, the channel families outside of our treatise are quite distant evolutionarily from the supercluster of TRP, Nav, and Cav channels (Yu, Yarov-Yarovoy, Gutman, & Catterall, 2005).

1.1 Place of TRP, Nav, and Cav Channels in the Flow of Excitation in Neural Circuits

What is the position of TRP, Nav, and Cav channels in the grand order of the living organism in a general sense? The schematic in Fig. 1 draws an analogy between an integrated electromechanical system and a neuronal circuit. Essential blocks of an electromechanical system are the sensor, amplifier, analog-to-digital converter, computer, digital-to-analog converter, and actuator. In the sensor—the input of the system—the energy of an environmental signal is absorbed and transformed, usually with great amplification, into voltage changes by a *transducer*—a device that can convert one form of energy into another. The resulting electrical signal is digitized and sent to the computer for interpretation, processing, and conditioning. Digital control commands are sent by the computer to drive peripheral devices. In the actuator, the commands are transformed back into an analog form. There, with the help of the output transducer, the commands trigger a mechanical action (Fig. 1A). Similarly, in a biological organism, specialized cells or cell compartments contain molecular receptors of various types that are finely adjusted to interact with environmental stimuli of certain modalities (chemical, mechanical, thermal, electromagnetic). The associated molecular machinery *transduces* and amplifies the stimulus, yielding amplitude- and frequency-modulated graded voltage responses, which are converted (digitized) into frequency-modulated trains of APs in the same or downstream neuron. Nervous system functions as a computer, while muscles and endocrine organs are analogous to peripheral (output) effectors, where digital

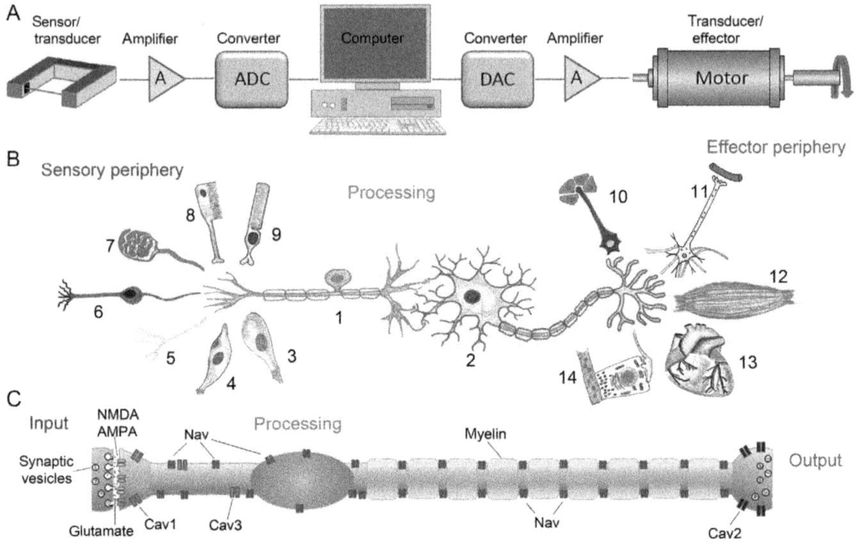

Figure 1 Flow of excitation in integrated artificial and biological systems. (A) Integrated electromechanical system consists of sensory periphery devices (a sensor with transducer, amplifier, analog-to-digital converter), central computing unit, and effector periphery devices (digital-to-analog converter, amplifier, transducer and motor). (B) Similarly, in the nervous systems, sensory periphery is represented by receptors (hair cell (3), taste receptor (4), free nerve ending (5), olfactory receptor (6), mechanosensory receptor (7), microvillar photoreceptor (8), rod photoreceptor (9)), computing is performed by neurons (sensory (1) and output (2) neurons depicted), while effector periphery includes various types of neurosecretory neurons (local (10) and global (11) neurosecretory neurons, adrenal chromaffin cell (14)) and myocytes (skeletal (12) and cardiac (13) muscles). (C) At the level of an individual neuron, input is provided by postsynaptic receptor channels, Nav channels are responsible for information processing and AP generation and propagation, and presynaptic Cav2 channels mediate neuronal output. Cav3 channels expressed in dendrites and soma condition Nav currents. Cav1 channels found in the postsynaptic regions provide still another, cell compartment-level output: depolarization of the terminal by NMDA/AMPA receptors opens Cav1 channels, with calcium influx initiating changes in gene expression, which can lead to LTP.

commands in form of AP trains are decoded into changes of the membrane potential, eventually governing contraction, secretion, and protein expression (Fig. 1B and C).

TRP and Cav channels operate as *biological transducers* at the *input* and *output* of neural circuits, correspondingly. TRP channels generate receptor potentials in many types of receptors, including mechanoreceptors, nociceptors, photoreceptors, temperature, taste and osmotic receptors, thus serving as the *endpoints* for the corresponding energy transduction mechanisms and pathways. In contrast, the function of Cav channels is to transform the signals carried by membrane depolarizations into the chemical signals. They do this by modulating calcium influx, which mediates regulation of enzymes, gene expression, excitation–contraction, and excitation–secretion coupling. Activation of Nav channels in neurons and muscles, on the other hand, is usually triggered by excitatory postsynaptic potentials, and can lead to generation of AP. In other words, *the main function of TRP channels is the initiation of excitatory electrical input into neural circuits; the role of Nav channels is limited to maintenance and regeneration of excitation in networks of neurons and myocytes; and Cav channels enable the neural circuit output*. This constitutes the paradigm, "flow-of-excitation for differential channel expression," which we present here. Crucially, to the best of our knowledge, no Nav or Cav channel serves as an input transducer in any receptor cell (with exception of electroreceptors), and no Nav or TRP channel operates as an immediate output transducer in any effector cell.

Why do the cells use Nav channels at all if Cav can provide depolarization alone? The conventional answer is that calcium is of much higher importance for the cell than just a depolarizing ion; in addition, phylogenetically it seems to be that calcium was used first for signaling in muscle and other effector cells and selected during evolution for this task earlier than action potentials appear (Yu et al., 2005). Accordingly, Nav channels are present in, e.g., yeast cells, but not in *C. elegans* (Bargmann, 1998). Calcium is a key signaling factor, which can drastically alter both the immediate and long-term cell function. Thus, using calcium influx mainly for membrane depolarization in neurons would obscure its unique signaling faculty. However, Cav channels can be solely responsible for depolarization in some effector cells, e.g., smooth muscles of blood vessels, where depolarization coincides with a global-regulated event.

Remarkably, the flow-of-excitation concept holds robust both at the systems level and at the level of individual elements of the neural circuit, and even at the level of cell compartments (Fig. 1B and C). A typical neuron receives electrical input through activation of ligand-gated receptor channels in the postsynaptic terminal. These channels can be excitatory or inhibitory. The excitatory cationic ones (e.g., glutamate receptor channels) function not just as analogs to TRP, but generally share TRP properties. In fact, TRPC channels can be found in postsynaptic membranes of some types of neurons, where they are functionally coupled to metabotropic glutamate receptors (Kim et al., 2003) and directly mediate synaptic transmission. Postsynaptic depolarization activates Nav channels and can trigger AP, which propagate to presynaptic terminals. There, they activate the locally expressed Cav2 channels, triggering synaptic vesicle release. Moreover, in striatal neurons the postsynaptic terminal itself can be envisioned as a local circuit consisting of the neurotransmitter-activated AMPA/kainate and NMDA receptors as an electrical input and the locally expressed Cav1 as an output. Although both AMPA and NMDA receptor channels pass Ca^{2+} ions in addition to the main Na^{+} current, they alone cannot launch Ca^{2+} and calmodulin-dependent events leading to the functional changes like long-term potentiation (LTP), which is believed to be related to memory imprinting. This process specifically requires Ca^{2+} influx through the closely associated Cav channels (Rajadhyaksha et al., 1999) and, likely, activation of calmodulin (CaM) used as a subunit by Cav channels but not by AMPA and NMDA receptors.

Similarly, in muscles, opening of acetylcholine receptors at the neuromuscular junction causes depolarizing sodium and calcium current, with consecutive opening of Nav channels and triggering of APs. This ensures both vigorous gating of Cav channels (Cav1.1), which initiates contraction, and rapid propagation of excitation along muscle fibers. Expression of Nav channels in myocytes appears to be a relatively recent evolutionary adaptation as in many ancient species and in smooth muscles of vertebrates APs in myocytes are provided by Cav1 channels alone (Berridge, 2008).

The last general aspect of the flow-of-excitation hypothesis to be discussed here is the characteristic pattern of physiological properties of the three groups of channels. The *input* excitatory cationic channels, whether it is TRP, CNG, ASIC, or ligand-gated channels, are generally characterized by relatively low ion selectivity, very low voltage dependence, and high variability in biophysical properties between different isoforms. The *output* channels are strongly selective for Ca^{2+}, highly voltage-dependent, and display a similarly high variability in properties. The Nav channels are selective for Na^{+}, highly sensitive to voltage, but their molecular similarity and biophysical properties do not merit their segregation into separate subfamilies. These differences can be easily understood in the context of channel function. In receptors, input channels do not receive voltage signals and so they do not need high voltage sensitivity, although many ligand-gated ion channels can be modulated by membrane potential. The mixed, predominantly sodium and calcium permeability of the input channels is required to generate membrane voltage response (both ion species take part in that), and, importantly, to adjust the functioning of the receptor cell through adaptation. The latter is usually triggered by calcium and involves several regulatory mechanisms. On the other hand, output Cav channels require voltage-dependence as they are activated by depolarization; they are regulated massively by local factors, adding to functional flexibility of the neural circuit output. Existence of many types of neural circuit inputs and outputs can explain why channels with wildly different properties are needed to adequately represent this diversity. In contrast, Nav channels are needed to propagate excitation and so most isoforms have very similar properties, although not without exceptions, as we show below.

Electroreceptors found in many marine species (Frings & Bradley, 2006) constitute an exception from the presented concept, as L-type Cav channels expressed in the apical segment of the electroreceptor seem to be the immediate sensors of the external signal, while L-type Cav channels in the basal part of the cell are responsible for synaptic transmission. However, since the stimulus and the response are of the same electrical modality, use of the same or a similar channel perhaps represents an optimal evolutionary development.

2. TRP CHANNELS

TRP channels were originally found in studies of visual transduction mechanisms in insect photoreceptors (Hardie, 2011). After the discovery of their role in vision (Hardie & Minke, 1992, 1993), their homologs were soon found both in yeast and algae and in all metazoan animals including humans (Wes et al., 1995; Zhu, Chu, Peyton, & Birnbaumer, 1995). It came much to a surprise to most of the scientific community that in vertebrates TRP channels are extremely common, possibly present in almost all cells and involved in a multitude of functions related to sensory input or cells' sensing of exogenous ligands. Their structure and genomic information shows that TRP channels are close relatives in evolutionary terms with voltage-activated Na^+ and Ca^{2+} channels (Yu et al., 2005), and they typically show high-to-moderate Ca^{2+} permeability, although there are exceptions. They are split into several families (six in mammals, seven in all), with several isoforms in each family, and they are central to most sensory functions (Clapham, Runnels, & Strubing, 2001; Frings & Bradley, 2006). In addition to their activation by various external stimuli in sensory cells, TRP channels are involved in many other neural and nonneural functions in the cardiovascular, gastrointestinal, endocrine, renal, and immune systems. Their activation and modulation mechanisms have shown to be extremely elusive, ranging from voltage sensitivity and mechanosensory and osmosensory properties to activation by both intra- and extracellular specific ligands and depletion of intracellular calcium stores. Typically, the heterologous expression of the channels exhibits different properties than in the original cell types, which is the case also with the "ancestor" channel, the TRP in *Drosophila* photoreceptors (Minke & Parnas, 2006). These properties render the TRP channels especially challenging for pharmacological endeavors, even when many of the bodily functions, where they have a role, would be natural targets for therapy.

2.1 Structure and Structural Varieties—Subfamilies of TRP Channels

As their relatives, the voltage-gated channels, the TRP channels have a tetrameric structure (Fig. 2). Each unit is modeled to be composed of six transmembrane segments (S1–6). A relatively long amino acid linker connecting S5 and S6 is called P-loop. It is partly situated within the membrane's outer leaflet and partly protrudes above the membrane. Channel pore is assembled from four P-loops with participation of amino acids of S5 and S6. In addition to transmembrane segments, the channels can have either an extensive or a small carboxyl (C) and amino (N) terminus amino acid chains. In heterologous expression studies, the subunits of the tetrameric channel protein may also be members of a different subfamily or isoform, although under natural conditions this has not been shown conclusively.

Altogether, seven TRP subfamilies, including ca. 30 isoforms, are recognized: TRPC (canonical), TRPA (ankyrin), TPRM (melastatin), TRPML (mucolipin), TRPV (vanilloid), TRPP (polycystin), and TRPN (also called NOMPC, no mechanoreceptor potential C). The latter has not been found in mammals. A common structural feature of TRP channels is a 25-amino acid motif, called TRP domain, containing TRP box on the C-terminal side of the sixth transmembrane segment. It appears to be involved in channel gating. Although a full structural itemization is not necessary here, it is good to know that some of the channels have clear structural hallmarks. The TRP domain and box are present in all TRPC channel genes, but not in all TRP channel genes. The N-terminal contains ankyrin repeats in TRPC, TRPA, and TRPV channels, but the channel isoforms belonging to the TRPC and TRPM subfamilies contain proline-rich regions in the region just C-terminal to S6. In the canonical channels (TRPC), C-terminal is very short but it is very long in TRPM (Clapham et al., 2001). Two subfamilies, TRPP and TRPML contain an ER retention motif in the C-terminal and their location in cells is likely to be the

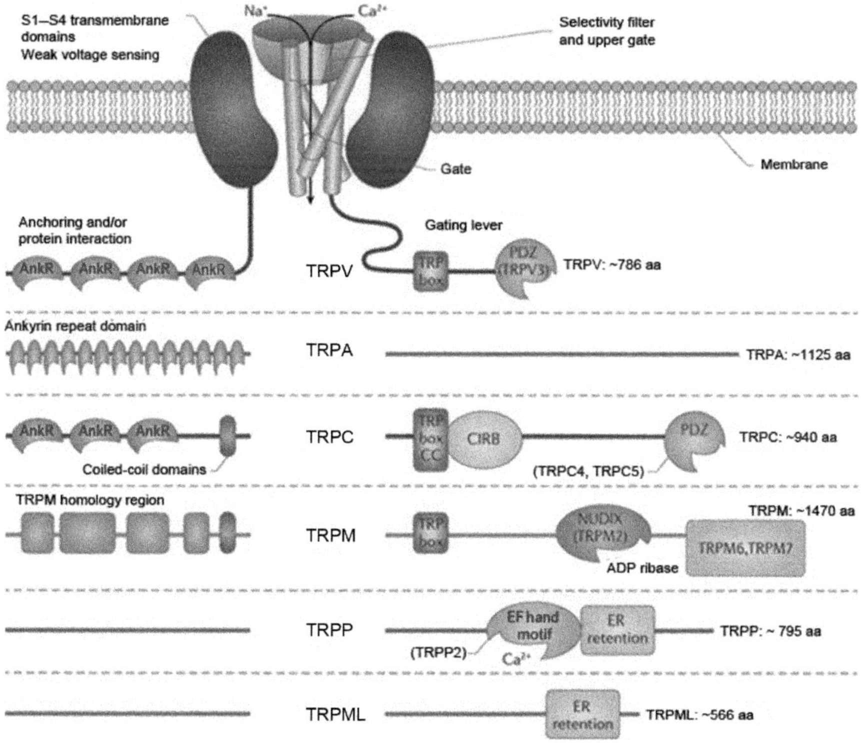

Figure 2 TRP channels. Different TRP families are characterized by distinct motifs in their N- and C-terminals. The ankyring repeats are found in TRPV, TRPA, and TRPC channels. The TRP box is present in TRPV, TRPM, and TRPC isoforms. TRPP and TRPML are characterized by endoplasmic reticulum (ER) retention domains indicating that they are expressed on the intracellular organelles; aa, amino acids; CIRB, CaM/IP_3 receptor-binding domain; NUDIX, nucleoside diphosphate-linked moiety X; PDZ, acronym for **P**SD95 (postsynaptic density protein 95), **D**LGA (*Drosophila* disc large tumor suppressor), and **Z**O1 (zonula occludens protein 1). *The image is reproduced, with permission, from Moran, McAlexander, Biro, and Szallasi (2011).*

membranes of intracellular organelles, ER, lysosomes, and the vesicles of the Golgi apparatus (Qian & Noben-Trauth, 2005; Venkatachalam, Hofmann, & Montell, 2006) (Fig. 2).

The originally discovered channels in insect phototransduction, TRP in photoreceptors, belong to the TRPC family and are thus considered as "canonical." Curiously, their mammalian homologs TRPC6 and TRPC7 have very similar properties and function, and are present in the intrinsically photosensitive retinal ganglion cells, providing input to the circadian control system in the brain. TRPC channels are substantially permeable for Ca^{2+}, with significant permeabilities for other cations. Because they mediate Ca^{2+} influx, they are involved in numerous excitation processes in cells, both in electrical terms and in activation of Ca^{2+}-dependent signaling pathways. By and large it can be said that, following the TRPC pattern, the TRP channels are cationic channels, where the exact permeability may vary largely between isoforms even within the same subfamily. Therefore the channel permeability is not a good way for their classification. The strength of voltage-dependence and the form of the current–voltage relation in cells varies widely from isoform to isoform, also because of other modulatory factors in cells. However, the structural features and localization at least partly define the activation and modulation mechanisms, which distinguish the subfamilies from each other.

2.2 Regulation and Activation Mechanisms

The ultimate goal of regulation of an ion channel is to maintain its momentary macroscopic conductance at the level optimal for the currently executed cellular function. To achieve this objective, cells utilize a complex set of approaches at a number of levels including ion channel gene transcription, posttranscriptional processing, translation, posttranslational modification, transportation of the channel protein to the plasma membrane, its assembly or incorporation into the membrane, interaction between various subunits, and regulation via signal transduction pathways.

In evolution, TRP channels have diversified enormously and there is mounting evidence that even homologous isoforms (or orthologs) show different modes of activation and/or modulation properties even within mammals (Kadowaki, 2015). Thus channel function may be different between e.g. mice and men, enormously complicating issues of all forms of therapeutics interventions related to TRPs. Of all the TRP channel activation mechanisms by far the best and most thoroughly investigated is the TRPC subfamily, especially the original TPR and TRPL channels from *Drosophila* photoreceptors. The activation of studied TRPC channels in mammals seems to follow the same pathway. The channels are activated by either DAG (diacylglycerol) or its derivatives or by changes in the tension of the lipid membrane produced by cleavage of DAG from the membrane lipid PIP_2, or by both simultaneously (Hardie & Franze, 2012; Hardie & Juusola, 2015). The latter is brought about by G_q protein-dependent activation of phospholipase C (PLC). What then leads to activation of G_q in different cell types may vary widely. In insect photoreceptors and in light-sensitive ganglion cells in mammalian retina, this is caused by light-activation of rhodopsin, but in most other cells, this is caused by other mechanisms. TRPCs 1, 4, and 5 and the vanilloid receptor subtype 1, in contrast, do not seem to be responsive to DAG (Hofmann et al., 1999; Venkatachalam, Zheng, & Gill, 2003). Otherwise, the link of many TRP channel types to the lipid signaling seems to be quite strong, and many other described mechanisms may in fact be secondary to the phospholipid processes, or at least modulated or enhanced by them (Qin, 2007). Another property that may be common to all TRP channels is the weak voltage dependence of activation. It seems that activating factors such as temperature (TRPV1, TRPM8, TRPV3), or the binding of various ligands (TRPV1, TRPV3, TRPM8, TRPM4), may act by shifting the voltage dependence of the channels to a physiologically relevant range (Nilius et al., 2005).

Some of the channel families show very specific ligand-binding properties, most famously the TRPV-family and especially the TRPV1 isoform, which can be recognized by the ability of its isoforms to bind and be activated by capsaicin (the "hot" ingredient in chilies) and its relatives (Caterina et al., 1997). Many TRP channels have been linked to the regulation or emptying of intracellular calcium stores and their potential role as CRAC (calcium release-activated) channels. This is linked again to the phospholipid signaling upstream of channel activation, where one of the products can be IP_3 (inositol 1,4,5-trisphosphate), for which receptors, acting as calcium channels, are localized in the ER calcium stores in many cell types. The CRAC function of some TRP channel isoforms (like TRPC4) may also be related to interaction with another channel protein, ORAI1 (Xu, Cioffi, Alexeyev, Rich, & Stevens, 2015).

2.3 Therapeutic Potential of TRP Channels

The function of TPR channels in the input pathways to neurons and other cells—in sensory processes themselves as well as in sensing different endo- and exogenous ligands—would seem to render them attractive targets for pharmaceutical interventions. This is strengthened by observing that the structural homology between subfamilies and sometimes even isoforms within the same subtype, is often small, enabling effective separation of therapeutic targets (see, e.g., Moran et al., 2011). In view of the highly varied pathologies associated with TRP channelopathies, it is clear that even when elucidating the details of the pathophysiological processes at cellular and molecular signaling level, we cannot gain very much general knowledge regarding the role of most of the TRP subfamilies, while specific mutations affect only a small range of those functions where the specific channel isoform is actually functional. This disheartening result limits the planning of any therapeutic interventions to solely specific subtypes of any TRP isoform-based disease. The effects of polymorphisms of TRP channels and their possible role in development of various clinical conditions are even further away from reality (Szallasi & McAlexander, 2015).

However, there is one clear case, where TRP channels have entered the pharmacological practice. That is the natural ability of the TRPV (especially TRPV1) channel to bind vanilloids, especially capsaicin and its derivatives. The stimulation of TRPV1 in skin and in several areas within the body (ligaments, joints, connective tissue membranes, etc.) produces pain and/or heat sensation. On this basis, it has seemed feasible to treat pain, especially chronic pain, by attempting to block the activation of TRPV1 channels (Szallasi & Sheta, 2012). Unfortunately, this also blocks heat sensation and the efforts have concentrated to treatment by desensitization on the skin by repeated topical application of capsaicin or by insertion of capsaicin analogs, like resiniferatoxin, into cancer areas (Brederson, Kym, & Szallasi, 2013; Cortright & Szallasi, 2009). Unfortunately, many attempts at pain relief via stimulation and desensitization of the TRPV1 (and other TRPV isoforms) using small-molecular capsaicin analogs have produced so much side-effects, like intense heat sensation, that they cannot be developed further (Lee et al., 2015). The treatment of pain condition via interacting with TRPA1 channels are all still in experimental phases.

3. VOLTAGE-GATED NA$^+$ CHANNELS

Nav channels are responsible for transmission of electrical signals in the neural circuits over large distances. In evolution, Nav channels first appear in the cnidarians, like the jellyfish, where they mediate rapid signal propagation over extended neural networks. In invertebrates, Nav channels are rarely found outside of the nervous system, while in chordates they are abundant in striated muscles, greatly outnumbering Cav channels (Marban, Yamagishi, & Tomaselli, 1998). In vertebrates, Nav underlie AP generation and propagation in neurons, myocytes, and, arguably, in certain types of glial cells. Sodium currents were discovered by Hodgkin and Huxley in their seminal work on squid giant axon (Hodgkin & Huxley, 1952a, 1952b, 1952c) and the channel was purified in 1978 on the basis of its affinity to tetrodotoxin (TTX), an extremely selective and potent blocker of several types of Nav channels (Agnew, Levinson, Brabson, & Raftery, 1978). Nav channels are characterized by extremely rapid activation (at the sub-millisecond timescale) and fast inactivation (within 1–2 ms), the properties crucial for AP generation. Other slower modes of inactivation were discovered later. Early works succeeded in development of a robust four-barrier mathematical model of Nav gating (Hille, 1975), determined ion selectivity and voltage-dependencies of gating, gained insight into the mechanism of fast inactivation, and discovered that local anesthetics exert their action by blocking Nav channels (Catterall, 2014). In medicine, Nav channels are the essential drug targets in anesthesiology, cardiology, and psychiatry.

3.1 Structure

Functional Nav channels usually consist of two or three subunits, the pore-forming α subunit (Navα) plus one or two smaller β subunits (Navβ), one of which is attached allosterically and another covalently. However, only the α subunit is needed for actual channel operation. Navα has a modular structure, consisting of four domains each containing six transmembrane helices or segments (S1–6). Domains are separated by extensive cytoplasmic linker sequences. The S4 helix of each domain contains several (4 to 7) repeats of positively charged amino acids followed by a couple of uncharged residues. This constitutes the voltage sensor; sliding displacement of all four voltage sensors upon the change in transmembrane potential provides energy for conformational transitions underlying channel opening (activation) or closing (deactivation). Like in TRP channels, channel pore is formed by S5 and S6. P-loops form the crucial aspects of the channel pore, including selectivity filter and the outer mouth (Catterall, 2014) (Fig. 3).

The four-domain, twenty-four-segment structure of the α subunit is stereotypical for many voltage-gated ion channel families, including Cav and the classical (Shaker- and Eag-related) potassium channels, as well as for the ligand-gated TRP. However, while the α subunits are comprised of four separate, often identical subunits six transmembrane segments each in channels belonging to the latter two classes, the amino acid sequences of different domains in Nav and Cav channels are not identical, with implications for channel gating, since each domain gates slightly differently. Navα contains numerous sites involved in modulation of the channel and interaction with regulatory molecules: N-linked glycosylation sites on the domain I P-loop (Nav channels are usually heavily glycosylated, this changes the local surface charge and, consequently, voltage-dependencies of gating (Bennett, Urcan,

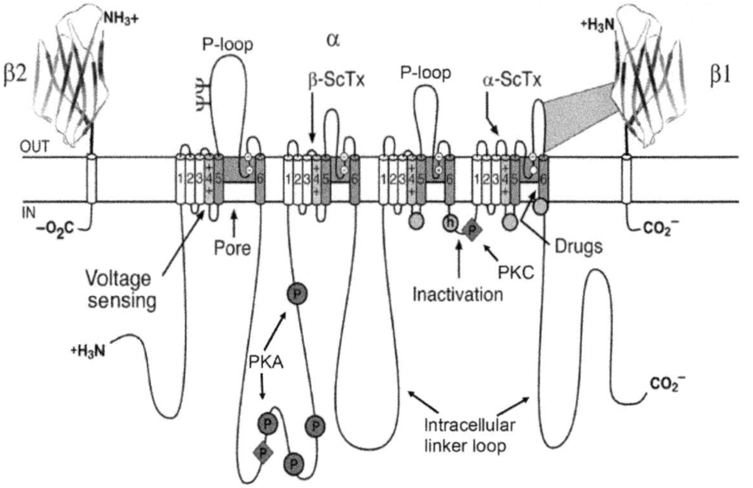

Figure 3 Nav channel with ancillary subunits. The α subunit consists of 24 transmembrane segments organized in four domains. The length of intracellular and P-loops is approximately proportional to the actual number of amino acids; yellow (light gray in the print version): voltage sensor segments; green (gray in the print version): pore-forming helices; Ψ: putative N-linked glycosylation sites; P in red (dark gray in the print version) circles and diamonds stands for phosphorylation sites by PKA and PKC, respectively; h in blue (light gray in the print version) circle: inactivation particle; blue (light gray in the print version) circles: sites involved in inactivation particle receptor. Sites for binding drugs (and TTX) and scorpion toxins are indicated. Immunoglobulin-like structure of extracellular domains of β subunits is shown. *The image is reproduced with changes, with permission, from Catterall (2014).*

Tinkle, Koszowski, & Levinson, 1997)), several PKA and PKC phosphorylation sites on the I–II intracellular linker, a PKC phosphorylation site on the III–IV linker, and β subunit interaction site on the domain IV P-loop. Importantly, the length of the I–II linker, and, therefore, the number of consensus phosphorylation sites, varies, with "long" isoforms usually found in neurons and myocardium, and "short" ones expressed in skeletal muscle (Marban et al., 1998).

3.2 Inactivation of Nav Channels

Inactivation is a crucial aspect of Nav function as evidenced by a number of disorders linked to defective inactivation. The classical "ball and chain" mechanism of rapid pore blockade, taking place in milliseconds and essential to secure high firing rate, was first discovered in Nav channels. Now it is known to be a feature of several dissimilar ion channels with rapid inactivation gating. The "ball" is formed by amino acids in the intracellular loop between domains III and IV, and its receptor site is formed by several residues on the intracellular end of transmembrane helices S4–6 in domains III and IV (Vassilev, Scheuer, & Catterall, 1988). The second inactivation mechanism, slow voltage-dependent inactivation (VDI, C-type inactivation), occurs on the scale of several hundred milliseconds to seconds and involves conformational changes in the selectivity filter (Catterall, 2014). It regulates Nav availability for consecutive depolarizations and adapts firing rates in activity-dependent manner (Vilin & Ruben, 2001).The third type of inactivation, closed-state inactivation (Bean, 1981), occurs in response to small depolarizations, resulting in movement of voltage sensors in domains III and IV only—not sufficient for channel opening but enough to trigger fast inactivation (Bahring & Covarrubias, 2011; Groome, Lehmann-Horn, & Holzherr, 2011).

However, it appears that as a result of failure of the fast inactivation mechanism, most if not all Nav isoforms can generate "persistent" non-inactivating currents at subthreshold voltages. Although very small, only few percents of the normal Nav current amplitude, persistent currents are functionally important for amplification of small depolarizing signals, which would otherwise fail to trigger response (Lampl, Schwindt, & Crill, 1998). In fact, some of the effect of local anesthetics can be attributed to inhibition of persistent Nav currents in sensory neurons (Baker, 2000; Devor, Wall, & Catalan, 1992; Eijkelkamp et al., 2012). In myocytes, persistent inward sodium current increases background Ca^{2+} levels and thus the tonic contraction level by stimulating the reverse transport via the sodium–calcium exchanger.

Still another aspect of Nav kinetics is the so-called "resurgent current" arising after ultra-fast inactivation or blockade (Cannon & Bean, 2010). Mechanistically, this phenomenon represents reactivation, a process opposite to inactivation, the rapid transition from the fast-inactivated to the open state. Normally, the pore part of the fast-inactivated channel stays in the conformation resembling an open state and undergoes deactivation-like transition after repolarization. However, rapid repolarization can destabilize the inactivation, causing the pore to shift to the open conformation. Since this occurs at repolarized potentials, most of the reopened channels close via deactivation, producing a kinetically distinct inward Na^+ current (Cannon & Bean, 2010). *In vivo*, resurgent currents are observed exclusively in neurons, and their appearance seems to be promoted by co-assembly with regulatory β subunits (Schiavon et al., 2006). Clinical importance of resurgent current arises from observations of resurgent current in some

Nav mutations causing severe neurological and cardiac channelopathies (Jarecki, Piekarz, Jackson, & Cummins, 2010).

3.3 Isoforms and Expression

In humans, nine genes encoding α subunits of Nav channels have been identified. All mammalian Nav channel isoforms are >50% identical in amino acid sequence (Catterall, Goldin, & Waxman, 2005). Nav1.1, Nav1.2, Nav1.3, and Nav1.7 are closely related. They have similar biophysical properties, are highly sensitive to TTX, and mainly expressed in central and peripheral neurons, specifically, in the initial segments of axons, in nodes of Ranvier and in dendrites. Nav1.4 is the skeletal muscle isoform. Nav1.5 is the cardiac sodium channel, although it is not found in the central part of sinoatrial node (SAN), the main heart pacemaker region (Park & Fishman, 2011); it is relatively resistant to TTX. Nav1.6 is expressed both in the central and peripheral nervous systems (CNS and PNS, respectively), primarily in the initial axon segments and nodes of Ranvie; it can underlie both fast transient and smaller persistent Na^+ currents. Nav1.8 and Nav1.9 are mainly expressed in the PNS, specifically in dorsal root ganglion neurons; they are involved in nociception, TTX-resistant, and upregulated by nerve growth factors. Nav1.8 is characterized by a relatively slow inactivation and rapid repriming. Nav1.9 has unique biophysical properties for a Nav channel: due to very slow kinetics it mediates a persistent current, which does not contribute to APs but regulates membrane excitability near the AP threshold, amplifying small depolarizations (Catterall et al., 2005; Jarecki et al., 2010). In addition, there is accumulating evidence that Nav channels are expressed and play important regulatory roles in many nonexcitable cells (for review, see Black & Waxman, 2013).

3.4 Regulation of Nav Channels

There are several intrinsic regulatory mechanisms of Nav channels, involving molecules directly interacting with Nav channels: β subunits, PKA, PKC, CaM, and CaMKII (CaM-dependent kinase II). Navβ subunits consist of a single transmembrane helix with a small C-terminal cytoplasmic tail and a large N-terminal immunoglobulin-like extracellular domain responsible for interaction with extracellular matrix and adhesion factors. Association with one or two β subunits changes Nav kinetics and can increase Nav trafficking to the membrane. Four Navβ genes have been identified. Studies in heterologous expression systems revealed dissimilar effects of different β subunits on Nav gating. Navβ1 subunit binds to Navα noncovalently and speeds up its inactivation. *Scn1b* gene shows alternative splicing, which still increases functional diversity of Navβ1. Navβ2 binds by disulfide bond and its effects on gating of different α subunits are disparate (Johnson & Bennett, 2006) but, unlike other β subunits, genetic defects in Navβ2 are not associated with severe pathologies. Nav co-assembled with Navβ3 show a hyperpolarizing shift in voltage-dependence of inactivation. In contrast, Navβ4 shifts the voltage-dependence of activation toward hyperpolarized voltages and does not affect inactivation (Namadurai et al., 2015).

Regulation of Nav channels by protein kinases is complex. The long neuronal isoforms have five consensus sites in the cytoplasmic domain I–II linker for phosphorylation by PKA, which reduces peak Na^+ currents without changes in kinetics. In contrast, the cardiac Nav1.5 channel has eight distinct consensus phosphorylation sites. In experiments on recombinant channels only two of these sites were phosphorylated by PKA, resulting in increased Nav current. Furthermore, the function of skeletal muscle Nav1.4 channels seems to be unaltered by PKA-mediated phosphorylation when studied in *Xenopus* oocytes (Smith & Goldin, 1992). PKC modulates Nav channels by phosphorylating a highly conserved serine in the III–IV intracellular linker, with diverse and isoform-specific functional outcomes (Marban et al., 1998).

Some regulatory mechanisms seem to be strictly isoform-specific. For example, Nav1.4 is selectively phosphorylated by myotonin protein kinase (Mounsey et al., 1995). Nav channels possess structural motifs for regulation by Ca^{2+}, CaM (a universal intracellular voltage sensor/effector protein), and CaMKII. Obviously, the existence of Ca^{2+}-dependent regulation is understandable given the overall molecular similarity between Nav and Cav channels (Yu et al., 2005) as well as close functional cooperation between Nav1.5 and Cav1.2 channels during cardiac AP. Indeed, these regulatory mechanisms are best studied in Nav1.5. The binding site for Ca^{2+} is the EF hand-like domain in the C-terminal located upstream from the CaM-binding IQ motif. The Nav channel IQ motif is highly similar to that in Cav channels (Chagot & Chazin, 2011). Calcium sensitivity mechanism also involves the intracellular III–IV linker, which is crucial for inactivation. It is thought that binding of Ca^{2+} to apo-CaM associated with C-terminal triggers conformational changes interfering with inactivation (Herren, Bers, & Grandi, 2013). CaMKII can directly phosphorylate Nav1.5; several candidate consensus sites are identified. Overall, the effects of Ca^{2+}-dependent regulation on Nav gating appear to be modest and expression system-specific

(Ben-Johny et al., 2014; Herren et al., 2013). In addition, Nav channels are regulated by methylation and S-nitrosylation (Gonzalez, Treuer, Sun, Stamler, & Hare, 2009). In the cardiac Nav1.5, methylation counters the effects of phosphorylation by PKA and vice versa (Beltran-Alvarez et al., 2015).

3.5 Nav Channels as Therapeutic Targets

Full suppression of a biomolecule is almost never a goal of therapeutic pharmacological intervention, for this would result in a chemical equivalent of genetic knockout, with potentially lethal consequences if the molecule in question is of such importance as Nav channels. Complete or near complete blockade of Nav channels—and, therefore, of neuronal and myocyte excitability—can swiftly result in death as evidenced by numerous cases of fatal TTX poisoning through consumption of poorly prepared Tetraodontiformes fish. Instead, pharmacological treatment normally aims at modulation of function, to suppress it partially in case of ion channel blockers. Nevertheless, Nav channels represent a remarkable exception from this general rule—and perhaps the only exception of its kind—as their complete inhibition in sensory nerves and nerve terminals in limited regions in the periphery is actively pursued in local anesthesia applications.

Nav channels are targeted by drugs from several large partially overlapping groups classified according to their effects—local anesthetics (LAs), antiarrhythmics class I, anticonvulsants, and analgesics. Mechanistically, Nav blockers bind to and block the same receptor site in the pore that is formed by amino acid residues in the S6 segments of I, II, and IV domains. They typically access the site from inside, which requires opening of the intracellular activation gate (Catterall, 2014). This dependence manifests as a frequency- or use-dependence, with greater channel inhibition observed at higher AP frequencies and depolarization voltages (Strichartz & Ritchie, 1987). This has immediate therapeutic implications as effects of highly frequency-dependent drugs, such as lidocaine, are stronger on hearts displaying abnormally high beating rates (tachycardia). However, frequency-dependence, while suggesting open-channel block, is also consistent with inactivated-channel block. The evidence suggests that LAs indeed have a higher affinity for inactivated channels than for open channels, and it is inactivated channels that specifically tend to accumulate during high-frequency heartbeat or neuronal firing. In addition, uncharged or hydrophobic Nav blockers can find their way to the binding site even when the channel is closed, resulting in the partial closed-channel block (Scholz, 2002; Strichartz & Ritchie, 1987).

In contrast, natural alkaloids such as TTX and saxitoxin bind to another site on the outer mouth of the pore accessible from outside (Chen & Chung, 2014). Several other toxins can bind and modulate Nav channels without pore blockade (six toxin-binding sites have been identified). This modulation results in either the inhibition or enhancement of Nav currents (Catterall et al., 2007).

3.5.1 Local Anesthetics

The first drug to be used as a local anesthetic (LA) was cocaine; however, its psychotropic and cardiovascular side-effects make it unsuitable for that role. The search for synthetic derivatives produced a number of aminoamide (e.g., lidocaine, bipuvacaine, ropivucaine) and aminoester (e.g., novocaine, tetracaine) LAs. Unlike cocaine, synthetic LAs have very low abuse potential and do not induce hypertension or arrhythmia at normal doses. Common applications of LAs include surgery (as a main or auxiliary method for anesthesia), management of acute (e.g., labor, traumas, postoperative) and chronic pain, as well as pain relief during instrumental diagnostic procedures (e.g., bronchoscopy). The number of application techniques available (surface, infiltration, peripheral nerve block, plexus block, epidural, spinal, intravenous regional, and body cavity anesthesia, etc.), together with the range of compounds with different properties in regard to potency and duration of block, allows virtually any practical configuration for the clinical use of LAs (Becker & Reed, 2012).

Aminoamide and aminoester compounds are weak bases usually formulated as hydrochloride salts. In tissues they dissociate according to their pKa, releasing free bases, which cross cellular membrane to reach the intracellular binding site. In the cytoplasm, after free bases are protonized again due to relatively low intracellular pH, they can bind and block Nav channels. This secondary ionization prevents escape of LA molecules from the cell interior. The increased extracellular acidity accompanying, e.g., inflammation in the vicinity of a wound can diminish membrane permeability and the efficacy of LAs. LAs, while capable of blocking conduction in all nerve fibers, in practice affect only sympathetic and sensory nerves mediating pain and touch. Clinically used concentrations of LAs are usually too low to block motor nerves.

Two more caveats need to be mentioned here. First, the clinically observed effects of LAs cannot be explained in their entirety by blockade of Nav channels. In particular, this follows from observations of systemic effects of TTX, a potent and highly selective blocker of most Nav isoforms.

Unlike lidocaine, TTX does not induce systemic sedation (Marcil et al., 2006). Lidocaine can block other essential ion channels (Meng, Xia, Liu, Bayliss, & Chen, 2011) and modulate the spinal glycinergic system (Muth-Selbach et al., 2009) at clinically relevant concentrations. Second, the antinociceptive effects of LAs used to treat neuropathic pain often manifest at concentrations that are too low for a real conductive sensory nerve block (Mao & Chen, 2000). However, although other putative mechanisms cannot be ruled out, this phenomenon can be explained by inhibition of the persistent Nav currents (Baker, 2000; Devor et al., 1992; Eijkelkamp et al., 2012).

3.5.2 Nav Blockers as Antiarrhythmic and Antianginal Drugs

Normal heartbeat is controlled by pacemaker myocytes in the sinoatrial node (SAN). These cells spontaneously generate APs, which propagate along the cardiac conduction system and in myocardium with complex spatio-temporal dynamics. SAN pacemaker activity is extensively regulated by neural and hormonal influences, which continuously adjust heart rate to physiological demands. Oscillatory self-excitation of central SAN myocytes is based on a special functional channelome, characterized by the absence of stable resting potential and fast depolarizing Nav currents, and by the presence of the slowly depolarizing (funny) Na^+ current (I_f) activated by hyperpolarization. Due to the absence of a rapid Nav current responsible for the phase 0 depolarization in the force-generating cardiomyocytes, SAN APs have a distinct slow onset. The upstroke of AP in SAN is provided by Ca^{2+} influx through L-type Cav channels, which open when membrane potential is gradually depolarized by I_f and T-type Cav channels. This situation is mirrored in the pacemaker myocytes of the atrioventricular node (AVN). Interestingly, the absence of phase 0 Nav current in central SAN myocytes does not mean the absence of Nav channels *per se*, since at the average resting potential of ca. −50 mV all Nav channels will be inactivated. Indeed, fast Nav current is found in the periphery of SAN, where myocytes have more negative resting potentials (Dobrev, 2009).

Dysregulation of cardiac function, manifesting in several types of abnormal heart rate and rhythm, is observed in a number of cardiovascular disorders and routinely treated with antiarrhythmic drugs. According to the Vaughan Williams classification (Vaughan Williams, 1975), antiarrhythmic drugs can be separated into five groups: Nav blockers (class I), β-blockers (class II), Kv channel blockers (class III), Cav blockers (class IV), and compounds with other mechanisms of action (class V). Depending on their effects on the effective refractory period (ERP) and AP duration (APD), class I compounds are clinically subdivided into three groups. Class IA drugs (quinidine, procainamide, disopyramide, etc.) increase ERP and APD, and moderately block phase 0 of AP. Class IB (lidocaine, tocainide, mexiletine) shorten ERP and APD and have the smallest effect on Nav current. Class IC drugs (flecainide, propafenone, moricizine) do not alter ERP but have the strongest blocking effect on Nav channels, typically causing a use-dependent (open-channel) block (Ramos & O'Leary, 2004), which makes them useful in treating atrial tachyarrhythmias and atrial fibrillation (Roden, 1998).

The main effect of class I drugs on cardiac function is blocking the phase 0 Nav current, with the ensuing slowing of propagation of depolarization in the *nonnodal tissue* of myocardium. This happens due to the capacitive nature of propagation of excitation in the interstitial fluid—the faster the depolarization, the further away in the neighboring cells it can trigger AP. Therefore, class I drugs have negative dromotropic effects, helping to suppress tachycardias caused by reentry mechanisms. It is important to recognize that pure Nav blockade can affect only phase 0 of AP and the small persistent Na^+ current (lidocaine and TTX can shorten APD in nonnodal cardiomyocytes; Wasserstrom & Salata, 1988). Thus the prolongation of ERP and APD, as well as other effects of class I drugs on cardiac excitability appear to stem from their action on *other targets*, such as Kv and other channels in case of quinidine, a prototypic class IA medicine and wide-spectrum ion channel blocker (for illustration of effects of quinidine on heartbeat in *Drosophila melanogaster* larvae, which lack Nav currents, see Frolov, Berim, & Singh, 2008). This relative nonselectivity can have both beneficial and deleterious effects. For instance, although in theory class I drugs should not alter SAN pacemaker functioning because SAN myocytes do not exhibit Nav currents, in practice some of them can inhibit the slowly depolarizing phase 4 current I_f and through this suppress abnormal pacemaker automaticity. On the other hand, increased APD, seen as prolonged QT interval in the ECG, may cause spatial dispersion of AP repolarization and precipitate a *torsades de pointes* (TdP) ventricular tachycardia, with dangerous consequences.

Ranolazine, a drug capable to block persistent inward sodium current I_{pNa}, has been recently approved for treatment of angina pectoris. I_{pNa} can reverse the sodium–calcium exchanger and increase calcium load, which contributes to impaired relaxation of the heart, causing deficient filling and increased myocardial stress during diastole and elevated end-diastolic arterial pressure. Block of I_{pNa} alleviates this and reduces oxygen requirements in ischemic hearts (Belardinelli, Shryock, & Fraser, 2006). However, ranolazine can also inhibit Kv channels and prolong the QT interval (Antzelevitch et al., 2004), which is again fraught with TdP and ventricular tachyarrhythmia. Another drug, riluzole, a Nav blocker used to treat epilepsy and amyotrophic lateral sclerosis, can similarly suppress I_{pNa} and protect against ischemia and reperfusion injury (Weiss, Benoist, White, Teng, & Saint, 2010).

3.5.3 Nav Blockers in Neurological Disordes

Nav channel blockers are prominently featured among antiepileptic drugs, because current theories of epileptic seizures are based on a concept of failed balance of excitation and inhibition in the neural networks (Treiman, 2001; Trevelyan, Muldoon, Merricks, Racca, & Staley, 2015). Partial blocking of neuronal Nav channels attenuates the release of excitatory neurotransmitters, particularly glutamate, helping to terminate or prevent epileptic seizures. Like most of antiarrhythmic compounds, anticonvulsant Nav blockers more often than not interact with other targets at therapeutic plasma concentrations, both in the brain and periphery, which makes it difficult to separate different modes of drug action. Nav blockers used for treatment of neurological disorders belong to different chemical groups, such as benzothiazoles (riluzole), carboxamides (carbamazepine), fatty acid derivatives (valproate), fructose derivatives (topiramate), hydantoins (phenytoin), and triazines (lamotrigine).

Phenytoin, carbamazepine, and lamotrigine block Nav channels in use-dependent manner, with slowing of recovery from inactivation, and thus reduce channel availability for consecutive APs. This ability to suppress abnormally high firing rates is utilized in treatment of generalized tonic–clonic seizures. A particular advantage of phenytoin is that its effects are strongly voltage-dependent, with greatest blockade observed at depolarized potentials. As a result, the drug suppresses neuronal hyperexcitation but shows little effect on normal neuronal firing and cognitive functions. In contrast to carbamazepine and phenytoin, lamotrigine is also used for treatment of absence seizures. However, since lamotrigine is a rather promiscuous channel blocker, it is not clear to what extent blocking of Nav channels contributes to this particular therapy. Similarly, topiramate and valproate, in addition to inhibition of Nav channels, act on many molecular targets, including other types of ion channels. They are widely prescribed to patients displaying partial and generalized tonic–clonic, absence, and myoclonic seizures. Both compounds can also selectively block persistent sodium currents, although the mechanism of topiramate action on Nav channels is not consistent with open channels block (Shank & Maryanoff, 2008). Riluzole, a blocker of presynaptic Nav channels and the persistent sodium current, is used as a primary neuroprotective agent in treatment of amyotrophic lateral sclerosis. Its neuroprotective effect derives from two mechanisms, a reduction of glutamate release due to presynaptic Nav inhibition, and from upregulation of glutamate uptake by astrocytes. Riluzole is also tested for treatment of angina pectoris and psychiatric diseases (Pittenger et al., 2008).

Nav blockers are used in therapy of other neurological conditions, such as migraine and neuropathic pain. Migraine is a widespread neurovascular disorder with significant genetic predisposition, manifesting in debilitating episodes of severe headache, probably triggered by a host of factors including but not limited to activation of nociceptors in meninges and blood vessels, central sensitization, abnormal modulation of central pain mechanisms, and neurogenic inflammation (Kalra & Elliott, 2007). Although administration of Nav blockers, such as valproate, topiramate, and lamotrigine, helps reducing the frequency of attacks, the use-dependent blockers phenytoin and carbamazepine turned out to be ineffective in clinical trials (Rogawski & Loscher, 2004), suggesting either an involvement of a Nav-unrelated mechanisms of therapeutic action for the three former drugs or the absence of pathogenic neuronal hyperexcitation amenable to Nav blockade during migraine attacks.

4. VOLTAGE-GATED CA^{2+} CHANNELS

Cav channels exist in all excitable and most nonexcitable cells and are the primary means of translation of electrical activity into nonelectrical output. They mediate excitation–contraction coupling in myocytes, excitation–transcription coupling in neurons and myocytes, and excitation–secretion coupling in neurons and secretory cells. The nervous systems would be essentially devoid of any output without Cav channels. The discovery of Cav channels can be traced to the seminal works by Paul Fatt, Bernard Katz, and Bernard Ginsborg on crustacean muscles (Fatt & Ginsborg, 1958; Fatt & Katz, 1953), in which AP does not depend on Nav channels as in the smooth muscles of vertebrates.

Intracellular concentration of free Ca^{2+}, compared to that of Na^+ and K^+, is very low—in the range of 20–300 nM in the resting cell. Considering that the typical extracellular concentration of Ca^{2+} is ca. 1–2 mM, the transmembrane driving force for Ca^{2+} is substantially larger than for any other major ion. However, since Cav currents are normally much smaller and load membrane capacitance much slower than Nav currents, APs mediated solely by Cav are usually less robust and have slower onset than those depending on Nav.

Opening of Cav channels can transiently increase intracellular Ca^{2+} concentration tremendously, by three to four orders of magnitude in the immediate vicinity of the channel. The resulting dramatic changes in free Ca^{2+} inside the cell show complex spatial and temporal patterns, depending on intracellular structures and cellular geometry. Ca^{2+} influx through Cav channels can alter many cellular functions, some of which show a steep dependence on Ca^{2+}. For example, neurotransmitter release in response to APs depends on the fourth power of extracellular Ca^{2+} which enters through Cav channels (Katz & Miledi, 1967). These wide-ranging regulatory influences confer a unique and extremely powerful role on these channels, and their disruption and dysregulation is linked to various cardiovascular and neurological disorders.

4.1 Structure and Function

The principal molecular determinant of Cav channels is the α1 subunit (Cavα1) (Fig. 4). Ten different Cavα1 identified in vertebrates share a common topology of twenty-four membrane-spanning helices (or "segments"), separated by intracellular linkers of various lengths into four domains six transmembrane segments each, strongly resembling the structure of Nav channels. P-loops contain highly conserved negatively charged amino acid residues, which contribute toward the permeation filter with high selectivity for calcium and similar ions (Cav channels are permeable to, e.g., strontium and barium ions but blocked by cadmium and some other metal ions). Alternative splicing reported for all Cavα1 genes further increases Cav channels' functional variability (Simms & Zamponi, 2014).

Both N- and C-terminals of Cavα1, as well as the intracellular linkers between domains contain numerous sites involved in regulation by auxiliary subunits and molecular regulatory factors. Functional Cav channels can be formed by either α1 subunit alone, or by co-assembly of α1 with several ancillary subunits: Cavβ, Cavα2δ, Cavγ, and CaM, with important implications for channel function. Two main physiological phenotypes, in regard to activation voltage threshold, are determined on the basis of such co-assembly: interaction of Cavα1 with Cavβ, Cavα2δ and CaM (and Cavγ

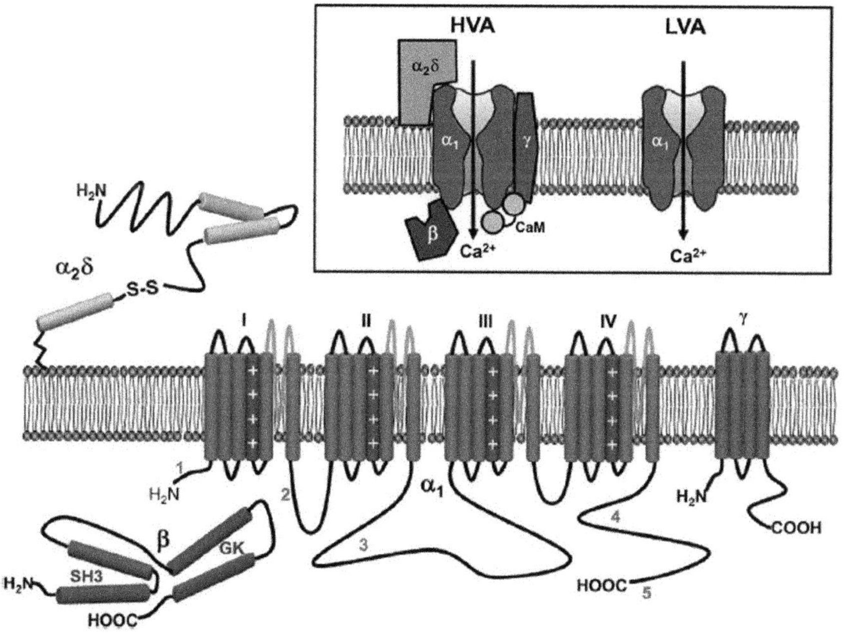

Figure 4 Cav channel with ancillary subunits. Transmembrane segments of the α subunit are organized in four domains (I–IV) like in the Nav α subunit (Fig. 3); re-entrant P-loops are shown in green (light gray in the print version). Red (dark gray in the print version) numbers indicate key protein interaction sites: (1) N-terminal CaM-association site in L-type channels; (2) Cavβ interaction site; (3) synprint site in Cav2 channels; (4) Cav1 and Cav2 channels' CaM-association sites PreIQ-IQ and IQ; (5) scaffolding protein interaction sites in Cav2 channels. Inset shows assemblies of the α with ancillary subunits in HVA and LVA Cav channels. *The image is adopted, with permission, from Simms and Zamponi (2014).*

in skeletal muscles) confers the so-called high voltage-activated (HVA) phenotype, while expression of Cavα1 alone bestows the low voltage-activated (LVA) phenotype. As the terms imply, HVA are activated by large depolarizations, while LVA-type channels are sensitive to relatively smaller depolarizing deviations from resting potential.

The intracellular linker between domains I and II houses a binding site for a Cavβ subunit—a soluble protein, which alters Cavα1 gating kinetics and increases channel trafficking to plasma membrane. Four β subunit genes are identified. Similarly, four genes are known to encode Cavα2δ subunits, which are bound to the external membrane layer. Increased expression of Cavα2δ augments Cav density, and, via an independent mechanism, increases the probability of synaptic vesicle release in neurons

(Hoppa, Lana, Margas, Dolphin, & Ryan, 2012). Cavγ subunits contain four transmembrane segments and are known to profoundly modulate the current density of skeletal muscle Cav. Out of seven proteins of this class, only γ1 and γ6 are recognized as authentic Cav subunits (Chen, Deng, Garcia, Sellers, & Best, 2007). Although CaM participates in many functions apart from regulation of Cav channels, its close physical and functional association with all HVA channels allows considering it as a genuine Cav subunit. This variability in ancillary subunits entails a much greater variability in biophysical properties of Cav channels than would be merited by variability in properties of the α1 subunits alone, contributing to fine-tuning of electrical output to the specific cellular function.

Inactivation of Cav channels is an essential mechanism preventing Ca^{2+} overload during prolonged depolarization. Two types of inactivation, VDI and calcium-dependent inactivation (CDI) are recognized, and, while the former constitutes an intrinsic channel property, CDI can be referred to as a regulatory mechanism. To various extents, both VDI and CDI can be observed in all Cav channels. In HVA channels, VDI is strongly modulated by Cavβ subunits. In contrast to rapid inactivation of Nav and some Kv channels, which involves actual channel block by an N-terminal "ball and chain" mechanism, Cav channel VDI is slow, and its main determinants appear to be the amino acids in S6 and in the domain I–II intracellular linker (Adams & Tanabe, 1997; Simms & Zamponi, 2014). It was proposed that conformational changes in S6, gradually developing during depolarization, expose a site to which the I–II linker can bind and render the channel pore impermeable (Simms & Zamponi, 2014; Stotz, Jarvis, & Zamponi, 2004). In contrast to HVA but like in Nav channels, VDI in Cav3 channels seems to be based on the domain III–IV linker region and C-terminus. Taking into account that Cav3 family is as distant molecularly from Nav channels as it is from HVA Ca^{2+} channels (Catterall, 2011), these results are not at all surprising.

4.2 Varieties and Expression

Based on molecular similarity, three major families of Cav channels are recognized. Distinct biophysical and protein interaction properties of each Cav isoform that arise from differences in amino acid sequence are closely matched to the specific functions they perform. Generally, Cav1 channels initiate contractions in different types of muscle cells, participate in modulation of gene expression in dendrites of neurons and mediate hormone secretion in endocrine cells. They are specifically used for synaptic transmission whenever persistent calcium influx is required, such as in the continuously active ribbon synapses (Baden, Euler, Weckström, & Lagnado, 2013). Cav2 channels are the main mediators of fast neurotransmitter release in the "regular" synapses within the nervous system, whereas Cav3 channels are involved in maintenance and modulation of rhythmical firing in pacemaker neurons and myocytes.

4.2.1 Cav1 Channels

Cav1 family consists of four HVA channels, also known as L-type channels (LTCC, "long-lasting"). Cav1 family is the best studied among Cav channels. LTCCs activate relatively slowly at a relatively high threshold, and also inactivate slowly. Vertebrate LTCCs can be blocked by 1,4-dihydropyridines (DHPs, e.g., nifedipine), phenylalkylamines (PAA, e.g., verapamil), and benzothiazepines (BT, e.g., diltiazem) and are insensitive, or only weakly sensitive, to the peptide toxin ω-conotoxin (Lewis, Dutertre, Vetter, & Christie, 2012; Triggle, 1990). Another hallmark of LTCCs is their sensitivity to augmentation by some DHPs, such as Bay K-8644 (Triggle, Langs, & Janis, 1989). L-type channels co-assemble with $\alpha2\delta$, $\beta3a$, and γ subunits.

Cav1.1 is predominantly found in the transverse tubules of skeletal muscles, where it is mechanically linked to the ryanodine receptor type 1 calcium release channel (RyR1) expressed on the cytoplasmic surface of the sarcoplasmic reticulum (SR). Conformational changes accompanying Cav1.1 opening directly activate RyR1, triggering Ca^{2+} release and initiating excitation–contraction coupling (ECC). Interestingly, in skeletal muscle calcium entry through Cav1.1 is not required for contraction, which effectively posits the channel as a voltage sensor serving the RyR1 channels. This peculiar organization seems to owe to the very slow channel pore opening after fast movement of voltage sensors. The time constant of Cav1.1 activation is too slow to respond adequately to milliseconds-long depolarizations initiating contraction; however, the voltage sensor movements have sufficient energy to activate the RyR1 channels.

Cav1.2 is the most abundant LTCC in the brain and also widespread in peripheral tissues (Hofmann, Flockerzi, Kahl, & Wegener, 2014). In hippocampal neurons, Cav1.2 (like other Cav1 channels) is often found in postsynaptic terminals. However, its role there is not electrogenic but regulatory, crucial for synaptic plasticity: opening of LTCC is essential for activation CaM/CaM kinase and Ras/MAPK pathways, and for consecutive phosphorylation and activation of CREB (cAMP response element-binding protein), a transcription factor promoting gene expression. This is essential for the development of LTP (Simms & Zamponi, 2014). This process strongly correlates with LTCC opening rather than with the accompanying general increase in intracellular calcium. Unusually, Cav1.2 gene produces not only the α1 subunit but also a polypeptide corresponding to the channel's C-terminus (calcium channel-associated transcript or CCAT), which independently acts as a transcription factor in brain neurons, modulating synthesis of many proteins and growth of neurites (Gomez-Ospina, Tsuruta, Barreto-Chang, Hu, & Dolmetsch, 2006). Cav1.2 expressed in the myocardium and smooth muscles of blood vessels plays a crucial role, mediating ECC and pacemaker function (Striessnig, Pinggera, Kaur, Bock, & Tuluc, 2014). However, in contrast to Cav1.1, Cav1.2 channels in cardiomyocytes are not tethered to the cardiac RyR type 2 Ca^{2+} release channels (RyR2), and an actual Ca^{2+} influx through Cav1.2 is needed to activate RyR2 and initiate a local calcium spark (Navedo & Santana, 2013). Still, Cav1.2 is directly coupled to vesicle release machinery in adrenal chromaffin cells, where it functions in concert with the third member of Cav1 family (Mahapatra et al., 2012). Cav1.1 and Cav1.2 channels in skeletal and cardiac muscles can be upregulated through the stress-response β-adrenergic receptor/cAMP/protein kinase A pathway, which can dramatically increase calcium influx and contractile force. Phosphorylation of Cav channels by PKA constitutes one of the mechanisms of voltage-dependent facilitation (Catterall, 2011). Cav1.2 is also regulated by cGMP-dependent protein kinase and protein kinase C associated with multiple G protein-coupled receptors (Kamp & Hell, 2000; Keef, Hume, & Zhong, 2001).

Cav1.3 is much less widespread in the nervous system than Cav1.2. These channels are found in dendrites of some neurons, and, crucially, in the presynaptic terminals of inner hair cells, where they mediate the exocytosis of glutamate and thus the signal transmission in ribbon synapses (Brandt, Striessnig, & Moser, 2003). Cav1.3 is particularly suited for this purpose as it is activated by relatively small depolarizations, at voltages as low as −70 mV—a rather unique property among the members of Cav1 and Cav2 families. In addition, the relatively weak CDI of Cav1.3 in these cells is instrumental for maintenance of the necessarily steady channel activity evoked by graded voltage responses. This particular property seems to be due to the local expression of CaM-like Ca^{2+}-binding proteins (CaBP4), which antagonize inactivation of Cav1.3 by CaM (Yang et al., 2006). In the periphery, Cav1.3 is involved in the SAN pacemaking

(Baig et al., 2011; Mesirca, Torrente, & Mangoni, 2015) and in catecholamine release from adrenal chromaffin cells (Simms & Zamponi, 2014). In murines, Cav1.3 carries almost all calcium current in the AVN (Marger et al., 2011). Also, Cav1.3 appears to serve as pacemaker in adrenal chromaffin cells and in some neurons (Comunanza et al., 2010; Olson et al., 2005).

Cav1.4 is expressed almost exclusively in retina and characterized by the smallest single-channel conductance among Cav channels (Doering, Hamid, Simms, McRory, & Zamponi, 2005). It is the main LTCC-mediating synaptic transmission in vertebrate rods. In the dark, photoreceptors in vertebrates are continuously depolarized and presynaptic Cav1.4 channels stay open, maintaining tonic glutamate release. Slow inactivation is an essential prerequisite for continuous Cav function, and Cav1.4 indeed inactivates very slowly (Doering et al., 2005). Like Cav1.3, this channel also conspicuously lacks CDI, although the mechanism is different: there is an autoinhibitory domain on the C-terminal, which competes with calcium-free CaM for CaM's binding site upstream. (However, CaBP4 that reduces inactivation of Cav1.3 is also found in presynaptic terminals of photoreceptors (Haeseleer et al., 2004).) This, together with relatively low voltage activation threshold, very small unitary conductance, and low open probability (the latter two features are useful for reducing channel noise in the visual system), makes Cav1.4 very well suited for maintaining the functioning of ribbon synapses. Cav1.4 is also found in the organs of the immune system, including thymus, spleen and T-cells. In T-cells, Cav1.4 channels are involved in homeostasis and signaling (McRory et al., 2004; Omilusik et al., 2011).

4.2.2 Cav2 Channels

Cav2 family has three members, which are the main presynaptic Cav channels initiating rapid synaptic transmission following depolarization of the presynaptic terminal. They are characterized by relatively low sensitivity to DHPs, PAAs and BTs. In vertebrates, Cav2 channels are mechanically connected to the vesicle release complex via the synaptic protein interaction (synprint) site located on the intracellular loop linking the second and third Cavα1 domains, through which the vesicle release complex proteins provide regulatory feedback, specifically by speeding up voltage-dependent inactivation (Simms & Zamponi, 2014). They are modulated by multiple G protein-coupled receptor pathways, which play a crucial role in controlling synaptic function both in the CNS and periphery (Gasparini, Kasyanov, Pietrobon,

Voronin, & Cherubini, 2001; Luebke, Dunlap, & Turner, 1993; Wu, Borst, & Sakmann, 1998). This modulation is also essential for voltage-dependent facilitation of Cav2 channels.

Cav2.1 gene produces high voltage-activated P- (first identified in the cerebellar Purkinje neurons) or Q-type channels, depending on splicing and co-assembly with ancillary subunits (Randall & Tsien, 1995). Pharmacologically, P/Q-type channels are defined by the sensitivity to ω-agatoxin IVA. Like LTCCs, Cav2.1 can promote CREB-mediated changes in gene transcription (Sutton, McRory, Guthrie, Murphy, & Snutch, 1999). However, in contrast to LTCCs, this requires a global elevation of Ca^{2+} and occurs in a much slower manner (Wheeler et al., 2012). P/Q-type channels are co-expressed with α2δ and β4a subunits.

Cav2.2 underlies the N-type (norepinephrine-related) currents sensitive to ω-conotoxins GVIA and MVIIA (Simms & Zamponi, 2014). It is involved in NFAT-mediated gene transcription (Hernandez-Ochoa, Contreras, Cseresnyes, & Schneider, 2007). Cav2.3 forms the so-called R-type channels ("R" stands for "resistant," that is the remaining current after blocking all other HVA channels), which can be selectively blocked by SNX-482 toxin. In the periphery, Cav2.3 is found in nociceptors and involved in pain control at the spinal and supraspinal levels (Fang et al., 2007; Saegusa et al., 2000; Sekiguchi, 1990). Due to structural differences, Cav2.3 is less sensitive to direct modulation by G proteins than Cav2.1 or Cav2.2 (Shekter, Taussig, Gillard, & Miller, 1997; Simen & Miller, 2000; Simms & Zamponi, 2014; Stephens, Canti, Page, & Dolphin, 1998). N- and R-type channels co-assemble with α2δ and β1b subunits.

4.2.3 Cav3 Channels

Three members of Cav3 family represent the LVA channels and are also known as T-type (tiny, transient) channels as they are activated by relatively small depolarizations and characterized by fast inactivation and similarly fast recovery from inactivation. They do not form stable complexes with ancillary subunits, possibly because Cav3 are only distantly related to other Cav families (Catterall, 2011). Pharmacologically, T-type channels are relatively resistant to DHPs and low micromolar concentrations of cadmium but sensitive to nickel ions (Simms & Zamponi, 2014). They are expressed ubiquitously and participate in many physiological processes including neuronal firing, synaptic transmission, hormone secretion, muscle contraction, myoblast fusion, and fertilization (Perez-Reyes, 2003). The C-terminal of Cav3 channels contains scaffolding protein interaction sites used to bind syntaxin and organize the channels in the active zones of synapses

(Weiss & Zamponi, 2012). This interaction may be crucial for the Cav3.2-mediated low-threshold exocytosis in chromaffin cells (Carabelli, Marcantoni, Comunanza, & Carbone, 2007; Weiss & Zamponi, 2012). There is increasing evidence on involvement of Cav3 channels in synaptic release (Egger, Svoboda, & Mainen, 2003; Jacus, Uebele, Renger, & Todorovic, 2012; Pan, Hu, Perring, & Andrade, 2001; Rozanski, Nath, Adams, & Stanley, 2013; Tang et al., 2011) and in neuronal pacemaker function (Perez-Reyes, 2003). Cav3 channels are also expressed in the cardiac conduction system, especially in the SAN. Because of their high voltage sensitivity and fast kinetics, T-type channels expressed in dendrites and soma of many types of neurons can strongly modulate the firing rate, a phenomenon known as "rebound bursting": T-type channels can rapidly recover from inactivation following a brief hyperpolarization, and then, upon next near-threshold depolarization, they produce a much larger current than before, depolarizing the cell and facilitating opening of Nav channels, which can then trigger a burst of APs instead of just one (Alvina, Ellis-Davies, & Khodakhah, 2009). Note, that this sort of electrogenic interaction between Cav and Nav does not contradict the flow-of-excitation hypothesis since T-type channels do not directly evoke Nav currents but rather condition it. Cav3 channels are regulated by G protein-coupled receptors and other factors (Catterall, 2011; Perez-Reyes, 2003; Sekiguchi & Kawabata, 2013).

4.3 Regulation

Arguably, the choice of the Cav isoform to be expressed is the most important "decision" made by the cell as specific determinants of molecular interactions between the channel and its regulatory factors are hardwired into its primary structure (Simms & Zamponi, 2014). Moreover, all pore-forming and ancillary subunits undergo extensive posttranscriptional alternative splicing, which confers additional functional specificity. This manifests in altered biophysical and pharmacological properties (Huang, Yu, & Soong, 2013; Liao et al., 2004). In this section, we outline the hitherto unmentioned direct regulatory mechanisms modulating Cav function.

The most powerful mechanisms of regulation of calcium influx through Cav channels are the CDI, calcium-dependent facilitation (CDF) and voltage-dependent facilitation (VDF). Due to differences in channel structures and molecular milieus *in situ*, these mechanisms manifest differently for specific Cav channels and in some cases are not noticeably present at all. This is particularly true for VDF, which seems to depend exclusively on upregulation of Cav2 channels via G protein-coupled receptor/PKA/PKC/PKG-mediated signaling pathways (Meza & Adams, 1998).

CDI is a powerful mechanism of Cav influx control. It can be visualized by comparing recordings performed using Ba^{2+} as a charge carrier (Ba^{2+} does not cause CDI) with recordings in the presence of Ca^{2+}. Although details of CDI mechanics differ for different Cav channels, it critically depends on activation of the Cav-associated CaM by the entering Ca^{2+} (Zuhlke & Reuter, 1998). CDI is greatly reduced upon introduction of a mutant CaM unable to bind Ca^{2+} or CaM inhibitors (Lee et al., 1999). Ca^{2+}-free CaM is thought to be normally bound as a subunit to the Cav channel C-terminal and, upon binding of Ca^{2+}, triggers conformational changes resulting in channel inactivation (Peterson, DeMaria, Adelman, & Yue, 1999; Qin, Olcese, Bransby, Lin, & Birnbaumer, 1999; Zuhlke, Pitt, Deisseroth, Tsien, & Reuter, 1999). Important differences exist between Cav1 and Cav2 channels in regard to CDI: while a local increase in Ca^{2+} concentration near the inner mouth of the pore is sufficient to elicit CDI in LTCCs, a much greater, global Ca^{2+} increase is needed to inactivate Cav2 channels (Catterall, 2011; Simms & Zamponi, 2014).

The relatively low Ca^{2+} sensitivity of CDI in Cav2 channels masks a highly Ca^{2+}-sensitive process with an opposite, facilitating effect on Cav2 conductance—calcium-dependent facilitation (CDF). Although both CDI and CDF are mediated by the same Ca^{2+} sensor CaM, they are spaced apart concentration-wise by virtue of having an additional binding site for CaM on the C-terminal of Cav2.1 immediately adjacent to the IQ motif involved in inactivation. This binding site is merely a modified IQ motif, with modification conferring a greatly diminished affinity to Ca^{2+} (Catterall, 2011; Simms & Zamponi, 2014). Interestingly, two lobes of CaM are differentially involved in CDI and CDF as mutational analysis revealed that the high-affinity C-terminal mainly promotes facilitation through interaction with IQ motif, while the low-affinity N-terminal mediates inactivation (DeMaria, Soong, Alseikhan, Alvania, & Yue, 2001; Erickson, Alseikhan, Peterson, & Yue, 2001).

In addition to the above-mentioned immediate regulatory mechanisms, Cav channels are regulated by a number of conditions and factors (reviewed in McDonald, Pelzer, Trautwein, & Pelzer, 1994). These include *S*-nitrosylation (Gonzalez et al., 2009), chronic treatment with Ca^{2+} channel antagonists or activators (Gengo et al., 1988), exposure to therapeutic agents and toxins, such as estrogen (Batra, 1987), reserpine (Powers & Colucci, 1985), lead (Rius, Govoni, & Trabucchi, 1986), etc. Regulation of Cav channels in turn affects a wide variety of cellular functions including electrical excitability, neuromuscular development, gene expression, etc.

4.4 Calcium Channels in Pharmacological Therapy

The discovery and subsequent introduction of LTCC blockers (LTCCB, for review, see Ertel & Godfraind, 2012) into medicine in late 1960s has revolutionized the treatment of cardiovascular disorders, such as hypertension, angina pectoris, and arrhythmias. Currently, a large number of different DHPs and several non-DHP Cav blockers are widely administered worldwide, and most of them target peripheral Cav1 channels. Discoveries of agents selectively modulating neuronal Cav channels expanded the application of Cav blockers (CCBs) to the treatment of chronic pain, epilepsy, and bipolar disorder. This pharmacological field continues to develop energetically, with new compounds highly selective toward specific types of Cav channels being discovered and new methods of delivery showing great promise for future therapeutics appearing (Kang et al., 2012).

An important caveat to understand in regard to the therapeutic use of CCBs is that in the majority of cases Cav channels are not targeted as causative factors of diseases nor considered even to be major participants in the pathogenesis, for which there is little evidence; rather, Cav channels are taken as well known and convenient targets for manipulation, as key elements of neural and myocyte excitability, in order to offset the functional influence of other, often obscure, pathogenic mechanisms. Unfortunately, our understanding of many of those pathogenic processes is still very limited, which leads commonly to a symptomatic treatment of common cardiovascular disorders. With humans involved, the opportunities to scrutinize the actual functional status of ion channels in smooth muscles of blood vessels and in myocardium are usually very limited.

4.4.1 L-Type Channel Blockers

Drugs entering the bloodstream through the conventional gastrointestinal and intravascular routes tend to redistribute and accumulate in tissues according to their specific binding affinities. However, the blood–brain barrier of healthy individuals usually restricts the availability of drugs to neural tissue in the CNS. With few exceptions, the action of LTCCBs is limited to the periphery. Furthermore, although LTCCBs can block all Cav1 channels, albeit with different affinities, the functional effects they induce can be attributed to the action on Cav1.2 and to the lesser extent on Cav1.3. Cav1.1 channels are exclusively expressed in skeletal muscles, and Ca^{2+}

influx through them is not necessary for contraction. Thus, open-channel blockade of Cav1.1 by LTCCBs is irrelevant for modulation of ECC. (It should be noted that LTCCBs specifically bind open and inactivated channels, directly blocking the pore. If LTCCBs would inhibit Cav1.1 channels via, e.g., a closed-channel block or by altering the voltage-dependence of activation, their effects on skeletal muscle contraction might become noticeable.) Cav1.4 channel, on the other hand, is almost exclusively expressed in retina, without significant presence in the periphery. Consistent with the effects of LTCCBs on Cav1.1, the drugs do not substantially affect catecholamine release from chromaffin cells in adrenal glands (O'Farrell, Ziogas, & Marley, 1997), where Cav1.2 channels mediate secretion by means of direct mechanical coupling with vesicle release machinery.

Frequency- or use-dependence is a crucial aspect of LTCCB action; similarly to Nav blockers, the LTCCB binding sites on the channel become accessible only after the channel opens, and remain accessible for the duration of open and inactivated states (Catterall & Swanson, 2015). This explains why channel block develops progressively with repetitive depolarizations and gets relieved during rest. Functionally, this can partly account for the differences in LTCCBs action on the depolarized, continuously contracting smooth muscles in arteries versus the relatively weakly contracting muscles in veins.

Six distinct mechanisms of LTCCB action in the periphery have been proposed. By blocking Cav1.2 channels, LTCCBs (1) decrease contractions in vascular smooth muscle of arteries, causing vasodilatation and reducing blood pressure (BP), and (2) reduce the force of contraction in the myocardium (negative inotropy). By blocking both Cav1.2 and Cav1.3 channels, (3) LTCCBs decrease the heart rate (negative chronotropy), (4) decrease the speed of conduction of excitation within the heart, particularly affecting the AVN (negative dromotropy), and (5) suppress the firing rates of aberrant and ectopic pacemaker sites in the heart. Finally, by blocking unspecified LTCCs in the adrenal cortex cells (Uebele, Nuss, Renger, & Connolly, 2004), they (6) inhibit secretion of aldosterone into the bloodstream, contributing to reduced retention of ions and water. Mechanisms 1 and 6 decrease BP and through this alleviate the cardiac afterload, which then decreases the contractile efforts required and lessens the myocardium oxygen demands. Mechanism 1 can directly improve oxygenation in the heart by dilating coronary arteries in case of vasospastic angina. Mechanisms 2 and 3 contribute to the same effect more directly, helping to ameliorate angina pectoris. Mechanisms 4 and 5 mediate the antiarrhythmic reaction of LTCCBs.

Three classes of LTCCBs have substantially different effects on LTCCs expressed in the arterial smooth muscle cells and myocardium. Due to reasons outlined above, DHPs (isradipine, nifedipine, etc.) have the strongest effect on the constantly contracting and depolarized arterial LTCCs and less pronounced effects on myocardial contractions, which are characterized by regular and relatively long periods when membrane potential stays at the very negative resting level (ca. −85 mV). Thus, DHPs are mainly used as antihypertensive agents, although long-acting DHPs are the drugs of choice in treatment of vasospastic angina. However, their powerful effects on systemic blood pressure can trigger baroreceptor-mediated reflectory tachycardia and increased inotropy through increased sympathetic stimulation thus offsetting their anti-anginal effects. Among the LTCCBs, verapamil and diltiazem are the only non-DHP compounds currently used in medicine. Verapamil preferentially blocks cardiac LTCCs with relatively small effect on the systemic BP. It is, therefore, used in therapy of angina and arrhythmias. Effects of diltiazem, on the other hand, are intermediate between DHPs and verapamil: it can reduce arterial pressure without pronounced reflectory cardiac stimulation.

4.4.2 Blockers and Modulators of Cav2 Channels

Several compounds selectively modulating Cav2 channels are currently used in medicine. However, none of them are as convenient as the classical LTCCBs. Ziconitide, a synthetic polypeptide analog of a ω-conotoxin, is used for intractable pain management in patients with, e.g., cancer. It is the only highly selective Cav2.2 channel blocker available (Lewis et al., 2012). However, the drug must be delivered directly into the spinal fluid, which drastically limits its practical applicability.

Gabapentinoid compounds, such as gabapentin and pregabalin, are widely used for treatment of chronic pain and epilepsy. Although they are structural analogs of the inhibitory neurotransmitter γ-aminobutyric acid (GABA), neither gabapentin nor pregabalin can bind GABA or benzodiazepine receptors directly. Instead, they modulate GABA synthesis and metabolism (Taylor, 1997). They also appear to downregulate the presynaptic Cav2.2 channel density in the membrane by binding to Cavα2δ subunits and interfering with reinsertion into membrane of Cav2.2 channels undergoing endocytosis. However, to what extent the Cav2.2 modulating mechanism contributes to the antiepileptic action of the drugs, is not known.

4.4.3 Nonspecific Cav Channel Inhibitors

A number of relatively nonselective drugs are also administered. This partly makes up for the absence of highly selective blockers for the most isoforms of Cav2 and Cav3 families. Examples include DHPs benidipine and efonidipine (Cav1/Cav3 channel blocker; Masumiya, Shijuku, Tanaka, & Shigenobu, 1998; Suh, Fujimoto, & Tseng, 1989), cilnidipine (Cav1/Cav2 channel blocker, suppresses sympathetic hyperactivity; Aritomi et al., 2015; Shiga et al., 2007). Other interesting drugs, structurally unrelated to the above-mentioned groups, include the blood–brain crossing drugs fluspirilene (used as an antipsychotic medicine; Spedding, Kenny, & Chatelain, 1995) and flunarizine (used in the treatment of migraine and epilepsy; Amery, 1983; Waszkielewicz et al., 2013).

5. CHANNELOPATHIES OF TRP, NAV, AND CAV CHANNELS

Pathologies caused by a structurally defective ion channel or by an abnormal channel function are called channelopathies. Although the distinction between the two aspects in this definition may appear trivial, it is in fact very substantial, since defects in the channel can originate only from mutations in the pore-forming or ancillary subunits, whereas abnormal channel function can be caused by a much greater host of causes encompassing all possible regulatory mechanisms pertaining to the channel. Because channelopathies are rare or extremely rare disorders, sometimes described just in few individuals, it is fair to state that societal impact of many channelopathies is substantially lower than their scientific importance. Observing, studying, and treating patients with channelopathies greatly advances our understanding of ion channel function *in vivo*. One may argue that pure pathological phenotypes presenting in channelopathies provide more essential information on the role of ion channels in physiology and pathophysiology than that obtained in meticulous studies of altered function of cellular channelomes and its consequences in diseases caused by unrelated factors—simply by the virtue of the absence of the additional causative factor and its other effects on the disease phenotype. Nevertheless, the cumulative societal effect of inherited or *de novo* channelopathies is quite large, as, for example, ca. 35% cases of sudden unexplained death and ca. 20% of sudden infant death syndrome cases can be attributed to channelopathies (Ackerman et al., 2011; Hund & Mohler, 2014; Klaver, Versluijs, & Wilders, 2011).

Numerous animal (especially mouse) models with mutations in virtually all voltage-gated channels have been developed and provide valuable information, including recapitulation of complex human disorder phenotypes. However, known differences between human and animal physiology make assessment of human relevance of such findings rather difficult; after all, even with the mice we have at least ~90 million years of divergent evolutionary history (Ferguson-Smith & Trifonov, 2007). Therefore, while describing channelopathies, we specifically focus on findings in humans. Furthermore, we do not attempt to provide a balanced and detailed account for each and every TRP, Nav, and Cav channelopathy, for neither it is possible within the scope of this chapter, nor it is necessary considering the number of reviews available on the topic. Instead, in this section we outline the major disorders, for which the sheer number of meticulously documented cases allows factoring out the phenotypic variability due to individual genetic, epigenetic and environmental moderation of clinical representation from the main mechanisms of the disorder (see examples in Kyndt et al., 2001). Thus, here we do not address channelopathies described just for one or few individual cases, unless such are the only disorders linked to a particular channel.

5.1 TRP Channelopathies

Several of the numerous TRP channels have been linked to the equally numerous disturbances of bodily function. This comes as no surprise, because TRP channels are present almost everywhere in the body. Detailed information of their involvement is available in physiology and pathophysiology of pain in various conditions. In contrast, their role in, e.g., cardiac hypertrophy and conduction disorders, development of diabetes mellitus, or oncogenesis is scarce and the results of investigations are not overtly informative (Smani, Shapovalov, Skryma, Prevarskaya, & Rosado, 2015). There are, however, clear cases, where the channels are somehow defective, the TRP channelopathies, either with gain-of-function or loss-of-function effects in specific organ systems. Also, the large diversity of the activation and modulation properties poses a big problem for understanding even the rather specific pathophysiology of channelopathies.

Of the several mutations in TPR channels a few may serve as examples of the diversity—and also of kind of unpredictability—of the described and characterized channelopathies. Of the TRPC subfamily the first channelopathy found was a mutation in the TRPC6 isoform (Reiser et al., 2005; Winn et al., 2005). It leads to malformation of the podocytes that form part of the permeability barrier in renal glomeruli and to a kidney malfunction called focal segmental glomerulosclerosis, with proteinuria and progressive kidney failure. The pathophysiological mechanism whereby defective TRPC6 channels cause this is not known in detail. The channel is present in the podocytes that form a part of the glomerular filter, and the mutant channel may have wrong association with other proteins involved or it may induce increased Ca^{2+} permeability in the podocytes and thereby increased apoptosis (Reiser et al., 2005; Walz, 2005).

The first TRPV class channelopathy is associated with point mutations in TRPV4 isoform, which causes brachyolmia, a debilitating stature and joints disease (Rock et al., 2008). This is not a surprise, since TRV4 channel is expressed strongly in osteocytes and chondrocytes. The mutant channels show increased constitutive activity, but it is not known how the resulting increased cationic conductance and changes in the membrane potential affect the development of bones and cartilage.

The first channelopathy in the TRPM subfamily, which was originally thought to be associated with melanin production and possibly melanoma progress, was found to cause congenital stationary night blindness (Audo et al., 2009). The decreased sensitivity for low intensity light is caused by expression of mutated channels in retinal bipolar neurons, which then have only small or nonexistent responses to small rod-based inputs (Peachey et al., 2012; Shen et al., 2009).

An amino acid substitution mutation in TRPA1, causing a shift in the gating properties of the channel and a dramatically increased inward cationic currents, is associated with familial episodic pain syndrome 1 (FEPS1), where various physiological stresses trigger intense upper body pain (Kremeyer et al., 2010).

Within the TRPML subfamily, 21 different mutations have been found in TRPML1, which all lead to disturbances in lysosomal trafficking, due to increased activation of the channels and loss of luminal Ca^{2+}. This is associated with a condition known as mucolipidosis type IV, characterized by widespread neuromuscular defects and retarded psychomotor development (Wakabayashi, Gustafson, Sidransky, & Goldin, 2011).

The best known channelopathy within the TRPP is represented by mutations in TRPP2, with more than 400 different types described in patients suffering from autosomal-dominant polycystic kidney disease (Chapin & Caplan, 2010). The pathophysiology of the cyst development is not known, but seems to involve disruption of several intracellular signaling pathways, possibly linked to calcium fluxes (Takiar & Caplan, 2011).

5.2 Channelopathies of Voltage-Gated Na^+ Channels

Due to the role Nav channels play in neuronal and cardiac excitability, Nav channelopathies are the best studied disorders of this disease class. They are caused by loss- or gain-of-function mutations in the α or β subunits and characterized by distinct sets of symptoms depending on the isoform involved, the pattern of expression of the defective tissue-specific splice variant, and the character of functional changes the mutation confers. A number of seizure, pain, neurodegenerative, muscle, cardiovascular, and immune disorders have been thoroughly documented and grouped into four clinical categories: brain, muscle, cardiac, and peripheral nerve sodium channelopathies.

The number of known distinct channelopathies varies for different Nav isoforms. For instance, while numerous entities have been described for Nav1.4 and Nav1.5 mutants, no clear link has been found between Nav1.3 and any genetic disorder. This can imply either the crucial importance of the Nav1.3 for the organism, or the total opposite, its relative insignificance. Indeed, studies of knockout mice have shown that although Nav1.3 channel is involved in nociception, its role seems to be minor (Nassar et al., 2006).

5.2.1 Nav1.1

Mutations in *SCN1A* are linked to several seizure disorders, such as severe myoclonic epilepsy of infancy (SMEI, also known as Dravet syndrome) and generalized epilepsy with febrile seizures plus type 2 (GEFS+ type 2). In fact, *SCN1A* is the most clinically relevant for seizure disorders, as it is the locus of the majority of identified epilepsy-associated mutations. The importance of Nav1.1 channel stems from the ubiquity of its expression in the CNS, particularly from its expression in inhibitory GABAergic neurons (Yu et al., 2006), dysregulation of which may disrupt the fine balance between excitation and inhibition in the brain. In addition, Nav1.1 mutations cause familial hemiplegic migraine type 3 (FHM3) and appear to be implicated in autism spectrum disorders (O'Roak et al., 2012).

Most of >700 identified mutations in *SCN1A* gene are nonsensical and result in a nonfunctional Nav1.1 channel protein. The accompanying clinical phenotype of the autosomal-dominant Dravet syndrome includes prolonged febrile and nonfebrile seizures during the first year of life, which can later develop into myoclonic or partial seizures, with concomitant psychomotor delay, ataxia, cognitive impairment, behavioral and motor deficits. Behavioral deficits may involve hyperactivity, impulsiveness, and signs of autism. The disorder is generally severe and can result in premature death. A milder clinical phenotype of GEFS+ type 2, characterized by various seizures usually ending by the age of six, is associated with >20 various nonfunctional, gain- or loss-of-function mutations in SCN1A. Clinical pictures and Nav1.1 functional defects found in SMEI and GEFS+ type 2 overlap substantially (for reviews, see Eijkelkamp et al., 2012; George, 2005; Ragsdale, 2008). The treatment of these disorders largely centers on the empirical selection of effective Nav blockers, which may show disparate effects depending on functional effects of actual mutations (Eijkelkamp et al., 2012).

Clinical manifestations of familial hemiplegic migraine type 3 include episodic onsets of migraine with unilateral weakness of the body, which can be accompanied by ataxia, coma, and even paralysis. Several mutations in *SCN1A* have been linked to the rare subtype 3 of FHM, observed in <5% of all FHM cases (Barros et al., 2014; Weller et al., 2014). One of the studied mutations, in the intracellular loop between domains III and IV, dramatically increases the rate of recovery from inactivation, which can lead to neuronal hyperexcitability (Dichgans et al., 2005).

5.2.2 Nav1.2

Mutations in *SCN2A* gene are found in patients suffering from for the most part benign seizure disorders, such as benign familial neonatal–infantile seizures (BFNIS) (Heron et al., 2002) and GEFS+ type 4. The effects of mutations on Nav1.2 permeation and kinetics are disparate. BFNIS is a mild condition, usually resolving by the age of 1 year. Spontaneous complete recovery from inherited disorders caused by mutations in *SCN2A* can be explained by developmental changes in the pattern of Nav1.2 expression: the channel is initially clustered in the immature nodes of Ranvier but later gets replaced there by Nav1.6 (Boiko et al., 2001). In adults, Nav1.2 is widely expressed in the CNS, specifically in the unmyelinated dendrites and somas of cortical and hippocampal neurons.

There is also some evidence regarding the role of Nav1.2 mutations in autism, which has heritability of ca. 90% and in one-third of cases is associated with seizures (Weiss et al., 2003).

5.2.3 Nav1.4

Mutations in the skeletal muscle Nav1.4 channels can cause several disorders, such as hyperkalemic and hypokalemic periodic paralyses, potassium-aggravated myotonia (PAM), and paramyotonia congenita (Jurkat-Rott, Holzherr, Fauler, & Lehmann-Horn, 2010). The character of each disorder is determined by the effect of each specific mutation on muscle excitability. For example, hypokalemic periodic paralysis type 2 (HOKPP2) constitutes ca. 20% of cases of HOKPP and is caused by mutations in the voltage sensor region, which partially neutralize the associated positive charge and result in aberrant gating pore currents and hypoexcitability of muscle fibers due to intracellular accumulation of Na^+ with consequent myocyte depolarization (Jurkat-Rott et al., 2010; Sokolov, Scheuer, & Catterall, 2010). Clinically, HOKPP2 manifests in adolescence and characterized by hypotonia, general or local muscle weakness, and by episodes of weakness or paralysis triggered by low serum K^+ (Venance et al., 2006). Because of the complexity of the relationship between the defective Nav1.4 channel and the disrupted myocyte function, the pathogenic mechanism of the disorder is not fully understood.

On the other hand, PAM and paramyotonia congenita are caused by mutations, which are effectively gain-of-function due to facilitated activation, suppressed inactivation, or amplified resurgent current. This confers muscle hyperexcitability, although for PAM and paramyotonia congenita, the hyperexcitability is characterized by distinct sets of clinical symptoms: while the former shows a work-out related and often painful muscle stiffness, the latter manifests as a cold-induced muscle stiffness (Jarecki et al., 2010; Jurkat-Rott et al., 2010).

Still a different picture is observed in patients with hyperkalemic periodic paralysis (HyperPP). The causative mutations alter Nav1.4 gating in such a manner that the channel does not inactivate completely and produces a persistent current. This current slightly depolarizes the myocyte, forcing a fraction of Kv channels to the open state. The ensuing steady K^+ efflux can increase the systemic extracellular K^+ concentration to dangerous levels fraught with cardiac complications.

5.2.4 Nav1.5

Despite of the crucial role of Nav1.5 in cardiac AP propagation, not all mutations are lethal despite, causing drastic changes in biophysical properties of the channel. A number of diseases with distinct clinical representations have been described, including but not limited to congenital long QT syndrome type 3 (LQTS3), idiopathic ventricular fibrillation type 1 (Brugada syndrome), isolated cardiac conduction system disease (ICCSD), and several other very rare conditions, which we would not address here, such as atrial standstill, congenital sick sinus syndrome, sudden infant death syndrome, and other forms of arrhythmias.

LQTS3 (named after a prolonged QT interval in the ECG) is observed in ca. 10% of patients with congenital LQTS. These are characterized by a rather unusual association of bradycardia with high frequency of occurrence of cardiac events during rest or sleep. Clinically, a cardiac episode in LQTS3 manifests in the ventricular TdP arrhythmia. This may lead to palpitations, syncope and, if ventricular fibrillation follows, to sudden death. The consensus understanding of LQTS3 pathogenesis is that the associated Nav1.5 mutations result in incomplete inactivation or accelerated recovery from inactivation, which manifests either in appearance of small persistent Na^+ current or in repetitive Nav activation during the same AP. Although the resulting excessive inward current is quite small in comparison with the peak Nav current, it is sufficient to offset the fine balance between depolarizing Nav/Cav and repolarizing Kv conductances. This leads to an increase in duration of cardiac AP in nonnodal cardiomyocytes and produces increased transmural dispersion of both repolarization and the refractory periods, resulting from delayed repolarization that amplifies the intrinsic electrical heterogeneity in the myocardium. This all predisposes the myocardium to early afterdepolarizations and re-entrant TdP events (Antzelevitch, Yan, & Shimizu, 1999).

In contrast to LQTS3, Brugada syndrome type 1 (BA1, ca. 20% of all BA cases; Kapplinger et al., 2010) is caused by loss-of-function Nav1.5 mutations through various mechanisms, from expression of nonfunctional channels or channels with defective gating to reduced density of functional channels due to decreased availability or disrupted trafficking. Clinically, BS1 is characterized by an abnormal ECG pattern and an increased risk of sudden death from idiopathic ventricular fibrillation. In latent cases, diagnostic administration of Nav blockers can help to reveal the pathology. It is generally accepted that, like in the case of LQTS3, cardiac pathology in BS1 results from exacerbation of transmural electric heterogeneities, albeit via a dissimilar mechanism. Different layers of myocardium generate APs of variable durations, with epicardial APs substantially shorter than endocardial APs due to greater expression of transient Kv current in the epicardium. Reduction of Nav current in BS1 patients further magnifies the duration gradient, increasing the probability of early afterdepolarizations.

Interestingly, many mutations with disparate effects on the Nav1.5 current, but not as prominent as to manifest in the form of LQTS3 or BS1, cause more benign ICCSD spectrum disorders (see examples in the review by George, 2005).

5.2.5 Nav1.6

Mutations in *SCN8A* gene are linked to infantile epileptic encephalopathy type 13, a severe disorder characterized by intractable seizures of various types, progressive encephalopathy, ataxia, developmental intellectual and psychomotor retardation or even regress. Although the effects of mutations on channel function in cellular essays are diverse, both loss- and gain-of-function, they cause neuronal hyperexcitation (Blanchard et al., 2015; de Kovel et al., 2014; Ohba et al., 2014; Veeramah et al., 2012).

5.2.6 Nav1.7

Dominant gain-of-function mutations in Nav1.7 channel cause primary erythromelalgia (PE) characterized by transient painful episodes, with episodic block of part of the circulation in the extremities, bilateral burning pain, hyperthermia, and reactive hyperemia of the hands, lower legs, and feet. Another rare disorder, paroxysmal extreme pain disorder (PEPD) or familial rectal pain syndrome, is characterized by the same set of symptoms, albeit involving not extremities but the mandibular, ocular, and rectal areas. The functioning of the CNS is not affected, which is consistent with the exclusive expression of Nav1.7 in the peripheral sensory systems. It appears that mutations causing PE and PEPD affect Nav1.7 gating differently: while in PE gain-of-function is caused by a depolarizing shift in voltage-dependence of activation and, often, slowing of deactivation, PEPD-associated mutations interfere with inactivation, resulting in persistent sodium current and, therefore, prolonged AP bursts.

In addition, gain-of-function mutations in Nav1.7 may account for as much as 30% of cases of painful peripheral small-fiber neuropathy, a widespread idiopathic neurological disorder involving pathology of nociceptive fibers of small diameter. All such mutant channel variants conferred hyperexcitability on dorsal root ganglion neurons (Faber et al., 2012).

Recessive loss-of-function mutations in Nav1.7 have been linked to congenital insensitivity to pain, a very rare disorder, with patients unable to feel pain. No other neurological or cognitive deficits are usually reported (Cox et al., 2006).

5.2.7 Nav1.8

Nav1.8 is also involved in sensation of pain; it is specifically responsible for nociception at low temperatures. The channel has unique biophysical properties: its inactivation is entirely temperature-insensitive, and low temperatures reduce Nav1.8 activation threshold, while at the same time increasing membrane resistance, resulting in relatively high membrane gain (Zimmermann et al., 2007). Consistently, studies on Nav1.8 pharmacological knockout mice confirmed complete loss of response to noxious cold and mechanical stimuli (Abrahamsen et al., 2008). Gain-of-function mutations in *SCN10A* gene were recently found to contribute to painful peripheral neuropathy (Faber et al., 2012).

Evidence accumulates on involvement of Nav1.8 in regulation of cardiac function. A recent study found mutations in *SCN10A* gene in 16% of patients with Brugada syndrome, implying a major role of this channel, second only to SCN5A. These patients displayed a substantially more severe clinical phenotype than BS patients without mutations in *SCN10A*. However, the mechanism through which the mutant channels exert they action is very unusual, involving direct physical association between Nav1.8 and the cardiac Nav1.5: when mutant Nav1.8 channels have been co-expressed with normal Nav1.5 channels, the Na^+ current has become almost twice as small than when Nav1.5 channels have been expressed alone. Moreover, coexpression of normal Nav1.8 and Nav1.5 channels resulted in nearly two-fold increase of the Nav1.5 (Hu et al., 2014). These surprising results underscore the possibility of complex interactions between different isoforms of the same channel class, which may open a new chapter in the study of channelopathies.

Like some Nav1.5 mutations, which do not result in clear LQTS or BS phenotypes and instead classified as cardiac conduction disease variants, several mutations in SCN10A gene have been linked to ICCSD (Chambers et al., 2010; Hu et al., 2014).

5.2.8 Nav1.9

Gain-of-function mutations in *SCN11A* have been found in a small fraction of patients with painful peripheral small-fiber neuropathy. Functional analysis of mutant Nav1.9 channels expressed in DRG neurons revealed a substantial −6.7 mV hyperpolarizing shift in half-activation potential and slowing of deactivation. This resulted in depolarization of resting potential and DRG hyperexcitability (Huang et al., 2014). Surprisingly, gain-of-function mutation phenotype was linked to congenital loss of pain perception. This excessive activity depolarizes nociceptors, impairing generation of APs (Leipold et al., 2013).

5.2.9 Nav β Subunit Channelopathies

Brugada syndrome type 5 is caused by mutations in *SCN1B* gene encoding β1 subunit of Nav channels. This subunit normally strongly upregulates Nav current. However, mutations can abolish this capacity (Watanabe et al., 2008). Similar aberrations in other β subunit genes are also linked to BS (Medeiros-Domingo et al., 2007). Mutations in *SCN1B* are also implicated in generalized epilepsy with febrile seizures plus, type 1 (Wallace et al., 1998). Mutations in all four all four Navβ subunit genes are linked to familial atrial fibrillation (Li et al., 2013).

5.3 Channelopathies of Voltage-Gated Ca^{2+} Channels

5.3.1 Cav1.1

Mutations in Cav1.1 channel gene *CACNA1S* have been linked to hypokalemic periodic paralysis type 1 (HOKPP1) and malignant hyperthermia susceptibility (MHS) syndromes.

Mutations in Cavα1 are implicated in ca. 75% of HOKPP1 cases. The mutations are typically in amino acid residues related to S4 helices in the domains II, III, or IV. The pathology is still poorly understood and muscle weakness is linked to lowered serum K^+ (Striessnig, Bolz, & Koschak, 2010); this is especially problematic, considering that Cav1.1 channels activate very slowly and Ca^{2+} influx through them is not needed for muscle contraction. Still, there is evidence that mutant Cav1.1 channels can possess a secondary pore mediating a small depolarizing cation current, and that in some mutants the voltage-dependence of Cav1.1 activation is altered dramatically (Jurkat-Rott et al., 2010; Venance et al., 2006).

Malignant hyperthermia susceptibility is a severe, potentially lethal muscle disorder, with attacks triggered by exposure to volatile anesthetics or depolarizing muscle relaxants, which in turn elicits an abnormally high influx of Ca^{2+} from SR. The uncontrollable contractures that follow lead to muscle hypermetabolism, ATP depletion, and excessive glycogenolysis. At the system level, hypercapnia, hypoxemia, lactic acidosis, and hyperthermia can be observed in severe cases. The accompanying rhabdomyolysis can cause renal failure and arrhythmia. The primary locus of MHS is the defective RyR1, but two causative mutations in Cav1.1 were also identified, one in the domain III–IV cytoplasmic linker and another in the domain I voltage sensor (Carpenter et al., 2009; Monnier, Procaccio, Stieglitz, & Lunardi, 1997; Striessnig et al., 2010). These mutations appear to alter voltage-dependence of Cav1.1 gating (Striessnig et al., 2010).

5.3.2 Cav1.2

Timothy syndrome (TS) and Brugada syndrome type 3 (BS3) are the two main clinical phenotypes associated with, correspondingly, gain- and loss-of-function mutations in the *CACNA1C* gene. In other studies of unrelated mutations in human Cav1.2 several *CACNA1C* SNPs were linked to bipolar disorder, recurrent depression, and schizophrenia (Simms & Zamponi, 2014).

Timothy syndrome is a severe autosomal-dominant disorder caused by gain-of-function mutations in *CACNA1C* gene. Clinically, it is characterized by physical abnormalities and neurological defects such as syndactyly, Long QT syndrome type 8 (LQTS8), morphological heart defects, and autism. Sufferers often die in childhood. Two forms, "classical" (or TS) and "atypical" (or TS2), manifest depending on the localization of mutation in one or another variant of the alternatively spliced *CACNA1C* exon 8/8a, which codes S6 of the domain I and the domain I–II intracellular linker loop. The alternatively spliced variants are expressed with certain tissue-specificity. This modulates the disease phenotype, since only one channel variant is compromised in each patient. Exon 8/8a is the major alternative splicing locus determining the tissue-specific penetration: exon 8a is predominantly expressed in smooth muscles, while exon 8 determines the main cardiac Cav1.2 variant (77%) (Splawski et al., 2004). All TS-associated mutations abolish VDI and CDI nearly completely, causing Ca^{2+} overload and prolonged depolarization, and this underlies the LQTS. Because of the tissue-specificity of defective channel variant expression, LQTS is less pronounced in TS than in TS2. Consistently with the prominent role of Cav1.2 in brain, neural development is abnormal in TS patients (Krey et al., 2013; Pietrobon, 2010). TS patients also occasionally report recurrent infections and bronchitis, which suggests that Cav1.2 participates in the functioning of the immune system.

Loss-of-function mutations in *CACNA1C* gene account for ca. 10% of all cases of Brugada syndrome (Hedley et al., 2009). Reduced L-type current in phase 1 of cardiac AP results in decreased depolarization and may manifest in short ST segment on ECG. Cardiac episodes can lead to ventricular fibrillation and sudden death (Hedley et al., 2009).

5.3.3 Cav1.3

Mutations in *CACNA1D* were linked to two novel pathological phenotypes in humans: SAN dysfunction and deafness syndrome (SANDD), and primary aldosteronism, seizures, and neurologic abnormalities (PASNA) syndrome. Also, increased Cav1.3 expression was found in brains of patients with early-stage Parkinson's disease (Hurley, Brandon, Gentleman, & Dexter, 2013).

The SANDD syndrome has been described for two consanguineous families suffering from deafness. The mutation is a single glycine insertion in an alternatively spliced locus near the channel pore. It renders Cav1.3 channels nonfunctional. Patients display profound deafness without vestibular dysfunction but with syncopes due to sporadic pronounced bradycardia (Baig et al., 2011).

Involvement of Cav1.3 channels in the PASNA syndrome, so far described for two children only, was determined following genetic analysis of patients with adrenal aldosterone-producing adenomas. Electrophysiological analysis of mutant channels revealed gain of function due to strong hyperpolarizing shift in voltage-dependence of activation. PASNA patients showed strong hypertension, renal, cardiac, and neurological dysfunction, including seizures and severe mental retardation (Scholl et al., 2013).

5.3.4 Cav1.4

The very discovery of the predominantly retinal Cav1.4 channel owes to the genetic analysis of incomplete congenital stationary night blindness type 2 (CSNB2), which is caused by mutations in the Cav1.4-coding gene *CACNA1F*. The disorder is not life-threatening. Patients present with a moderately low visual acuity, myopia, nystagmus, and night blindness of varying severity. Since the gene is situated on X-chromosome, in practice only males suffer. Around 40 different mutations in *CACNA1F* with disparate effects on conductance were linked to the disorder (Burtscher et al., 2014). These observations emphasize the importance of proper matching of the presynaptic channel voltage activation range to the dynamic range of graded voltage responses it has to transmit.

5.3.5 Cav2.1

Cav2.1 is the major channel responsible for fast synaptic transmission in the nervous system, and thus mutations in the *CACNA1A* gene are linked to several pathologies—unlike mutations in two genes encoding other members of Cav2 family, Cav2.2 and Cav2.3, for which no clinical disorders have been identified so far, presumably because of the mildness of associated phenotypes. Here, we address three rare neurological autosomal-dominant disorders: familial hemiplegic migraine type 1 (FHM1), episodic ataxia type 2 (EA2), and spinocerebellar ataxia type 6 (SCA6) (for detailed review, see Pietrobon, 2010).

Mutations in Cav2.1 account for approximately half of cases of familial hemiplegic migraine. The clinical symptoms of FHM1 can overlap with EA2, SCA6, benign familial infantile epilepsy, and alternating hemiplegia of childhood. One of the proposed mechanisms of migraine pathogenesis is the spreading cortical depression, a slowly propagating wave of strong neuronal and glial depolarization, which can be triggered in animal models by localized cortical stimulation (Pietrobon, 2010). Most but not all mutations causing FHM1 are of gain-of-function type; their effects on channel permeation and kinetics are disparate, ranging from altered channel kinetics to reduced inhibition of Cav2.1 by G proteins.

Episodic ataxia is characterized by sporadic stress-provoked episodes of cerebellar dysfunction, clinically manifesting in paroxysms, ataxia, dysarthria, and nystagmus. The attacks are usually much longer in EA2 patients than in EA1 patients with defects in Kv1.1 channels. The disease usually begins in childhood and often progresses to include cerebellar syndrome with partial cerebellar atrophy and severe ataxia. Cognitive development impairment and epileptic seizures are also observed. More than 50 loss-of-function mutations in *CACNA1A* gene have been found in connection with EA2, including insertions, duplications, deletions, truncations, frame shifts, etc. In some cases, expression of the truncated and misfolded mutant protein triggers degradation of normal Cav2.1 channels produced by the intact gene copy. This helps to explain the dominant-negative character of the disorder (Mezghrani et al., 2008).

Spinocerebellar ataxia type 6 is a late-onset ataxia syndrome additionally characterized by positional vertigo and downbeating nystagmus. Morphologically, a marked degeneration in the superior vermis region of cerebellum is observed with a severe loss of Purkinje cells. The disease progresses slowly. It is unique among other Cav channelopathies in the sense that it is caused by usually a very limited (21–30 repeats) prolongation of the polyglutamine sequence at the C-terminal in some of the splicing variants, with neuronal defects developing in line with pathological mechanisms found in other polyglutamine disorders, which are as a rule characterized by much longer polyglutamine repeats. Studies of recombinant channels with varied numbers of glutamine repeats in heterologous expression systems have yielded rather inconclusive results and the mechanisms through which the mutant channels confer neurodegeneration *in situ* are not yet understood (Giunti, Mantuano, Frontali, & Veneziano, 2015).

5.3.6 Cav3 Channels

Currently, no clearly defined channelopathies caused by T-type channels have been described. However, there are links between Cav3 channels and neurological diseases in humans, although the direct link between the channel and the pathological process is lacking. Polymorphisms in Cav3.1 were found in families with juvenile myoclonic epilepsy (Singh et al., 2007). A large number of mutations in Cav3.2 have been found in patients with idiopathic generalized epilepsies. Electrophysiological analysis of 11 variants expressed in HEK-293 cells revealed that nine variants slightly upregulate the channels, thus representing gain-of-function mutations. The authors suggested that these variants can possibly predispose an individual to epilepsy but are not causative factors *per se* (Heron et al., 2007). However, it has been shown recently that one of the mutant Cav3.2 variants, when expressed in neurons can increase neuronal firing and thus seizure susceptibility both directly by reducing the rebound bursting threshold and indirectly through altered gene expression (Eckle et al., 2014). Finally, all three Cav3 channels may be involved in some forms of autistic phenotypes (Simms & Zamponi, 2014).

5.3.7 Cav Ancillary Subunit Channelopathies

Mutations in *CACNB2* gene encoding Cavβ2 subunit cause Brugada syndrome type 4 (Hedley et al., 2009; Kanter et al., 2012). Cavβ2 is predominantly expressed in the heart and associates with Cav1.2 channel, serving as its chaperone. Experiments in CHO cells involving coexpression of the recombinant mutant Cavβ2 from patients with BS with normal α subunit resulted in a markedly reduced L-type current. The defective current, however, was not due to compromised trafficking, probably resulting from interference of the mutation with the stimulating influence of Cavβ2 on Cav1.2 (Antzelevitch et al., 2007). Other mutations in *CACNB2* have been linked to cardiac conduction disease (Hu et al., 2010), hypertension, and bipolar disorder (Lee et al., 2011; Soldatov, 2015).

Mutations in *CACNB4*, a gene encoding Cavβ4 subunit, are linked to juvenile myoclonic epilepsy (JME) and episodic ataxia type 5 (EA5). The mutant subunits increase Cav2.1 conductance by ca. 30–40%, leading to neuronal hyperexcitability. Phenotypes of both disorders are partly overlapping: JME patients present with episodes of muscle twitches and seizures, and EA5 patients show recurrent episodes of ataxia with vertigo, with nystagmus and dysarthria observed between attacks (Escayg et al., 2000).

A homozygous cytosine to adenine transversion at nucleotide 2406 in *CACNA2D4* gene encoding Cavα2δ4 subunit, which introduces a premature stop-codon resulting in truncation of an approximately one-third of the protein, has been linked to slowly progressing retinal cone dystrophy (Wycisk et al., 2006), indicating that mutant α2δ subunit can compromise the normal function of Cav1.4 in photoreceptors.

5.4 Acquired Channelopathies

It has long been established that ion channel function can be altered in many nongenetic, acquired disorders as a result of action of a whole host of factors, such as toxins, drugs, hormones, immune system, various regulatory mechanisms, etc. Unlike primary genetic ion channel defects, these secondary channel-related pathologies affect a substantial fraction of population and are of great epidemiological importance. For instance, changes in density or function of Cav channels have been associated with arrhythmia (Carnes et al., 2007), Lambert–Eaton myasthenic syndrome (Kleopa, 2011), Alzheimer's disease (Tan, Deng, & Qing, 2012), and Parkinson's disease (Dragicevic, Schiemann, & Liss, 2015). Specifically, increased S-nitrosylation, a part of normal NO-based signaling (Gonzalez et al., 2009), contributes to the reduced L-type current in patients with atrial fibrillation (Carnes et al., 2007). Similarly, Nav function is altered in such acquired pathologies as angina, heart failure, and arrhythmias. For instance, dysregulation of cardiac Nav1.5 channels by CaMKII in oxidative stress caused by ischemia can lead to a decreased I_{Na} and increased persistent Na^+ current, with consequent cardiac conduction defects and arrhythmias replicating appropriate clinical phenotypes of Nav1.5 channelopathies, from LQTS to Brugada syndrome (Herren et al., 2013).

However, only a few disorders with altered ion channel function at the center of the pathology can be considered as "acquired channelopathies."

Candidate disorders include drug-induced channelopathies and autoimmune ion channel disorders. The former group can be further separated into disorders caused by unintended targeting of ion channels by prescribed medications (e.g., drug-induced LQTS; Kannankeril, Roden, & Darbar, 2010), overdose of channel blockers or channel-regulating agents (i.e., beta-blockers) (Shepherd, 2006), or by toxin poisoning (Zimmer, 2010). Of these, acute channelopathies due to drug overdose have the greatest epidemiological importance. In practice, none of these drug-induced channelopathies are classified as such; consecutively, their treatment strategies focus on the removal of the offending agent and life support.

Over the last decade, a number of disorders caused by autoimmune attacks on neuronal ion channels in the CNS and PNS have been described (Vincent, 2009), possibly explaining some postinfectious syndromes affecting skin and muscle, joints, and the nervous and cardiovascular systems. These conditions are known as paraneoplastic neurological disorders, as they are predominantly accompany neoplastic diseases, and less often occur after infections or allergic episodes, which apparently implicates misbalance of the immune system. The immune system can attack ion channels expressed in the healthy cells via different mechanisms, including complement-mediated damage and direct channel block, or facilitate abnormal channel degradation (Vincent, 2009). Autoimmune channelopathies has been described for several voltage- and ligand-gated channels, but, to the best of our knowledge, only two of them, Lambert–Eaton myasthenic syndrome (LEMS) and cerebellar ataxia with antibodies to Cav channels, fall within the scope of this chapter. So far, no distinct acquired autoimmune pathology has been linked to Nav channels.

In most cases, LEMS accompanies small cell lung cancer and is coupled to other paraneoplastic neurological syndromes. Importantly, paraneoplastic LEMS usually manifests well before the actual tumor is uncovered, which has obvious implications for survival. LEMS is characterized by progressive proximal weakness, especially in the legs, fatigability, and muscle stiffness. LEMS is a neuromuscular disorder caused by a decreased presynaptic Cav current, which results in reduced synaptic release of acetylcholine, smaller compound AP amplitudes and weak contractions. Repetitive or high-frequency nerve stimulation results in buildup of calcium in the presynaptic terminal, increased acetylcholine release and restored muscle force, which explains alleviation of muscle weakness following exercise. Reduced Cav channel density and altered localization patterns are caused by IgG antibodies to Cav2 channels. Patients usually respond well to immunomodulating therapy (Kleopa, 2011).

Cerebellar ataxia is a very common paraneoplastic syndrome. Clinically, neurological defects include ataxia, diplopia, dysarthria, and dysphagia (Mason et al., 1997). Morphologically, cerebellar inflammation is observed, followed by extensive loss of Purkinje cells. Antibodies to Cav channels are detected in less than a half of patients. These antibodies bind to extracellular portions of Cav channels and block their function, which is thought to contribute significantly to the clinical picture (Kleopa, 2011).

5.5 Considerations on the Treatment of Channelopathies

In this section, we briefly address the general therapeutic approaches to channelopathies. Not all channelopathies require therapy; for the detailed information on the treatment of specific types of channelopathies the reader is advised to consult the many available reviews.

Like the treatment of any other disease where causative pathogenic factor is known, the treatment of channelopathies includes etiological interventions, which aim at the causative mechanism, symptomatic therapy to terminate/prevent life-threatening episodes and improve the quality of life, and at life-style and dietary adjustments. The goal of etiological treatment is to reverse the physiological effects of the mutation, ideally at the same level where they occur. In case of gain-of-function mutations, this involves administration of channel blockers or modulators, if possible, with selective action toward the compromised channel isoform and with specific tissue affinity. Such therapy is complicated by the need to find a fine balance between inhibiting excessive functions and inacceptable deterioration of the physiological functions. For instance, the goal of administration of Nav blockers in the treatment of potassium-aggravated myotonia is to reduce the frequency and duration of spontaneous AP bursts in muscles without interfering with voluntary movements. Use-dependent Nav blockers such as mexiletine and flecainide, which have little effect on normal firing but effectively block Nav channels accumulating in the inactivated state, are well suited for this purpose (Jurkat-Rott et al., 2010). This is an example of etiological therapy targeting the very source of functional disruption.

The situation is less straightforward when loss-of-function mutations or haploid insufficiency (a loss of one of the two alleles of the gene) is involved. Although in theory the drug therapy could be aimed at optimization of the residual channel function, in practice pharmacological interventions of this precision are still way beyond the state-of-the-art. In such situations, a more viable approach can be to suppress the opposing influences. For instance, quinidine, a promiscuous channel blocker, appears to be quite useful in treatment of the Brugada syndrome, effectively preventing TdP and ventricular fibrillation, presumably via a blockade of I_{to}, which directly opposes I_{Na} (Belhassen, Glick, & Viskin, 2004). In other cases, however, the mechanisms underlying beneficial drug effects remain obscure. One example is carbonic anhydrase inhibitors used in therapy of hypokalemic periodic paralysis. The drugs reduce the frequency of attacks through an unknown mechanism independent of carbonic anhydrase inhibition (Venance et al., 2006). Management of hypokalemic periodic paralysis also involves measures to supplement K^+ and increase its retention in the plasma, which helps to prevent attacks by reducing the paradoxical membrane depolarization. This constitutes an example of etiological therapy not targeting the source of pathology but interfering with the essential pathogenic mechanism (Jurkat-Rott et al., 2010).

Channelopathies of cardiac Nav and Cav channels associated with potentially fatal arrhythmias are often treated by implantation of a cardiac pacemaker or cardioverter defibrillator, which can terminate the onset of ventricular arrhythmias (for review, see Stephenson & Berul, 2007). Unfortunately, similarly effective instrumental etiologic therapy methods are not yet available for severe neurological channelopathies.

A very effective long-term, cross-generational medical genetics approach to treat familial channelopathies involves genetic counseling, screening, preimplantation, and early prenatal diagnostics, with termination of pregnancies recommended in cases when fetuses carry a defective gene with strong predisposition to a severe disease. This essentially eugenic approach is effective regardless the mode of inheritance, that is for both dominant and recessive types of disease transmission. Near eradication of the devastating neurodegenerative Tay–Sachs disease, an autosomal recessive disorder caused by a deficiency of β-hexosaminidase and mainly endemic among Jewish Ashkenazi populations, represents a compelling example of effective genetic screening as primary prevention (Lew, Burnett, Proos, & Delatycki, 2015). Unfortunately, in the absence of mass prenatal screening drawing upon the comprehensive libraries of mutant alleles, similar prevention of *de novo* channelopathies and other genetic diseases is not possible.

6. HARNESSING THE FLOW OF EXCITATION IN NEURAL CIRCUITS

The flow-of-excitation concept postulates that there is an orderly distribution of excitatory channel types in the neural circuits according to their function: the input is provided by ligand-gated cationic conductances with relatively low ionic selectivity expressed in receptors, the propagation of excitation depends on neuronal Nav channels, and the output is mediated by Cav channels, which initiate excitation–contraction or excitation–secretion coupling. This pattern—in the form of both temporal succession of excitation and spatial organization of the expressed conductances—is also evident at the level of individual neurons, which are the elementary building blocks of neural circuits. However, while stimulation of sensory receptors at the input of neural circuits results in depolarization (with the exception of vertebrate photoreceptors), the situation *within* neural circuits is different. There are two types of electrical inputs into neurons—excitatory depolarizing ones mediated by cationic conductances essentially similar to TRP channels and inhibitory ones promoting hyperpolarization and predominantly based on ligand-gated chloride channels, such as ionotropic GABA receptors. If the flow-of-excitation concept is expanded to include all varieties of neuronal input conductances, then the following question can be raised: what is the optimal ion channel target for pharmacological therapy aimed at different parts of the neural circuits? In other words, what channels should be targeted if the neural input, AP propagation, or neural output, is to be suppressed?

The pharmacological therapy approach focuses almost exclusively on boosting the physiological inhibition or suppressing excitation in neural circuits. As a result, medical practice deals with inhibitors of excitatory ion channels or stimulators of inhibitory channels, and almost never with compounds stimulating excitatory conductances. Thus, this is not a trivial question, since, as discussed above, each component of the neural circuit (receptor, neuron, and myocyte) often expresses the full suite of different channels mediating the flow of excitation, and each consecutive stage is dependent on the preceding one. Thus, at least in theory, synaptic release in the axon can be controlled by altering the neuronal input in the dendritic tree, and strength of muscle contractions can be regulated by changing the firing rate of the innervating motor neuron. One way to scrutinize this question is to inspect the appropriate pharmacological therapies.

For attenuation of a nervous system input, e.g., a sensory function, it is necessary to target the appropriate receptor channels or Nav channels involved in signal propagation from the receptor cell or nerve input terminal to the CNS. In therapeutic practice, this almost exclusively concerns suppression of nociception, neuropathic pain and itching (Raap, Stander, &

Metz, 2011). In such cases, contemporary treatment of choice involves administration of nonspecific local anesthetics and Nav blockers. However, recent advances in research of TRP channels mediating nociception and development of isoform-selective Nav1.7–9 blockers promise near-future breakthroughs in the therapy of neuropathic disorders.

Next, to alter AP generation and propagation within the neural networks, both neuronal input conductances and Nav channels can be targeted. Indeed, in the treatment of epilepsy spectrum disorders, among the first-line agents (valproate, lamotrigine, topiramate, carbamazepine, oxcarbazepine, phenytoin, and ethosuximide), all except ethosuximide can block Nav channels, while two drugs, valproate and topiramate, can augment GABAergic activity (Goldenberg, 2010). Valproate can also block excitatory postsynaptic NMDA receptors and reduce the release of the neurotransmitter gamma-hydroxybutyrate (Maitre, 1997). In contrast, ethosuximide suppresses activity of T-type Ca^{2+} channels involved in firing rate regulation. Of the second line drugs with known mechanisms of action, most compounds alter the function of GABAergic neurons and also block Nav channels (Goldenberg, 2010). Similarly, pharmacological regulation (i.e., invariably slowing) of AP propagation within myocardium crucially depends on block of Nav channels (class I antiarrhythmics) everywhere except in the nodal tissue lacking functional Nav conductance. There, Cav blockers (class III antiarrhythmics) are used to decrease the speed of propagation of excitation and suppress the firing rates of aberrant and ectopic pacemaker sites.

Lastly, to attenuate the output function *per se*, e.g., the strength of contraction in myocytes or the amount of secreted hormone (Uebele et al., 2004), block of Cav channels appears to be the optimal therapeutic strategy. Examination of compounds used for treatment of angina pectoris and hypertension reveals that there are many Cav blockers but only one Nav blocker—ranolazine.

These practical examples show that in order to alter propagation of excitation, its immediate substrate, Nav channels, should be acted upon, whereas to diminish Ca^{2+}-dependent output it is best to suppress Cav function. This supports the opinion that for therapeutic interference to be most effective and specific, it has to target the direct structural determinant of the function in question while avoiding intermediates, whose function tends to be too complicated in the actual physiological framework, involving feedback and feed-forward regulatory processes that are difficult to account for by outside intervention.

7. CONCLUSION AND PERSPECTIVES

Intensive research in the fields of ion channel biophysics, mechanisms of channel-drug interactions, and pharmaceutical development of isoform-specific agents promises future breakthroughs in the treatment of ion channel-related disorders. Numerous clinical trials involving selective channel modulators are either under way or already completed, and many more compounds are being discovered and tested in animal studies (for review, see Bagal et al., 2014; Eijkelkamp et al., 2012; Moran et al., 2011), although not without drawbacks: for instance, search for isoform-specific TRP blockers did not yet produce compounds suitable for general medical practice (Moran et al., 2011), obviously in part caused by the multifunction nature of even single TRP isoforms. The basic and clinical research is facilitated by the advent of high-throughput automated patch-clamp, which allows screening of drug-channel interactions at an unprecedented rate, by several orders of magnitude surpassing that of the traditional electrophysiological analysis (Picones, 2015).

It is worth mentioning that apart from the conventional pharmaceutical research, advances are made in the development of peptides and antibodies capable to block ion channels. Toxins and antibodies are usually highly potent and much more specific than traditional drugs, due to their larger interface for interaction with their targets. Unfortunately, challenges accompanying research of toxins derived from animal venom are such that, despite a great number of natural and synthetic toxins and their derivatives assessed in laboratories prior to clinical trials, there are as yet very few examples of successful introduction of biological toxins into medicine. These include captopril, a derivative of *Bothrops jararaca* snake venom, which inhibits angiotensin converting enzyme; the already mentioned Cav2.2 blocker ziconotide, the synthetic version of the peptide MVIIA from venom of cone snail *Conus magus*; hirudin, the anticoagulant peptide from the medicinal leech *Hirudo medicinalis*; and microbial botulinum toxins (Harvey, 2014). On the other hand, polyclonal antibodies against extracellular epitopes of ion channels can be very effective in suppression of channel function. Indeed, such antibodies against TRPV1 (Klionsky et al., 2006), TRPM3 (Naylor, Milligan, Zeng, Jones, & Beech, 2008), Nav1.5 (Chioni et al., 2005), and Cav2.1 (Liao et al., 2008) have been created and successfully tested in laboratory. Nevertheless, no anti-ion channel antibody has been approved so far for use in humans.

In conclusion, therapeutic strategies targeting ion channels are likely to develop with an increased pace in the future. To make this happen, it needs an equally increasing solid base of fundamental research into the molecular, cell, and whole-body physiology. This is, first, because even with "absolute specificity," when able to change the function of one molecular isoform only, the same isoform is likely to be found participating in many functions in the same cells or in other tissues. Second, the physiological processes are so complex, with many bodily functions and signaling pathways intertwined in so many ways, that understanding them fully becomes of primary importance, when trying to cure any particular defective process causing the disease. This also means that pharmacologists, electrophysiologists, molecular biologists, and others working on basic physiological processes necessarily have to work together, because the intractable problems remaining in ion channel pharmacotherapy cannot be solved within any single discipline alone.

ACKNOWLEDGMENTS

The authors thank Irina Ignatova for Fig. 1B drawings.

REFERENCES

Abrahamsen, B., Zhao, J., Asante, C. O., Cendan, C. M., Marsh, S., Martinez-Barbera, J. P., et al. (2008). The cell and molecular basis of mechanical, cold, and inflammatory pain. *Science*, *321*(5889), 702–705.

Ackerman, M. J., Priori, S. G., Willems, S., Berul, C., Brugada, R., Calkins, H., et al. (2011). HRS/EHRA expert consensus statement on the state of genetic testing for the channelopathies and cardiomyopathies. This document was developed as a partnership between the Heart Rhythm Society (HRS) and the European Heart Rhythm Association (EHRA), *Heart Rhythm*, *13*(8), 1077–1109.

Adams, B., & Tanabe, T. (1997). Structural regions of the cardiac Ca channel alpha subunit involved in Ca-dependent inactivation. *The Journal of General Physiology*, *110*(4), 379–389.

Agnew, W. S., Levinson, S. R., Brabson, J. S., & Raftery, M. A. (1978). Purification of the tetrodotoxin-binding component associated with the voltage-sensitive sodium channel from *Electrophorus electricus* electroplax membranes. *Proceedings of the National Academy of Sciences of the United States of America*, *75*(6), 2606–2610.

Alvina, K., Ellis-Davies, G., & Khodakhah, K. (2009). T-type calcium channels mediate rebound firing in intact deep cerebellar neurons. *Neuroscience*, *158*(2), 635–641.

Amery, W. K. (1983). Flunarizine, a calcium channel blocker: A new prophylactic drug in migraine. *Headache*, *23*(2), 70–74.

Antzelevitch, C., Belardinelli, L., Zygmunt, A. C., Burashnikov, A., Di Diego, J. M., Fish, J. M., et al. (2004). Electrophysiological effects of ranolazine, a novel antianginal agent with antiarrhythmic properties. *Circulation*, *110*(8), 904–910.

Antzelevitch, C., Pollevick, G. D., Cordeiro, J. M., Casis, O., Sanguinetti, M. C., Aizawa, Y., et al. (2007). Loss-of-function mutations in the cardiac calcium channel underlie a new clinical entity characterized by ST-segment elevation, short QT intervals, and sudden cardiac death. *Circulation*, *115*(4), 442–449.

Antzelevitch, C., Yan, G. X., & Shimizu, W. (1999). Transmural dispersion of repolarization and arrhythmogenicity: The Brugada syndrome versus the long QT syndrome. *Journal of Electrocardiology*, *32*(Suppl.), 158–165.

Aritomi, S., Harada, E., Sugino, K., Nishimura, M., Nakamura, T., & Takahara, A. (2015). Comparison of the cardioprotective and renoprotective effects of the L/N-type calcium channel blocker, cilnidipine, in adriamycin-treated spontaneously-hypertensive rats. *Clinical and Experimental Pharmacology & Physiology*, *42*(4), 344–352.

Audo, I., Kohl, S., Leroy, B. P., Munier, F. L., Guillonneau, X., Mohand-Said, S., et al. (2009). TRPM1 is mutated in patients with autosomal-recessive complete congenital stationary night blindness. *American Journal of Human Genetics*, *85*(5), 720–729.

Baden, T., Euler, T., Weckström, M., & Lagnado, L. (2013). Spikes and ribbon synapses in early vision. *Trends in Neurosciences*, *36*(8), 480–488.

Bagal, S. K., Chapman, M. L., Marron, B. E., Prime, R., Storer, R. I., & Swain, N. A. (2014). Recent progress in sodium channel modulators for pain. *Bioorganic & Medicinal Chemistry Letters*, *24*(16), 3690–3699.

Bahring, R., & Covarrubias, M. (2011). Mechanisms of closed-state inactivation in voltage-gated ion channels. *The Journal of Physiology*, *589*(Pt. 3), 461–479.

Baig, S. M., Koschak, A., Lieb, A., Gebhart, M., Dafinger, C., Nurnberg, G., et al. (2011). Loss of Ca(v)1.3 (CACNA1D) function in a human channelopathy with bradycardia and congenital deafness. *Nature Neuroscience*, *14*(1), 77–84.

Baker, M. D. (2000). Selective block of late Na(+) current by local anaesthetics in rat large sensory neurones. *British Journal of Pharmacology*, *129*(8), 1617–1626.

Bargmann, C. I. (1998). Neurobiology of the *Caenorhabditis elegans* genome. *Science*, *282*(5396), 2028–2033.

Barros, J., Ferreira, A., Brandao, A. F., Lemos, C., Correia, F., Damasio, J., et al. (2014). Familial hemiplegic migraine due to L263V SCN1A mutation: Discordance for epilepsy between two kindreds from Douro valley. *Cephalalgia*, *34*(12), 1015–1020.

Batra, S. (1987). Increase by oestrogen of calcium entry and calcium channel density in uterine smooth muscle. *British Journal of Pharmacology*, *92*(2), 389–392.

Bean, B. P. (1981). Sodium channel inactivation in the crayfish giant axon. Must channels open before inactivating? *Biophysical Journal*, *35*(3), 595–614.

Becker, D. E., & Reed, K. L. (2012). Local anesthetics: Review of pharmacological considerations. *Anesthesia Progress*, *59*(2), 90–101. quiz 102–103.

Belardinelli, L., Shryock, J. C., & Fraser, H. (2006). Inhibition of the late sodium current as a potential cardioprotective principle: Effects of the late sodium current inhibitor ranolazine. *Heart*, *92*(Suppl. 4), iv6–iv14.

Belhassen, B., Glick, A., & Viskin, S. (2004). Efficacy of quinidine in high-risk patients with Brugada syndrome. *Circulation*, *110*(13), 1731–1737.

Beltran-Alvarez, P., Feixas, F., Osuna, S., Diaz-Hernandez, R., Brugada, R., & Pagans, S. (2015). Interplay between R513 methylation and S516 phosphorylation of the cardiac voltage-gated sodium channel. *Amino Acids*, *47*(2), 429–434.

Ben-Johny, M., Yang, P. S., Niu, J., Yang, W., Joshi-Mukherjee, R., & Yue, D. T. (2014). Conservation of Ca2+/calmodulin regulation across Na and Ca2+ channels. *Cell*, *157*(7), 1657–1670.

Bennett, E., Urcan, M. S., Tinkle, S. S., Koszowski, A. G., & Levinson, S. R. (1997). Contribution of sialic acid to the voltage dependence of sodium channel gating. A possible electrostatic mechanism. *The Journal of General Physiology*, *109*(3), 327–343.

Berridge, M. J. (2008). Smooth muscle cell calcium activation mechanisms. *The Journal of Physiology*, *586*(Pt. 21), 5047–5061.

Black, J. A., & Waxman, S. G. (2013). Noncanonical roles of voltage-gated sodium channels. *Neuron, 80*(2), 280–291.

Blanchard, M. G., Willemsen, M. H., Walker, J. B., Dib-Hajj, S. D., Waxman, S. G., Jongmans, M. C., et al. (2015). De novo gain-of-function and loss-of-function mutations of SCN8A in patients with intellectual disabilities and epilepsy. *Journal of Medical Genetics, 52*(5), 330–337.

Boiko, T., Rasband, M. N., Levinson, S. R., Caldwell, J. H., Mandel, G., Trimmer, J. S., et al. (2001). Compact myelin dictates the differential targeting of two sodium channel isoforms in the same axon. *Neuron, 30*(1), 91–104.

Brandt, A., Striessnig, J., & Moser, T. (2003). CaV1.3 channels are essential for development and presynaptic activity of cochlear inner hair cells. *The Journal of Neuroscience, 23*(34), 10832–10840.

Brederson, J. D., Kym, P. R., & Szallasi, A. (2013). Targeting TRP channels for pain relief. *European Journal of Pharmacology, 716*(1–3), 61–76.

Burtscher, V., Schicker, K., Novikova, E., Pohn, B., Stockner, T., Kugler, C., et al. (2014). Spectrum of Cav1.4 dysfunction in congenital stationary night blindness type 2. *Biochimica et Biophysica Acta, 1838*(8), 2053–2065.

Cannon, S. C., & Bean, B. P. (2010). Sodium channels gone wild: Resurgent current from neuronal and muscle channelopathies. *The Journal of Clinical Investigation, 120*(1), 80–83.

Carabelli, V., Marcantoni, A., Comunanza, V., & Carbone, E. (2007). Fast exocytosis mediated by T- and L-type channels in chromaffin cells: Distinct voltage-dependence but similar Ca2+-dependence. *European Biophysics Journal, 36*(7), 753–762.

Carnes, C. A., Janssen, P. M., Ruehr, M. L., Nakayama, H., Nakayama, T., Haase, H., et al. (2007). Atrial glutathione content, calcium current, and contractility. *The Journal of Biological Chemistry, 282*(38), 28063–28073.

Carpenter, D., Ringrose, C., Leo, V., Morris, A., Robinson, R. L., Halsall, P. J., et al. (2009). The role of CACNA1S in predisposition to malignant hyperthermia. *BMC Medical Genetics, 10*, 104.

Caterina, M. J., Schumacher, M. A., Tominaga, M., Rosen, T. A., Levine, J. D., & Julius, D. (1997). The capsaicin receptor: A heat-activated ion channel in the pain pathway. *Nature, 389*(6653), 816–824.

Catterall, W. A. (2011). Voltage-gated calcium channels. *Cold Spring Harbor Perspectives in Biology, 3*(8), a003947.

Catterall, W. A. (2014). Structure and function of voltage-gated sodium channels at atomic resolution. *Experimental Physiology, 99*(1), 35–51.

Catterall, W. A., Cestele, S., Yarov-Yarovoy, V., Yu, F. H., Konoki, K., & Scheuer, T. (2007). Voltage-gated ion channels and gating modifier toxins. *Toxicon, 49*(2), 124–141.

Catterall, W. A., Goldin, A. L., & Waxman, S. G. (2005). International Union of Pharmacology. XLVII. Nomenclature and structure-function relationships of voltage-gated sodium channels. *Pharmacological Reviews, 57*(4), 397–409.

Catterall, W. A., & Swanson, T. M. (2015). Structural basis for pharmacology of voltage-gated sodium and calcium channels. *Molecular Pharmacology, 88*(1), 141–150.

Chagot, B., & Chazin, W. J. (2011). Solution NMR structure of Apo-calmodulin in complex with the IQ motif of human cardiac sodium channel NaV1.5. *Journal of Molecular Biology, 406*(1), 106–119.

Chambers, J. C., Zhao, J., Terracciano, C. M., Bezzina, C. R., Zhang, W., Kaba, R., et al. (2010). Genetic variation in SCN10A influences cardiac conduction. *Nature Genetics, 42*(2), 149–152.

Chapin, H. C., & Caplan, M. J. (2010). The cell biology of polycystic kidney disease. *The Journal of Cell Biology, 191*(4), 701–710.

Chen, R., & Chung, S. H. (2014). Mechanism of tetrodotoxin block and resistance in sodium channels. *Biochemical and Biophysical Research Communications*, *446*(1), 370–374.

Chen, R. S., Deng, T. C., Garcia, T., Sellers, Z. M., & Best, P. M. (2007). Calcium channel gamma subunits: A functionally diverse protein family. *Cell Biochemistry and Biophysics*, *47*(2), 178–186.

Chioni, A. M., Fraser, S. P., Pani, F., Foran, P., Wilkin, G. P., Diss, J. K., et al. (2005). A novel polyclonal antibody specific for the Na(v)1.5 voltage-gated Na(+) channel 'neonatal' splice form. *Journal of Neuroscience Methods*, *147*(2), 88–98.

Clapham, D. E., Runnels, L. W., & Strubing, C. (2001). The TRP ion channel family. *Nature Reviews. Neuroscience*, *2*(6), 387–396.

Comunanza, V., Marcantoni, A., Vandael, D. H., Mahapatra, S., Gavello, D., Carabelli, V., et al. (2010). CaV1.3 as pacemaker channels in adrenal chromaffin cells: Specific role on exo- and endocytosis? *Channels (Austin, Tex.)*, *4*(6), 440–446.

Cortright, D. N., & Szallasi, A. (2009). TRP channels and pain. *Current Pharmaceutical Design*, *15*(15), 1736–1749.

Cox, J. J., Reimann, F., Nicholas, A. K., Thornton, G., Roberts, E., Springell, K., et al. (2006). An SCN9A channelopathy causes congenital inability to experience pain. *Nature*, *444*(7121), 894–898.

de Kovel, C. G., Meisler, M. H., Brilstra, E. H., van Berkestijn, F. M., van 't Slot, R., van Lieshout, S., et al. (2014). Characterization of a de novo SCN8A mutation in a patient with epileptic encephalopathy. *Epilepsy Research*, *108*(9), 1511–1518.

DeMaria, C. D., Soong, T. W., Alseikhan, B. A., Alvania, R. S., & Yue, D. T. (2001). Calmodulin bifurcates the local Ca2+ signal that modulates P/Q-type Ca2+ channels. *Nature*, *411*(6836), 484–489.

Devor, M., Wall, P. D., & Catalan, N. (1992). Systemic lidocaine silences ectopic neuroma and DRG discharge without blocking nerve conduction. *Pain*, *48*(2), 261–268.

Dichgans, M., Freilinger, T., Eckstein, G., Babini, E., Lorenz-Depiereux, B., Biskup, S., et al. (2005). Mutation in the neuronal voltage-gated sodium channel SCN1A in familial hemiplegic migraine. *Lancet*, *366*(9483), 371–377.

Dobrev, D. (2009). Ion channel portrait of the human sinus node: Useful for a better understanding of sinus node function and dysfunction in humans? *Circulation*, *119*(12), 1556–1558.

Doering, C. J., Hamid, J., Simms, B., McRory, J. E., & Zamponi, G. W. (2005). Cav1.4 encodes a calcium channel with low open probability and unitary conductance. *Biophysical Journal*, *89*(5), 3042–3048.

Dragicevic, E., Schiemann, J., & Liss, B. (2015). Dopamine midbrain neurons in health and Parkinson's disease: Emerging roles of voltage-gated calcium channels and ATP-sensitive potassium channels. *Neuroscience*, *284*, 798–814.

Eckle, V. S., Shcheglovitov, A., Vitko, I., Dey, D., Yap, C. C., Winckler, B., et al. (2014). Mechanisms by which a CACNA1H mutation in epilepsy patients increases seizure susceptibility. *The Journal of Physiology*, *592*(Pt. 4), 795–809.

Egger, V., Svoboda, K., & Mainen, Z. F. (2003). Mechanisms of lateral inhibition in the olfactory bulb: Efficiency and modulation of spike-evoked calcium influx into granule cells. *The Journal of Neuroscience*, *23*(20), 7551–7558.

Eijkelkamp, N., Linley, J. E., Baker, M. D., Minett, M. S., Cregg, R., Werdehausen, R., et al. (2012). Neurological perspectives on voltage-gated sodium channels. *Brain*, *135*(9), 2585–2612.

Erickson, M. G., Alseikhan, B. A., Peterson, B. Z., & Yue, D. T. (2001). Preassociation of calmodulin with voltage-gated Ca(2+) channels revealed by FRET in single living cells. *Neuron*, *31*(6), 973–985.

Ertel, E., & Godfraind, T. (2012). *Calcium channel blockers*. Basel: Birkhäuser.

Escayg, A., De Waard, M., Lee, D. D., Bichet, D., Wolf, P., Mayer, T., et al. (2000). Coding and noncoding variation of the human calcium-channel beta4-subunit gene CACNB4

in patients with idiopathic generalized epilepsy and episodic ataxia. *American Journal of Human Genetics*, *66*(5), 1531–1539.
Faber, C. G., Hoeijmakers, J. G., Ahn, H. S., Cheng, X., Han, C., Choi, J. S., et al. (2012). Gain of function Nanu1.7 mutations in idiopathic small fiber neuropathy. *Annals of Neurology*, *71*(1), 26–39.
Faber, C. G., Lauria, G., Merkies, I. S., Cheng, X., Han, C., Ahn, H. S., et al. (2012). Gain-of-function Nav1.8 mutations in painful neuropathy. *Proceedings of the National Academy of Sciences of the United States of America*, *109*(47), 19444–19449.
Fang, Z., Park, C. K., Li, H. Y., Kim, H. Y., Park, S. H., Jung, S. J., et al. (2007). Molecular basis of Ca(v)2.3 calcium channels in rat nociceptive neurons. *The Journal of Biological Chemistry*, *282*(7), 4757–4764.
Fatt, P., & Ginsborg, B. L. (1958). The ionic requirements for the production of action potentials in crustacean muscle fibres. *The Journal of Physiology*, *142*(3), 516–543.
Fatt, P., & Katz, B. (1953). The electrical properties of crustacean muscle fibres. *The Journal of Physiology*, *120*(1–2), 171–204.
Ferguson-Smith, M. A., & Trifonov, V. (2007). Mammalian karyotype evolution. *Nature Reviews. Genetics*, *8*(12), 950–962.
Frings, S., & Bradley, J. (2006). *Transduction channels in sensory cells.* Weinheim, Germany: Wiley.
Frolov, R. V., Berim, I. G., & Singh, S. (2008). Inhibition of delayed rectifier potassium channels and induction of arrhythmia: A novel effect of celecoxib and the mechanism underlying it. *The Journal of Biological Chemistry*, *283*(3), 1518–1524.
Gasparini, S., Kasyanov, A. M., Pietrobon, D., Voronin, L. L., & Cherubini, E. (2001). Presynaptic R-type calcium channels contribute to fast excitatory synaptic transmission in the rat hippocampus. *The Journal of Neuroscience*, *21*(22), 8715–8721.
Gengo, P., Skattebol, A., Moran, J. F., Gallant, S., Hawthorn, M., & Triggle, D. J. (1988). Regulation by chronic drug administration of neuronal and cardiac calcium channel, beta-adrenoceptor and muscarinic receptor levels. *Biochemical Pharmacology*, *37*(4), 627–633.
George, A. L., Jr. (2005). Inherited disorders of voltage-gated sodium channels. *The Journal of Clinical Investigation*, *115*(8), 1990–1999.
Giunti, P., Mantuano, E., Frontali, M., & Veneziano, L. (2015). Molecular mechanism of spinocerebellar ataxia type 6: Glutamine repeat disorder, channelopathy and transcriptional dysregulation. The multifaceted aspects of a single mutation. *Frontiers in Cellular Neuroscience*, *9*, 36.
Goldenberg, M. M. (2010). Overview of drugs used for epilepsy and seizures: Etiology, diagnosis, and treatment. *P & T*, *35*(7), 392–415.
Gomez-Ospina, N., Tsuruta, F., Barreto-Chang, O., Hu, L., & Dolmetsch, R. (2006). The C terminus of the L-type voltage-gated calcium channel Ca(V)1.2 encodes a transcription factor. *Cell*, *127*(3), 591–606.
Gonzalez, D. R., Treuer, A., Sun, Q. A., Stamler, J. S., & Hare, J. M. (2009). *S*-Nitrosylation of cardiac ion channels. *Journal of Cardiovascular Pharmacology*, *54*(3), 188–195.
Groome, J., Lehmann-Horn, F., & Holzherr, B. (2011). Open- and closed-state fast inactivation in sodium channels: Differential effects of a site-3 anemone toxin. *Channels (Austin, Tex.)*, *5*(1), 65–78.
Haeseleer, F., Imanishi, Y., Maeda, T., Possin, D. E., Maeda, A., Lee, A., et al. (2004). Essential role of Ca2+-binding protein 4, a Cav1.4 channel regulator, in photoreceptor synaptic function. *Nature Neuroscience*, 7(10), 1079–1087.
Hardie, R. C. (2011). A brief history of trp: Commentary and personal perspective. *Pflügers Archiv*, *461*(5), 493–498.
Hardie, R. C., & Franze, K. (2012). Photomechanical responses in Drosophila photoreceptors. *Science*, *338*(6104), 260–263.

Hardie, R. C., & Juusola, M. (2015). Phototransduction in Drosophila. *Current Opinion in Neurobiology, 34*, 37–45.

Hardie, R. C., & Minke, B. (1992). The trp gene is essential for a light-activated Ca2+ channel in Drosophila photoreceptors. *Neuron, 8*(4), 643–651.

Hardie, R. C., & Minke, B. (1993). Novel Ca2+ channels underlying transduction in Drosophila photoreceptors: Implications for phosphoinositide-mediated Ca2+ mobilization. *Trends in Neurosciences, 16*(9), 371–376.

Harvey, A. L. (2014). Toxins and drug discovery. *Toxicon, 92*, 193–200.

Hedley, P. L., Jorgensen, P., Schlamowitz, S., Moolman-Smook, J., Kanters, J. K., Corfield, V. A., et al. (2009). The genetic basis of Brugada syndrome: A mutation update. *Human Mutation, 30*(9), 1256–1266.

Hernandez-Ochoa, E. O., Contreras, M., Cseresnyes, Z., & Schneider, M. F. (2007). Ca2+ signal summation and NFATc1 nuclear translocation in sympathetic ganglion neurons during repetitive action potentials. *Cell Calcium, 41*(6), 559–571.

Heron, S. E., Crossland, K. M., Andermann, E., Phillips, H. A., Hall, A. J., Bleasel, A., et al. (2002). Sodium-channel defects in benign familial neonatal-infantile seizures. *Lancet, 360*(9336), 851–852.

Heron, S. E., Khosravani, H., Varela, D., Bladen, C., Williams, T. C., Newman, M. R., et al. (2007). Extended spectrum of idiopathic generalized epilepsies associated with CACNA1H functional variants. *Annals of Neurology, 62*(6), 560–568.

Herren, A. W., Bers, D. M., & Grandi, E. (2013). Post-translational modifications of the cardiac Na channel: Contribution of CaMKII-dependent phosphorylation to acquired arrhythmias. *American Journal of Physiology. Heart and Circulatory Physiology, 305*(4), H431–H445.

Hille, B. (1975). Ionic selectivity, saturation, and block in sodium channels. A four-barrier model. *The Journal of General Physiology, 66*(5), 535–560.

Hodgkin, A. L., & Huxley, A. F. (1952a). The components of membrane conductance in the giant axon of Loligo. *The Journal of Physiology, 116*(4), 473–496.

Hodgkin, A. L., & Huxley, A. F. (1952b). Currents carried by sodium and potassium ions through the membrane of the giant axon of Loligo. *The Journal of Physiology, 116*(4), 449–472.

Hodgkin, A. L., & Huxley, A. F. (1952c). The dual effect of membrane potential on sodium conductance in the giant axon of Loligo. *The Journal of Physiology, 116*(4), 497–506.

Hofmann, F., Flockerzi, V., Kahl, S., & Wegener, J. W. (2014). L-type CaV1.2 calcium channels: from in vitro findings to in vivo function. *Physiological Reviews, 94*(1), 303–326.

Hofmann, T., Obukhov, A. G., Schaefer, M., Harteneck, C., Gudermann, T., & Schultz, G. (1999). Direct activation of human TRPC6 and TRPC3 channels by diacylglycerol. *Nature, 397*(6716), 259–263.

Hoppa, M. B., Lana, B., Margas, W., Dolphin, A. C., & Ryan, T. A. (2012). alpha2delta expression sets presynaptic calcium channel abundance and release probability. *Nature, 486*(7401), 122–125.

Hu, D., Barajas-Martinez, H., Nesterenko, V. V., Pfeiffer, R., Guerchicoff, A., Cordeiro, J. M., et al. (2010). Dual variation in SCN5A and CACNB2b underlies the development of cardiac conduction disease without Brugada syndrome. *Pacing and Clinical Electrophysiology, 33*(3), 274–285.

Hu, D., Barajas-Martinez, H., Pfeiffer, R., Dezi, F., Pfeiffer, J., Buch, T., et al. (2014). Mutations in SCN10A are responsible for a large fraction of cases of Brugada syndrome. *Journal of the American College of Cardiology, 64*(1), 66–79.

Huang, J., Han, C., Estacion, M., Vasylyev, D., Hoeijmakers, J. G., Gerrits, M. M., et al. (2014). Gain-of-function mutations in sodium channel Na(v)1.9 in painful neuropathy. *Brain, 137*(Pt. 6), 1627–1642.

Huang, H., Yu, D., & Soong, T. W. (2013). C-terminal alternative splicing of CaV1.3 channels distinctively modulates their dihydropyridine sensitivity. *Molecular Pharmacology*, *84*(4), 643–653.

Hund, T. J., & Mohler, P. J. (2014). Nav channel complex heterogeneity: New targets for the treatment of arrhythmia? *Circulation*, *130*(2), 132–134.

Hurley, M. J., Brandon, B., Gentleman, S. M., & Dexter, D. T. (2013). Parkinson's disease is associated with altered expression of CaV1 channels and calcium-binding proteins. *Brain*, *136*(Pt. 7), 2077–2097.

Jacus, M. O., Uebele, V. N., Renger, J. J., & Todorovic, S. M. (2012). Presynaptic Cav3.2 channels regulate excitatory neurotransmission in nociceptive dorsal horn neurons. *The Journal of Neuroscience*, *32*(27), 9374–9382.

Jarecki, B. W., Piekarz, A. D., Jackson, J. O., 2nd, & Cummins, T. R. (2010). Human voltage-gated sodium channel mutations that cause inherited neuronal and muscle channelopathies increase resurgent sodium currents. *The Journal of Clinical Investigation*, *120*(1), 369–378.

Johnson, D., & Bennett, E. S. (2006). Isoform-specific effects of the beta2 subunit on voltage-gated sodium channel gating. *The Journal of Biological Chemistry*, *281*(36), 25875–25881.

Jurkat-Rott, K., Holzherr, B., Fauler, M., & Lehmann-Horn, F. (2010). Sodium channelopathies of skeletal muscle result from gain or loss of function. *Pflügers Archiv*, *460*(2), 239–248.

Kadowaki, T. (2015). Evolutionary dynamics of metazoan TRP channels. *Pflügers Archiv*, *467*(10), 2043–2053.

Kalra, A. A., & Elliott, D. (2007). Acute migraine: Current treatment and emerging therapies. *Therapeutics and Clinical Risk Management*, *3*(3), 449–459.

Kamp, T. J., & Hell, J. W. (2000). Regulation of cardiac L-type calcium channels by protein kinase A and protein kinase C. *Circulation Research*, *87*(12), 1095–1102.

Kang, S., Cooper, G., Dunne, S. F., Dusel, B., Luan, C. H., Surmeier, D. J., et al. (2012). CaV1.3-selective L-type calcium channel antagonists as potential new therapeutics for Parkinson's disease. *Nature Communications*, *3*, 1146.

Kannankeril, P., Roden, D. M., & Darbar, D. (2010). Drug-induced long QT syndrome. *Pharmacological Reviews*, *62*(4), 760–781.

Kanter, R. J., Pfeiffer, R., Hu, D., Barajas-Martinez, H., Carboni, M. P., & Antzelevitch, C. (2012). Brugada-like syndrome in infancy presenting with rapid ventricular tachycardia and intraventricular conduction delay. *Circulation*, *125*(1), 14–22.

Kapplinger, J. D., Tester, D. J., Alders, M., Benito, B., Berthet, M., Brugada, J., et al. (2010). An international compendium of mutations in the SCN5A-encoded cardiac sodium channel in patients referred for Brugada syndrome genetic testing. *Heart Rhythm*, 7(1), 33–46.

Katz, B., & Miledi, R. (1967). The release of acetylcholine from nerve endings by graded electric pulses. *Proceedings of the Royal Society of London B: Biological Sciences*, *167*(1006), 23–38.

Keef, K. D., Hume, J. R., & Zhong, J. (2001). Regulation of cardiac and smooth muscle Ca(2+) channels (Ca(V)1.2a, b) by protein kinases. *American Journal of Physiology. Cell Physiology*, *281*(6), C1743–C1756.

Kim, S. J., Kim, Y. S., Yuan, J. P., Petralia, R. S., Worley, P. F., & Linden, D. J. (2003). Activation of the TRPC1 cation channel by metabotropic glutamate receptor mGluR1. *Nature*, *426*(6964), 285–291.

Klaver, E. C., Versluijs, G. M., & Wilders, R. (2011). Cardiac ion channel mutations in the sudden infant death syndrome. *International Journal of Cardiology*, *152*(2), 162–170.

Kleopa, K. A. (2011). Autoimmune channelopathies of the nervous system. *Current Neuropharmacology*, *9*(3), 458–467.

Klionsky, L., Tamir, R., Holzinger, B., Bi, X., Talvenheimo, J., Kim, H., et al. (2006). A polyclonal antibody to the prepore loop of transient receptor potential vanilloid type 1

blocks channel activation. *The Journal of Pharmacology and Experimental Therapeutics, 319*(1), 192–198.
Kremeyer, B., Lopera, F., Cox, J. J., Momin, A., Rugiero, F., Marsh, S., et al. (2010). A gain-of-function mutation in TRPA1 causes familial episodic pain syndrome. *Neuron, 66*(5), 671–680.
Krey, J. F., Pasca, S. P., Shcheglovitov, A., Yazawa, M., Schwemberger, R., Rasmusson, R., et al. (2013). Timothy syndrome is associated with activity-dependent dendritic retraction in rodent and human neurons. *Nature Neuroscience, 16*(2), 201–209.
Kyndt, F., Probst, V., Potet, F., Demolombe, S., Chevallier, J. C., Baro, I., et al. (2001). Novel SCN5A mutation leading either to isolated cardiac conduction defect or Brugada syndrome in a large French family. *Circulation, 104*(25), 3081–3086.
Lampl, I., Schwindt, P., & Crill, W. (1998). Reduction of cortical pyramidal neuron excitability by the action of phenytoin on persistent Na+ current. *The Journal of Pharmacology and Experimental Therapeutics, 284*(1), 228–237.
Lee, M. T., Chen, C. H., Lee, C. S., Chen, C. C., Chong, M. Y., Ouyang, W. C., et al. (2011). Genome-wide association study of bipolar I disorder in the Han Chinese population. *Molecular Psychiatry, 16*(5), 548–556.
Lee, Y., Hong, S., Cui, M., Sharma, P. K., Lee, J., & Choi, S. (2015). Transient receptor potential vanilloid type 1 antagonists: A patent review (2011–2014). *Expert Opinion on Therapeutic Patents, 25*(3), 291–318.
Lee, A., Wong, S. T., Gallagher, D., Li, B., Storm, D. R., Scheuer, T., et al. (1999). Ca2+/calmodulin binds to and modulates P/Q-type calcium channels. *Nature, 399*(6732), 155–159.
Leipold, E., Liebmann, L., Korenke, G. C., Heinrich, T., Giesselmann, S., Baets, J., et al. (2013). A de novo gain-of-function mutation in SCN11A causes loss of pain perception. *Nature Genetics, 45*(11), 1399–1404.
Lew, R. M., Burnett, L., Proos, A. L., & Delatycki, M. B. (2015). Tay-Sachs disease: Current perspectives from Australia. *The Application of Clinical Genetics, 8*, 19–25.
Lewis, R. J., Dutertre, S., Vetter, I., & Christie, M. J. (2012). Conus venom peptide pharmacology. *Pharmacological Reviews, 64*(2), 259–298.
Li, R. G., Wang, Q., Xu, Y. J., Zhang, M., Qu, X. K., Liu, X., et al. (2013). Mutations of the SCN4B-encoded sodium channel beta4 subunit in familial atrial fibrillation. *International Journal of Molecular Medicine, 32*(1), 144–150.
Liao, Y. J., Safa, P., Chen, Y. R., Sobel, R. A., Boyden, E. S., & Tsien, R. W. (2008). Anti-Ca2+ channel antibody attenuates Ca2+ currents and mimics cerebellar ataxia in vivo. *Proceedings of the National Academy of Sciences of the United States of America, 105*(7), 2705–2710.
Liao, P., Yu, D., Lu, S., Tang, Z., Liang, M. C., Zeng, S., et al. (2004). Smooth muscle-selective alternatively spliced exon generates functional variation in Cav1.2 calcium channels. *The Journal of Biological Chemistry, 279*(48), 50329–50335.
Luebke, J. I., Dunlap, K., & Turner, T. J. (1993). Multiple calcium channel types control glutamatergic synaptic transmission in the hippocampus. *Neuron, 11*(5), 895–902.
Mahapatra, S., Calorio, C., Vandael, D. H., Marcantoni, A., Carabelli, V., & Carbone, E. (2012). Calcium channel types contributing to chromaffin cell excitability, exocytosis and endocytosis. *Cell Calcium, 51*(3–4), 321–330.
Maitre, M. (1997). The gamma-hydroxybutyrate signalling system in brain: Organization and functional implications. *Progress in Neurobiology, 51*(3), 337–361.
Mao, J., & Chen, L. L. (2000). Systemic lidocaine for neuropathic pain relief. *Pain, 87*(1), 7–17.
Marban, E., Yamagishi, T., & Tomaselli, G. F. (1998). Structure and function of voltage-gated sodium channels. *The Journal of Physiology, 508*(Pt. 3), 647–657.
Marcil, J., Walczak, J. S., Guindon, J., Ngoc, A. H., Lu, S., & Beaulieu, P. (2006). Antinociceptive effects of tetrodotoxin (TTX) in rodents. *British Journal of Anaesthesia, 96*(6), 761–768.

Marger, L., Mesirca, P., Alig, J., Torrente, A., Dubel, S., Engeland, B., et al. (2011). Functional roles of Ca(v)1.3, Ca(v)3.1 and HCN channels in automaticity of mouse atrioventricular cells: Insights into the atrioventricular pacemaker mechanism. *Channels (Austin, Tex.)*, *5*(3), 251–261.

Mason, W. P., Graus, F., Lang, B., Honnorat, J., Delattre, J. Y., Valldeoriola, F., et al. (1997). Small-cell lung cancer, paraneoplastic cerebellar degeneration and the Lambert-Eaton myasthenic syndrome. *Brain*, *120*(Pt. 8), 1279–1300.

Masumiya, H., Shijuku, T., Tanaka, H., & Shigenobu, K. (1998). Inhibition of myocardial L- and T-type Ca2+ currents by efonidipine: Possible mechanism for its chronotropic effect. *European Journal of Pharmacology*, *349*(2–3), 351–357.

McDonald, T. F., Pelzer, S., Trautwein, W., & Pelzer, D. J. (1994). Regulation and modulation of calcium channels in cardiac, skeletal, and smooth muscle cells. *Physiological Reviews*, *74*(2), 365–507.

McRory, J. E., Hamid, J., Doering, C. J., Garcia, E., Parker, R., Hamming, K., et al. (2004). The CACNA1F gene encodes an L-type calcium channel with unique biophysical properties and tissue distribution. *The Journal of Neuroscience*, *24*(7), 1707–1718.

Medeiros-Domingo, A., Kaku, T., Tester, D. J., Iturralde-Torres, P., Itty, A., Ye, B., et al. (2007). SCN4B-encoded sodium channel beta4 subunit in congenital long-QT syndrome. *Circulation*, *116*(2), 134–142.

Meng, Q. T., Xia, Z. Y., Liu, J., Bayliss, D. A., & Chen, X. (2011). Local anesthetic inhibits hyperpolarization-activated cationic currents. *Molecular Pharmacology*, *79*(5), 866–873.

Mesirca, P., Torrente, A. G., & Mangoni, M. E. (2015). Functional role of voltage gated Ca(2+) channels in heart automaticity. *Frontiers in Physiology*, *6*, 19.

Meza, U., & Adams, B. (1998). G-protein-dependent facilitation of neuronal alpha1A, alpha1B, and alpha1E Ca channels. *The Journal of Neuroscience*, *18*(14), 5240–5252.

Mezghrani, A., Monteil, A., Watschinger, K., Sinnegger-Brauns, M. J., Barrere, C., Bourinet, E., et al. (2008). A destructive interaction mechanism accounts for dominant-negative effects of misfolded mutants of voltage-gated calcium channels. *The Journal of Neuroscience*, *28*(17), 4501–4511.

Minke, B., & Parnas, M. (2006). Insights on TRP channels from in vivo studies in Drosophila. *Annual Review of Physiology*, *68*, 649–684.

Monnier, N., Procaccio, V., Stieglitz, P., & Lunardi, J. (1997). Malignant-hyperthermia susceptibility is associated with a mutation of the alpha 1-subunit of the human dihydropyridine-sensitive L-type voltage-dependent calcium-channel receptor in skeletal muscle. *American Journal of Human Genetics*, *60*(6), 1316–1325.

Moran, M. M., McAlexander, M. A., Biro, T., & Szallasi, A. (2011). Transient receptor potential channels as therapeutic targets. *Nature Reviews. Drug Discovery*, *10*(8), 601–620.

Mounsey, J. P., Xu, P., John, J. E., 3rd, Horne, L. T., Gilbert, J., Roses, A. D., et al. (1995). Modulation of skeletal muscle sodium channels by human myotonin protein kinase. *The Journal of Clinical Investigation*, *95*(5), 2379–2384.

Muth-Selbach, U., Hermanns, H., Stegmann, J. U., Kollosche, K., Freynhagen, R., Bauer, I., et al. (2009). Antinociceptive effects of systemic lidocaine: Involvement of the spinal glycinergic system. *European Journal of Pharmacology*, *613*(1–3), 68–73.

Namadurai, S., Yereddi, N. R., Cusdin, F. S., Huang, C. L., Chirgadze, D. Y., & Jackson, A. P. (2015). A new look at sodium channel beta subunits. *Open Biology*, *5*(1), 140192.

Nassar, M. A., Baker, M. D., Levato, A., Ingram, R., Mallucci, G., McMahon, S. B., et al. (2006). Nerve injury induces robust allodynia and ectopic discharges in Nav1.3 null mutant mice. *Molecular Pain*, *2*, 33.

Navedo, M. F., & Santana, L. F. (2013). CaV1.2 sparklets in heart and vascular smooth muscle. *Journal of Molecular and Cellular Cardiology*, *58*, 67–76.

Naylor, J., Milligan, C. J., Zeng, F., Jones, C., & Beech, D. J. (2008). Production of a specific extracellular inhibitor of TRPM3 channels. *British Journal of Pharmacology, 155*(4), 567–573.

Nilius, B., Talavera, K., Owsianik, G., Prenen, J., Droogmans, G., & Voets, T. (2005). Gating of TRP channels: A voltage connection? *The Journal of Physiology, 567*(Pt. 1), 35–44.

O'Farrell, M., Ziogas, J., & Marley, P. D. (1997). Effects of N- and L-type calcium channel antagonists and (+/−)-Bay K8644 on nerve-induced catecholamine secretion from bovine perfused adrenal glands. *British Journal of Pharmacology, 121*(3), 381–388.

Ohba, C., Kato, M., Takahashi, S., Lerman-Sagie, T., Lev, D., Terashima, H., et al. (2014). Early onset epileptic encephalopathy caused by de novo SCN8A mutations. *Epilepsia, 55*(7), 994–1000.

Olson, P. A., Tkatch, T., Hernandez-Lopez, S., Ulrich, S., Ilijic, E., Mugnaini, E., et al. (2005). G-protein-coupled receptor modulation of striatal CaV1.3 L-type Ca2+ channels is dependent on a Shank-binding domain. *The Journal of Neuroscience, 25*(5), 1050–1062.

Omilusik, K., Priatel, John J., Chen, X., Wang, Yiwen T., Xu, H., Choi, Kyung B., et al. (2011). The CaV1.4 calcium channel is a critical regulator of T cell receptor signaling and naive T cell homeostasis. *Immunity, 35*(3), 349–360.

O'Roak, B. J., Vives, L., Girirajan, S., Karakoc, E., Krumm, N., Coe, B. P., et al. (2012). Sporadic autism exomes reveal a highly interconnected protein network of de novo mutations. *Nature, 485*(7397), 246–250.

Pan, Z. H., Hu, H. J., Perring, P., & Andrade, R. (2001). T-type Ca(2+) channels mediate neurotransmitter release in retinal bipolar cells. *Neuron, 32*(1), 89–98.

Park, D. S., & Fishman, G. I. (2011). The cardiac conduction system. *Circulation, 123*(8), 904–915.

Peachey, N. S., Pearring, J. N., Bojang, P., Jr., Hirschtritt, M. E., Sturgill-Short, G., Ray, T. A., et al. (2012). Depolarizing bipolar cell dysfunction due to a Trpm1 point mutation. *Journal of Neurophysiology, 108*(9), 2442–2451.

Perez-Reyes, E. (2003). Molecular physiology of low-voltage-activated t-type calcium channels. *Physiological Reviews, 83*(1), 117–161.

Peterson, B. Z., DeMaria, C. D., Adelman, J. P., & Yue, D. T. (1999). Calmodulin is the Ca2 + sensor for Ca2+-dependent inactivation of L-type calcium channels. *Neuron, 22*(3), 549–558.

Picones, A. (2015). Ion channels as medicinal targets of biological toxins: The impact of automated patch-clamp electrophysiology. *Current Topics in Medicinal Chemistry, 15*(7), 631–637.

Pietrobon, D. (2010). CaV2.1 channelopathies. *Pflügers Archiv, 460*(2), 375–393.

Pittenger, C., Coric, V., Banasr, M., Bloch, M., Krystal, J. H., & Sanacora, G. (2008). Riluzole in the treatment of mood and anxiety disorders. *CNS Drugs, 22*(9), 761–786.

Powers, R. E., & Colucci, W. S. (1985). An increase in putative voltage dependent calcium channel number following reserpine treatment. *Biochemical and Biophysical Research Communications, 132*(2), 844–849.

Qian, F., & Noben-Trauth, K. (2005). Cellular and molecular function of mucolipins (TRPML) and polycystin 2 (TRPP2). *Pflügers Archiv, 451*(1), 277–285.

Qin, F. (2007). Regulation of TRP ion channels by phosphatidylinositol-4,5-bisphosphate. *Handbook of Experimental Pharmacology, 179*, 509–525.

Qin, N., Olcese, R., Bransby, M., Lin, T., & Birnbaumer, L. (1999). Ca2+-induced inhibition of the cardiac Ca2+ channel depends on calmodulin. *Proceedings of the National Academy of Sciences of the United States of America, 96*(5), 2435–2438.

Raap, U., Stander, S., & Metz, M. (2011). Pathophysiology of itch and new treatments. *Current Opinion in Allergy and Clinical Immunology, 11*(5), 420–427.

Ragsdale, D. S. (2008). How do mutant Nav1.1 sodium channels cause epilepsy? *Brain Research Reviews*, *58*(1), 149–159.

Rajadhyaksha, A., Barczak, A., Macias, W., Leveque, J. C., Lewis, S. E., & Konradi, C. (1999). L-type Ca(2+) channels are essential for glutamate-mediated CREB phosphorylation and c-fos gene expression in striatal neurons. *The Journal of Neuroscience*, *19*(15), 6348–6359.

Ramos, E., & O'Leary, M. E. (2004). State-dependent trapping of flecainide in the cardiac sodium channel. *The Journal of Physiology*, *560*(Pt. 1), 37–49.

Randall, A., & Tsien, R. W. (1995). Pharmacological dissection of multiple types of Ca2+ channel currents in rat cerebellar granule neurons. *The Journal of Neuroscience*, *15*(4), 2995–3012.

Reiser, J., Polu, K. R., Moller, C. C., Kenlan, P., Altintas, M. M., Wei, C., et al. (2005). TRPC6 is a glomerular slit diaphragm-associated channel required for normal renal function. *Nature Genetics*, *37*(7), 739–744.

Rius, R., Govoni, S., & Trabucchi, M. (1986). Regional modification of brain calcium antagonist binding after in vivo chronic lead exposure. *Toxicology*, *40*(2), 191–197.

Rock, M. J., Prenen, J., Funari, V. A., Funari, T. L., Merriman, B., Nelson, S. F., et al. (2008). Gain-of-function mutations in TRPV4 cause autosomal dominant brachyolmia. *Nature Genetics*, *40*(8), 999–1003.

Roden, D. M. (1998). Mechanisms and management of proarrhythmia. *The American Journal of Cardiology*, *82*(4A), 49I–57I.

Rogawski, M. A., & Loscher, W. (2004). The neurobiology of antiepileptic drugs for the treatment of nonepileptic conditions. *Nature Medicine*, *10*(7), 685–692.

Rozanski, G. M., Nath, A. R., Adams, M. E., & Stanley, E. F. (2013). Low voltage-activated calcium channels gate transmitter release at the dorsal root ganglion sandwich synapse. *The Journal of Physiology*, *591*(Pt. 22), 5575–5583.

Saegusa, H., Kurihara, T., Zong, S., Minowa, O., Kazuno, A., Han, W., et al. (2000). Altered pain responses in mice lacking alpha 1E subunit of the voltage-dependent Ca2+ channel. *Proceedings of the National Academy of Sciences of the United States of America*, *97*(11), 6132–6137.

Schiavon, E., Sacco, T., Cassulini, R. R., Gurrola, G., Tempia, F., Possani, L. D., et al. (2006). Resurgent current and voltage sensor trapping enhanced activation by a beta-scorpion toxin solely in Nav1.6 channel. Significance in mice Purkinje neurons. *The Journal of Biological Chemistry*, *281*(29), 20326–20337.

Scholl, U. I., Goh, G., Stolting, G., de Oliveira, R. C., Choi, M., Overton, J. D., et al. (2013). Somatic and germline CACNA1D calcium channel mutations in aldosterone-producing adenomas and primary aldosteronism. *Nature Genetics*, *45*(9), 1050–1054.

Scholz, A. (2002). Mechanisms of (local) anaesthetics on voltage-gated sodium and other ion channels. *British Journal of Anaesthesia*, *89*(1), 52–61.

Sekiguchi, S. (1990). Proffered opinion forum medical decision making, impact on laboratory tests. *Rinsho Byori*, *38*(5), 601–603.

Sekiguchi, F., & Kawabata, A. (2013). T-type calcium channels: Functional regulation and implication in pain signaling. *Journal of Pharmacological Sciences*, *122*(4), 244–250.

Shank, R. P., & Maryanoff, B. E. (2008). Molecular pharmacodynamics, clinical therapeutics, and pharmacokinetics of topiramate. *CNS Neuroscience & Therapeutics*, *14*(2), 120–142.

Shekter, L. R., Taussig, R., Gillard, S. E., & Miller, R. J. (1997). Regulation of human neuronal calcium channels by G protein betagamma subunits expressed in human embryonic kidney 293 cells. *Molecular Pharmacology*, *52*(2), 282–291.

Shen, Y., Heimel, J. A., Kamermans, M., Peachey, N. S., Gregg, R. G., & Nawy, S. (2009). A transient receptor potential-like channel mediates synaptic transmission in rod bipolar cells. *The Journal of Neuroscience*, *29*(19), 6088–6093.

Shepherd, G. (2006). Treatment of poisoning caused by beta-adrenergic and calcium-channel blockers. *American Journal of Health-System Pharmacy, 63*(19), 1828–1835.

Shiga, T., Yamada, Y., Matsuda, N., Tanaka, T., Urae, A., Hashiguchi, M., et al. (2007). Influence of cilnidipine or nisoldipine on sympathetic activity in healthy male subjects. *Heart and Vessels, 22*(6), 404–409.

Simen, A. A., & Miller, R. J. (2000). Involvement of regions in domain I in the opioid receptor sensitivity of alpha1B Ca(2+) channels. *Molecular Pharmacology, 57*(5), 1064–1074.

Simms, B. A., & Zamponi, G. W. (2014). Neuronal voltage-gated calcium channels: Structure, function, and dysfunction. *Neuron, 82*(1), 24–45.

Singh, B., Monteil, A., Bidaud, I., Sugimoto, Y., Suzuki, T., Hamano, S., et al. (2007). Mutational analysis of CACNA1G in idiopathic generalized epilepsy. Mutation in brief #962. Online. *Human Mutatation, 28*(5), 524–525.

Smani, T., Shapovalov, G., Skryma, R., Prevarskaya, N., & Rosado, J. A. (2015). Functional and physiopathological implications of TRP channels. *Biochimica et Biophysica Acta, 1853*(8), 1772–1782.

Smith, R. D., & Goldin, A. L. (1992). Protein kinase A phosphorylation enhances sodium channel currents in Xenopus oocytes. *The American Journal of Physiology, 263*(3 Pt. 1), C660–C666.

Sokolov, S., Scheuer, T., & Catterall, W. A. (2010). Ion permeation and block of the gating pore in the voltage sensor of NaV1.4 channels with hypokalemic periodic paralysis mutations. *The Journal of General Physiology, 136*(2), 225–236.

Soldatov, N. M. (2015). CACNB2: An emerging pharmacological target for hypertension, heart failure, arrhythmia and mental disorders. *Current Molecular Pharmacology, 8*, 32–42.

Spedding, M., Kenny, B., & Chatelain, P. (1995). New drug binding sites in Ca2+ channels. *Trends in Pharmacological Sciences, 16*(4), 139–142.

Splawski, I., Timothy, K. W., Sharpe, L. M., Decher, N., Kumar, P., Bloise, R., et al. (2004). Ca(V)1.2 calcium channel dysfunction causes a multisystem disorder including arrhythmia and autism. *Cell, 119*(1), 19–31.

Stephens, G. J., Canti, C., Page, K. M., & Dolphin, A. C. (1998). Role of domain I of neuronal Ca2+ channel alpha1 subunits in G protein modulation. *The Journal of Physiology, 509*(Pt. 1), 163–169.

Stephenson, E. A., & Berul, C. I. (2007). Electrophysiological interventions for inherited arrhythmia syndromes. *Circulation, 116*(9), 1062–1080.

Stotz, S. C., Jarvis, S. E., & Zamponi, G. W. (2004). Functional roles of cytoplasmic loops and pore lining transmembrane helices in the voltage-dependent inactivation of HVA calcium channels. *The Journal of Physiology, 554*(Pt. 2), 263–273.

Strichartz, G. R., & Ritchie, J. M. (1987). The action of local anesthetics on ion channels of excitable tissues. In G. Strichartz (Ed.), *Local anesthetics. Vol. 81.* (pp. 21–52). Berlin, Heidelberg: Springer.

Striessnig, J., Bolz, H. J., & Koschak, A. (2010). Channelopathies in Cav1.1, Cav1.3, and Cav1.4 voltage-gated L-type Ca2+ channels. *Pflügers Archiv, 460*(2), 361–374.

Striessnig, J., Pinggera, A., Kaur, G., Bock, G., & Tuluc, P. (2014). L-type Ca channels in heart and brain. *Wiley Interdisciplinary Reviews. Membrane Transport and Signalling, 3*(2), 15–38.

Suh, H. H., Fujimoto, J. M., & Tseng, L. L. (1989). Differential mechanisms mediating beta-endorphin- and morphine-induced analgesia in mice. *European Journal of Pharmacology, 168*(1), 61–70.

Sutton, K. G., McRory, J. E., Guthrie, H., Murphy, T. H., & Snutch, T. P. (1999). P/Q-type calcium channels mediate the activity-dependent feedback of syntaxin-1A. *Nature, 401*(6755), 800–804.

Szallasi, A., & McAlexander, M. A. (2015). *TRP channels as therapeutic targets: From basic science to clinical use.* Saint Louis, MO: Elsevier Science. from, http://alltitles.ebrary.com/Doc?id=11043766.

Szallasi, A., & Sheta, M. (2012). Targeting TRPV1 for pain relief: Limits, losers and laurels. *Expert Opinion on Investigational Drugs*, *21*(9), 1351–1369.
Takiar, V., & Caplan, M. J. (2011). Polycystic kidney disease: Pathogenesis and potential therapies. *Biochimica et Biophysica Acta*, *1812*(10), 1337–1343.
Tan, Y., Deng, Y., & Qing, H. (2012). Calcium channel blockers and Alzheimer's disease. *Neural Regeneration Research*, 7(2), 137–140.
Tang, A. H., Karson, M. A., Nagode, D. A., McIntosh, J. M., Uebele, V. N., Renger, J. J., et al. (2011). Nerve terminal nicotinic acetylcholine receptors initiate quantal GABA release from perisomatic interneurons by activating axonal T-type (Cav3) Ca(2)(+) channels and Ca(2)(+) release from stores. *The Journal of Neuroscience*, *31*(38), 13546–13561.
Taylor, C. P. (1997). Mechanisms of action of gabapentin. *Revue Neurologique (Paris)*, *153*(Suppl. 1), S39–S45.
Treiman, D. M. (2001). GABAergic mechanisms in epilepsy. *Epilepsia*, *42*(Suppl. 3), 8–12.
Trevelyan, A. J., Muldoon, S. F., Merricks, E. M., Racca, C., & Staley, K. J. (2015). The role of inhibition in epileptic networks. *Journal of Clinical Neurophysiology*, *32*(3), 227–234.
Triggle, D. J. (1990). Calcium, calcium channels, and calcium channel antagonists. *Canadian Journal of Physiology and Pharmacology*, *68*(11), 1474–1481.
Triggle, D. J., Langs, D. A., & Janis, R. A. (1989). Ca2+ channel ligands: Structure-function relationships of the 1,4-dihydropyridines. *Medicinal Research Reviews*, *9*(2), 123–180.
Uebele, V. N., Nuss, C. E., Renger, J. J., & Connolly, T. M. (2004). Role of voltage-gated calcium channels in potassium-stimulated aldosterone secretion from rat adrenal zona glomerulosa cells. *The Journal of Steroid Biochemistry and Molecular Biology*, *92*(3), 209–218.
Vassilev, P. M., Scheuer, T., & Catterall, W. A. (1988). Identification of an intracellular peptide segment involved in sodium channel inactivation. *Science*, *241*(4873), 1658–1661.
Vaughan Williams, E. M. (1975). Classification of antidysrhythmic drugs. *Pharmacology & Therapeutics. Part B*, *1*(1), 115–138.
Veeramah, K. R., O'Brien, J. E., Meisler, M. H., Cheng, X., Dib-Hajj, S. D., Waxman, S. G., et al. (2012). De novo pathogenic SCN8A mutation identified by whole-genome sequencing of a family quartet affected by infantile epileptic encephalopathy and SUDEP. *American Journal of Human Genetics*, *90*(3), 502–510.
Venance, S. L., Cannon, S. C., Fialho, D., Fontaine, B., Hanna, M. G., Ptacek, L. J., et al. (2006). The primary periodic paralyses: Diagnosis, pathogenesis and treatment. *Brain*, *129*(Pt. 1), 8–17.
Venkatachalam, K., Hofmann, T., & Montell, C. (2006). Lysosomal localization of TRPML3 depends on TRPML2 and the mucolipidosis-associated protein TRPML1. *The Journal of Biological Chemistry*, *281*(25), 17517–17527.
Venkatachalam, K., Zheng, F., & Gill, D. L. (2003). Regulation of canonical transient receptor potential (TRPC) channel function by diacylglycerol and protein kinase C. *The Journal of Biological Chemistry*, *278*(31), 29031–29040.
Vilin, Y. Y., & Ruben, P. C. (2001). Slow inactivation in voltage-gated sodium channels: Molecular substrates and contributions to channelopathies. *Cell Biochemistry and Biophysics*, *35*(2), 171–190.
Vincent, A. (2009). Autoimmune channelopathies: New antibody-mediated disorders of the central nervous system. *F1000 Biology Reports*, *1*, 61.
Wakabayashi, K., Gustafson, A. M., Sidransky, E., & Goldin, E. (2011). Mucolipidosis type IV: An update. *Molecular Genetics and Metabolism*, *104*(3), 206–213.
Wallace, R. H., Wang, D. W., Singh, R., Scheffer, I. E., George, A. L., Jr., Phillips, H. A., et al. (1998). Febrile seizures and generalized epilepsy associated with a mutation in the Na+-channel beta1 subunit gene SCN1B. *Nature Genetics*, *19*(4), 366–370.
Walz, G. (2005). Slit or pore? A mutation of the ion channel TRPC6 causes FSGS. *Nephrology, Dialysis, Transplantation*, *20*(9), 1777–1779.

Wasserstrom, J. A., & Salata, J. J. (1988). Basis for tetrodotoxin and lidocaine effects on action potentials in dog ventricular myocytes. *The American Journal of Physiology*, *254*(6 Pt. 2), H1157–H1166.

Waszkielewicz, A. M., Gunia, A., Szkaradek, N., Sloczynska, K., Krupinska, S., & Marona, H. (2013). Ion channels as drug targets in central nervous system disorders. *Current Medicinal Chemistry*, *20*(10), 1241–1285.

Watanabe, H., Koopmann, T. T., Le Scouarnec, S., Yang, T., Ingram, C. R., Schott, J. J., et al. (2008). Sodium channel beta1 subunit mutations associated with Brugada syndrome and cardiac conduction disease in humans. *The Journal of Clinical Investigation*, *118*(6), 2260–2268.

Weiss, S., Benoist, D., White, E., Teng, W., & Saint, D. A. (2010). Riluzole protects against cardiac ischaemia and reperfusion damage via block of the persistent sodium current. *British Journal of Pharmacology*, *160*(5), 1072–1082.

Weiss, L. A., Escayg, A., Kearney, J. A., Trudeau, M., MacDonald, B. T., Mori, M., et al. (2003). Sodium channels SCN1A, SCN2A and SCN3A in familial autism. *Molecular Psychiatry*, *8*(2), 186–194.

Weiss, N., & Zamponi, G. W. (2012). Regulation of voltage-gated calcium channels by synaptic proteins. *Advances in Experimental Medicine and Biology*, *740*, 759–775.

Weller, C. M., Pelzer, N., de Vries, B., Lopez, M. A., De Fabregues, O., Pascual, J., et al. (2014). Two novel SCN1A mutations identified in families with familial hemiplegic migraine. *Cephalalgia*, *34*(13), 1062–1069.

Wes, P. D., Chevesich, J., Jeromin, A., Rosenberg, C., Stetten, G., & Montell, C. (1995). TRPC1, a human homolog of a Drosophila store-operated channel. *Proceedings of the National Academy of Sciences of the United States of America*, *92*(21), 9652–9656.

Wheeler, D. G., Groth, R. D., Ma, H., Barrett, C. F., Owen, S. F., Safa, P., et al. (2012). Ca(V)1 and Ca(V)2 channels engage distinct modes of Ca(2+) signaling to control CREB-dependent gene expression. *Cell*, *149*(5), 1112–1124.

Winn, M. P., Conlon, P. J., Lynn, K. L., Farrington, M. K., Creazzo, T., Hawkins, A. F., et al. (2005). A mutation in the TRPC6 cation channel causes familial focal segmental glomerulosclerosis. *Science*, *308*(5729), 1801–1804.

Wu, L. G., Borst, J. G., & Sakmann, B. (1998). R-type Ca2+ currents evoke transmitter release at a rat central synapse. *Proceedings of the National Academy of Sciences of the United States of America*, *95*(8), 4720–4725.

Wycisk, K. A., Zeitz, C., Feil, S., Wittmer, M., Forster, U., Neidhardt, J., et al. (2006). Mutation in the auxiliary calcium-channel subunit CACNA2D4 causes autosomal recessive cone dystrophy. *American Journal of Human Genetics*, *79*(5), 973–977.

Xu, N., Cioffi, D. L., Alexeyev, M., Rich, T. C., & Stevens, T. (2015). Sodium entry through endothelial store-operated calcium entry channels: Regulation by Orai1. *American Journal of Physiology. Cell Physiology*, *308*(4), C277–C288.

Yang, P. S., Alseikhan, B. A., Hiel, H., Grant, L., Mori, M. X., Yang, W., et al. (2006). Switching of Ca2+-dependent inactivation of Ca(v)1.3 channels by calcium binding proteins of auditory hair cells. *The Journal of Neuroscience*, *26*(42), 10677–10689.

Yu, F. H., Mantegazza, M., Westenbroek, R. E., Robbins, C. A., Kalume, F., Burton, K. A., et al. (2006). Reduced sodium current in GABAergic interneurons in a mouse model of severe myoclonic epilepsy in infancy. *Nature Neuroscience*, *9*(9), 1142–1149.

Yu, F. H., Yarov-Yarovoy, V., Gutman, G. A., & Catterall, W. A. (2005). Overview of molecular relationships in the voltage-gated ion channel superfamily. *Pharmacological Reviews*, *57*(4), 387–395.

Zhu, X., Chu, P. B., Peyton, M., & Birnbaumer, L. (1995). Molecular cloning of a widely expressed human homologue for the Drosophila trp gene. *FEBS Letters*, *373*(3), 193–198.

Zimmer, T. (2010). Effects of tetrodotoxin on the mammalian cardiovascular system. *Marine Drugs*, *8*(3), 741–762.
Zimmermann, K., Leffler, A., Babes, A., Cendan, C. M., Carr, R. W., Kobayashi, J., et al. (2007). Sensory neuron sodium channel Nav1.8 is essential for pain at low temperatures. *Nature*, *447*(7146), 855–858.
Zuhlke, R. D., Pitt, G. S., Deisseroth, K., Tsien, R. W., & Reuter, H. (1999). Calmodulin supports both inactivation and facilitation of L-type calcium channels. *Nature*, *399*(6732), 159–162.
Zuhlke, R. D., & Reuter, H. (1998). Ca2+-sensitive inactivation of L-type Ca2+ channels depends on multiple cytoplasmic amino acid sequences of the alpha1C subunit. *Proceedings of the National Academy of Sciences of the United States of America*, *95*(6), 3287–3294.

CHAPTER THREE

Ion Channels in Neurological Disorders

Pravir Kumar*,†,1, Dhiraj Kumar*, Saurabh Kumar Jha*, Niraj Kumar Jha*, Rashmi K. Ambasta*

*Molecular Neuroscience and Functional Genomics Laboratory, Delhi Technological University (Formerly DCE), Delhi, India
†Department of Neurology, Adjunct faculty, Tufts University School of Medicine, Boston, Massachusetts, USA
[1]Corresponding author: e-mail address: Pravir.Kumar@tufts.edu

Contents

Abstract

The convergent endeavors of the neuroscientist to establish a link between clinical neurology, genetics, loss of function of an important protein, and channelopathies behind neurological disorders are quite intriguing. Growing evidence reveals the impact of ion channels dysfunctioning in neurodegenerative disorders (NDDs). Many neurological/neuromuscular disorders, viz, Alzheimer's disease, Parkinson's disease, Huntington's disease, multiple sclerosis, amyotrophic lateral sclerosis, and age-related disorders are caused due to altered function or mutation in ion channels. To maintain cell homeostasis, ion channels are playing a crucial role which is a large transmembrane protein. Further, these channels are important as it determines the membrane potential and playing

Advances in Protein Chemistry and Structural Biology, Volume 103
ISSN 1876-1623
http://dx.doi.org/10.1016/bs.apcsb.2015.10.006

critically in the secretion of neurotransmitter. Behind NDDs, losses of pathological proteins and defective ion channels have been reported and are found to aggravate the disease symptoms. Moreover, ion channel dysfunctions are eliciting a range of symptoms, including memory loss, movement disabilities, neuromuscular sprains, and strokes. Since the possible mechanistic role played by aberrant ion channels, their receptor and associated factors in neurodegeneration remained elusive; therefore, it is a challenging task for the neuroscientist to implement the therapeutics for targeting NDDs. This chapter reviews the potential role of the ion channels in membrane physiology and brain homeostasis, where ion channels and their associated factors have been characterized with their functional consequences in neurological diseases. Moreover, mechanistic role of perturbed ion channels has been identified in various NDDs, and finally, ion channel modulators have been investigated for their therapeutic intervention in treating common NDDs.

1. INTRODUCTION

Ion channels are the important constituents of neurons, which is responsible for triggering nerve impulse and synaptic transmission (neurotransmitter's release). These channels are divided into two major classes, viz, (i) voltage-gated (Na^+, K^+, Ca^{2+}, Cl^-) and (ii) ligand-gated (nicotinic acetylcholine receptors (nAChRs), γ-amino butyric acid (GABA), *N*-methyl-D-aspartate receptors (NMDARs), ryanodine receptors (RyRs)) that are involved in impulse transmission across the synapses. These channels play a crucial role in various intracellular and extracellular processes of the brain, for instance, cell communication, cell adhesion, and cell migration. Moreover, it also allows the brain to receive and route information via producing electrical signals in order to maintain normal brain homeostasis (Kurachi & North, 2004). However, research over the past few decades has identified various genetic defects or aberrations in channel-forming genes, which are responsible for numerous neurological outcomes, such as memory disorders, movement disorders, and neuromuscular disorders. Therefore, such type of diseases that occurred due to faulty ion channels has been classified as neurological channelopathies, which include myotonia, congenital myasthenic syndromes, malignant hyperthermia, and periodic paralysis (Cooper & Jan, 1999). Furthermore, there are several neurodegenerative disorders (NDDs), including Alzheimer's disease (AD), Parkinson's disease (PD), Huntington's disease (HD), multiple sclerosis (MS), and Amyotrophic lateral sclerosis (ALS) that exhibit defective ion channels, but their related mechanism is quite elusive. In order to solve this mystery, this chapter explores the pertinent role of ion channels in neurons and their associated NDDs. Further, therapeutic potential of well-known channel blockers and modulators has been explored, depicting their role in disease prevention, which might be a promising therapeutic agent for the treatment of the most common NDDs.

1.1 Intricacy of Ion Channels in Membrane Physiology

Ion channel protein creates a passage for ions like sodium, potassium, calcium, and chloride to travel across the impermeable lipid bilayer. Further, it is involved in regulating various physiological processes, for instance, electrical conduction in neuronal cell, muscular contraction, and release of neurotransmitters (Lodish et al., 2000). Ion channels are known to govern three prominent roles in regulating membrane physiology that is (i) setting up membrane potential of cells where movement of ions across the membrane constitutes a potential gradient that determines resting potentials of membrane and generates action potentials. For instance, at the beginning of action potential, sodium channels rapidly open and potassium channels close which causes depolarization; however, within a few milliseconds, sodium channels close, and calcium channels open to sustain a depolarized state. Later on, calcium channels get inactivated and potassium channels open that again causes membrane repolarization (Barnett & Larkman, 2007), (ii) transmission of electrical signals, i.e., opening and closing of channels constitute electrical signals in excitable tissues like neurons and muscles, (iii) maintain electrolytic balance across cell membrane to regulate cell volume (Strange, 2004), and (iv) generation of regulatory signals in the cell. These ion channels enable ion flow to constitute electric signals that trigger intracellular signaling cascades linked with neurotransmitter release, hormonal secretion, muscular contraction, and altered gene expression (Gouaux & Mackinnon, 2005). Thus, ion channel research improves our understanding of cellular activity and organ function in normal versus diseased condition. Apart from ion channels, other proteins like transporters and pumps also play a significant role in regulating the membrane physiology.

1.2 Role of Ion Channels in the Brain Homeostasis

To maintain physiological equilibrium in the brain, proper functioning of ion channel is essential, for instance, maintaining ionic balance across postsynaptic neuronal membrane. Moreover, an ion influx through the channels regulates a number of processes that include neuronal growth and differentiation, synaptic plasticity, neurotransmitter release, neuronal gene expression, neuronal motility, and excitability. Further, cellular responses to humoral or synaptic inputs are determined by multiple ion channels that are expressed by a specific neuron (Kurachi & North, 2004). The ion channel activity is frequently modulated by hormones and neurotransmitters that are involved in central regulation of metabolism in the brain (Sohn, 2013). Upon oxygen deprivation in the brain, glutamate (an excitatory neurotransmitter) causes the cationic influx through ion channels thereby triggering excitotoxicity due to anoxic depolarization (Madry, Haglerød, & Attwell, 2010; Weilinger et al., 2013). Importantly, during ischemia, excessive release of glutamate from the presynaptic terminal leads to overactivation of glutamate receptors (NMDA and AMPA) that causes an excessive influx of calcium ions. Further, these receptors activate pannexin-1 channels that get hyperactivated and release ATP outside. This extracellular overburden of glutamate and ATP consecutively activates various complexes that trigger apoptotic and necrotic cascades eventually leading to neuronal injury or death (Abbracchio, Burnstock, Verkhratsky, & Zimmermann, 2009). Recent reports advocate the dysregulation of energy and glucose homeostasis due to defective ion channel subunits in specific neuronal populations (Liu et al., 2012; Parton et al., 2007). Moreover, ion channels are also known to control cell migration and organization in central nervous system (Ariano et al., 2006; Komuro & Kumada, 2005), where ion channels are involved in the regulation of signaling processes; therefore, any disturbances in the structure and function of channels would have detrimental effects on nervous system. Elevated intracellular ion level disturbs the homeostasis and lead to the development of NDDs. The prominent cause for ion imbalance is the defective ion channels encoded by mutated genes. For instance, calcium ion channel mutations govern the onset of spinocerebellar ataxia, episodic ataxia type 2, X-linked congenital stationary night blindness, and familial hemiplegic migraine (Pietrobon, 2002). Furthermore, mutated sodium, potassium, and calcium channels, or acetylcholine- and glycine-gated channels are known to cause epilepsy, Lambert–Eaton myasthenic syndrome, hyperekplexia, schizophrenia, AD, and PD (Kim, 2014).

1.3 Impact of Channels on Blood–Brain Barrier

The blood–brain barrier (BBB) is the semipermeable barrier of endothelial cell lining that separates the brain from circulating blood and hence plays a critical defense system for the brain against blood-borne toxins, proteins, and cells, which are the prime requisite for the healthy functioning of the brain. It is comprised of four cellular components that include endothelial cells, astrocytic end feet, microglial cells, and pericytes. These cells interact among themselves directly or indirectly to selectively regulate the passage of molecules and cells in the brain (Wolburg & Lippoldt, 2002) that maintains the neurovascular cell homeostasis (Abbott, Ronnback, & Hansson, 2006). Moreover, the BBB functionalities have been maintained by different cell junctions at BBB sites that are tight junctions, adherent junctions, and gap junctions. Tight junction enables cell sealing via cell sheet formation, whereas adherent junction allows the cell attachment with neighboring cells or with the extracellular matrix. Likewise, gap junction facilitates the transmission of chemical or electrical signals among interacting cells (Omidi & Gumbleton, 2005). Moreover, the restrictive function of BBB is provided by tight junction, while selective permeability is provided by the intricate transport machineries present at the membrane of brain capillary endothelial cells that are responsive to the autocrine and paracrine signals (Rubin & Staddon, 1999). Therefore, any disturbances in BBB integrity would govern the onset of numerous neurological disorders in response to brain inflammation caused by infiltrating leukocytes; cytokines from blood and thus activated glial cells affect the extracellular milieu around neurons. For instance, dysfunctioning of BBB caused infiltration of the tumor necrosis factor-α in the brain which has been found to aggravate the progression of PD (Qin et al., 2007). Similarly, damaged biotrafficking across BBB has also been associated with numerous neurological disorders such as epilepsies, strokes, traumatic brain injury, AD, and PD (Friedman, Kaufer, & Heinemann, 2009; Lo, Dalkara, & Moskowitz, 2003; Zlokovic, 2008).

1.4 How Channels Affect Gap Junctions, Release of Ions, and Homeostasis?

Multicellular organisms maintain their tissue/organ homeostasis with the help of cell communication machinery, by which it responds differently with the change in environmental conditions. Such cell communication machinery includes gap junctions, which are intercellular channels clustered in specialized regions of plasma membrane (Meşe, Richard, & White, 2006). These gap junctional channels mediate cytoplasmic connections between two cells and provide a medium for the transport of small metabolites like glucose, ions like K^+/Ca^{2+} and secondary messengers like cAMP, cGMP, and IP3 (inositol 1,4,5-triphosphate) (Kanno & Loewenstein, 1964;

Lawrence, Beers, & Gilula, 1978). Moreover, gap junctional communication is responsible for controlling numerous physiological processes, which include cell motility, cell growth or proliferation, cell differentiation, cell synchronization, and metabolic coordination of vascular cell (Vinken et al., 2006; White & Paul, 1999). Furthermore, in a complex tissue like nervous tissue, gap junctions mediate electrical coupling and recovery upon neuronal injury. These gap junctional intercellular channels are encoded by two families of proteins, pannexins and connexins, which are expressed in highly active neuronal cells, astrocytes, and oligodendrocytes. In mammals, pannexins are reported to form hemichannels while connexins to form both gap junction channels and hemichannels. For instance, connexin proteins in animals are comprised of 12 subunits, which are assembled as a hemichannel, i.e., 6 connexin subunits in each connected cell (Goodenough, Paul, & Jesaitis, 1988). Further, gap junctional channels are responsible for cytoplasmic connection between cells thereby coordinating their electrical and metabolic activity, whereas hemichannels interconnect intra- and extracellular compartments of cells thus providing a diffusion pathway for ions and small molecules (Figueroa et al., 2014). Mutations and structural aberrations in connexins have been reported in various diseases such as hereditary epilepsy, neoplasms, bacterial/parasitic infections and also in AD (Mylvaganam, Ramani, Krawczyk, & Carlen, 2014; Quintanilla, Orellana, & Von Bernhardi, 2012; Vega et al., 2013). However, gap junctional channels protect neurons via spatial buffering of neurotoxic substrates, but under stressful conditions during neuronal injury, microglial cells get overactivated and release proinflammatory cytokines (TNF and IL-1) that hamper gap junctional homeostasis thereby hindering astroglial communication in the brain (Orellana et al., 2009). Continual glial activation in response to toxic proteins expressed in numerous neurodegenerative diseases induce glial hemichannels and gap junctions that contribute to neurodegeneration during inflammation and thus disrupt the brain homeostasis (Orellana et al., 2009).

1.5 What Are the Different Channels That Cause Ion Disturbance in Neurological Disorders?

The drastic increase in the incidence of neurological disorders associated with aberrant ion channels had been reported in past 20 years. There are two types of channels: ligand- and voltage-gated that are associated with ion disturbance in various neurological disorders (Kumar, Ambasta, & Kumar, 2014). For instance, mutated nAChRs, ionotropic glutamate receptors, γ-amino butyric acid type A receptors (GABAARs), and P2X receptors (P2XRs) dysfunctions are associated with epilepsy, Alzheimer's, and PD progression (Lemoine et al., 2012). Likewise, defective sodium channels SCN2A, SCN8A, and SCN9A have been found to cause ataxia, epilepsy, seizures, periodic paralysis, and sensitivity to pain (Eijkelkamp et al., 2012; Meisler & Kearney, 2005). Similarly, potassium channel mutations are linked to the pathogenesis of dyskinesias, seizures, epilepsy, and ataxias (Kim, 2014). Further calcium ion channel aberrations are observed in the patients affected with ataxia, paralysis, and migraine (George, 2004). Furthermore, these defective ion channels have been reported to govern the pathophysiology of a range of diverse neurological disorders, which has been summarized in Table 1, and their molecular architecture has been depicted in Fig. 1.

2. ABERRANT CHANNELS IN NDDs

2.1 Alzheimer's Disease

AD is a form of dementia mainly characterized by cognitive dysfunction, memory loss, and neuronal death. The focal hallmark of this disease relies on diverse factors, including Aβ accumulation, tau hyper phosphorylation, mutations in the catalytic domain of γ secretase (Zhao & Zhao, 2013). However, abnormal levels of intracellular ions (Ca^{2+}, Na^+, K^+, and Cl^-) have been reported to cause ionic imbalance that is also associated with AD pathogenesis (Fig. 2A). In 1992, Hardy and Higgins first reported that Aβ peptides perturb Ca^{2+} homeostasis in neurons and increase intracellular Ca^{2+}, which was later confirmed by Mattson and his colleagues (Hardy & Higgins, 1992). Presently, several successive studies laid the foundation for a novel idea: Aβ peptides are in part, toxic to neurons because they form aberrant ion channels in neuronal membranes and thereby disrupt neuronal homeostasis (Itkin et al., 2011). Moreover, accumulated Aβ peptides have the potential to disrupt the activity of numerous calcium conducting ion channels including, voltage-gated calcium channels (VGCCs): L-, N-, P/Q-, and R-type calcium channels (Kim & Rhim, 2011; Neelands, King, & Macdonaldm, 2000; Nimmrich & Gross, 2012) and ligand-gated calcium channels: NMDARs, nAChRs, and RyRs channels (Danysz & Parsons, 2012; Oulès et al., 2012; Parri & Dineley, 2010). Normally, an influx of Ca^{2+} ions are tightly regulated that evokes neurotransmitters release such as glutamate from presynaptic terminal and trigger downstream signals to govern various cellular processes—synaptogenesis, synaptic transmission,

Table 1 Ion Channels and Their Associated Neurological Diseases

Channels	Involved Genes	Associated Proteins/Subunits	Functions	Neurological Diseases	References
Ca^{2+} channels	CACNA1D	$Ca_v1.3$ (α-1D subunit)	It provides pacemaking activity to DA neuron; any alteration in their functions leads to high Ca^{2+} load to neurons	PD	Dragicevic et al. (2014)
	CACNA1F	$Ca_v1.4$(α-1F subunit)	It provides cell's ability to generate and transmit electrical signals. Moreover, it is also involved in transportation of Ca^{2+} across cell membranes	Aland Island eye disease and cone–rod dystrophy	Boycott, Pearce, and Bech-Hansen (2000)
	CACNA1A	$Ca_v2.1$(α-1A subunit)	It provides P/Q-type current in Purkinje and granule cells and also to presynaptic terminals	FHM, SCA6, and EA2	Bruun et al. (2015)
	CACNA1C	$Ca_v1.2$(α-1C subunit)	It provides pacemaking activity to DA neuron; any alteration in their functions leads to high Ca^{2+} load to neuron. Moreover, it is also associated with toxic Aβ	AD and schizophrenia	Koran, Hohman, and Thornton-Wells (2014), Takahashi, Glatt, Uchiyama, Faraone, and Tsuang (2015)

	CACNA1H	$Ca_v3.2$ (α-1H subunit)	It makes neuronal cells prone to disease by directly altering neuronal electrical properties and indirectly by changing gene expression	IGE and CAE	Eckle et al. (2014)
	CACNB4	β-4 Subunit	It plays a vital role in Ca^{2+} channel function by modulating G protein inhibition, increasing Ca^{2+} influx, controlling the α-1 subunit membrane targeting, and shifting the voltage dependence of activation and inactivation	IGE, JME, and EA5	Delgado-Escueta, Koeleman, Bailey, Medina, and Durón (2013) and Etemad, Campiglio, Obermair, and Flucher (2014)
	HVA (Ca^+)	–	It changes the neuronal Ca^{2+} homeostasis and also altered the Na^+ and K^+ channel function	HD	Cepeda, Wu, André, Cummings, and Levine (2006)
	BEST1	β-4 subunit	It is believed to function as chloride channels that may also serve as regulators of intracellular Ca^{2+} signaling	Bestrophinopathy, vitelliform macular dystrophy, and autosomal recessive retinitis pigmentosa	Lin et al. (2015) and Marmorstein et al. (2015)
	RYR1	–	The RyRs are a family of Ca^{2+} release channels, found on intracellular Ca^{2+} storage/release organelles. It involves in intracellular Ca^{2+} release	MHS and CCD	Murayama et al. (2015)

Continued

Table 1 Ion Channels and Their Associated Neurological Diseases—cont'd

Channels	Involved Genes	Associated Proteins/ Subunits	Functions	Neurological Diseases	References
	TRPA1	Subfamily A, member 1	It is believed to function as a mechanical and chemical stress sensor. However, it is involved in signal transduction and growth control	FEPS	Kremeyer et al. (2010)
	TRPM1	Subfamily M, member 1	It mediates Ca^{2+} entry into the cells	CSNB1C	Bellone et al. (2013)
	TRPV4	Subfamily V, member 4	In the brain, it is involved in the regulation of systemic osmotic pressure	congenital dSMA and CMT2C	(Evangelista et al. (2015)
K^+ channels	KCNA1	$K_v1.1$	It involves in repolarization of axon and also interferes with the demyelinating process of axons	EA1 and MS	Lassche et al. (2014) and Wacker et al. (2012)
	KCNA2	$K_v1.2$	It interferes with demyelinating process of axons	MS	Wacker et al. (2012)

	KCNA3	$K_v1.3$	It is associated with LPS-activated microglia as well as toxic Aβ. However, it also interferes with the demyelinating process of axons	AD and MS	Chung, Lee, Joe, and Uhm (2001) and Rus et al. (2005)
	KCNA5	$K_v1.5$	It function is associated with toxic Aβ as well as involved in regulatory process of insulin secretion	AD	Chung et al. (2001)
	KCNC3	$K_v3.3$	It mediates the voltage-dependent K^+ permeability of excitable membranes	SCA13	Irie, Matsuzaki, Sekino, and Hirai (2014)
	KCNQ2	$K_v7.2$	It involves in repolarization activity of axons	BFNS and EIEE7	Ambrosino et al. (2015) and Gürsoy and Erçal (2015)
	KCNQ3	$K_v7.3$	It involves in repolarization activity of axons	BFNS	Miceli et al. (2015)
	KCNQ4	$K_v7.4$	It plays a crucial role in the regulation of neuronal excitability in association with KCNQ3	DFNB2A	Gao, Yechikov, Vázquez, Chen, and Nie (2013)

Continued

Table 1 Ion Channels and Their Associated Neurological Diseases—cont'd

Channels	Involved Genes	Associated Proteins/ Subunits	Functions	Neurological Diseases	References
	KCNJ3	Kir3.1	It associates with Kir3.2 to form GIRK-1 in neurons that direct to degeneration of DA neurons	PD	Lüscher and Slesinger (2010)
	KCNJ2	Kir2.1	It is having a greater tendency to allow K^+ to flow into a cell rather than out of a cell. Further, it participates in establishing action potential and excitability of neuronal and muscle cells	Andersen–Tawil syndrome	Ördög et al. (2015)
	KCNJ6	Kir3.2	It causes increasing influx of Na^+ as a substitute of K^+ that lead to cause developmental loss of cells in the cerebellum and SNc of DA neurons	PD	Hartfield et al. (2014)
	KCNJ10	Kir4.1	It is characterized by having a greater tendency to allow K^+ influx into the cells	HD, IGE, and EAST syndrome	Cross et al. (2013) and Tong et al. (2014)
	KCNMA1	KCa1.1	It involves in controlling of smooth muscle tone and neuronal excitability	GEPD	Lee and Cui (2009)

	KCNT1	KCa4.1	It is having diverse functions, such as regulating neurotransmitter release, insulin secretion, neuronal excitability, and cell volume	NFLE5 and EIEE14	Møller et al. (2015) and Ohba et al. (2015)
	K-ATP channels	–	It makes membrane potential hyperpolarized and also reduces SN DA activity	PD	Schiemann et al. (2012)
Na^+ channels	SCN1A	$Na_v1.1$(α-1 subunit)	It increases Na^+ influx in somatodendritic cells	GEFS+, SMEI, Dravet syndrome, and FHM3	Gürsoy and Erçal (2015) and Helbig (2015)
	SCN1B	β-1 Subunit	It modulates the function of α-subunit of Na_v channels	GEFS+	Helbig (2015)
	SCN2A	$Na_v1.2$ (α-subunit)	It increases fast Na^+ influx in axons and also involved in axonal degeneration	MS and GEFS+	Black, Newcombe, Trapp, and Waxman (2007) and Nakayama (2009)
	SCN5A	$Na_v1.5$ (α-subunit)	It is involved in axonal degeneration and promotes Ca^{2+} influx in a neuron that interferes with demyelinating neurons	MS, cerebellar ataxia, and EAE	Black, Newcombe, and Waxman (2010)
	SCN8A	$Na_v1.6$ (α-subunit)			Black et al. (2007)
	SCN10A	$Na_v1.8$ (α-subunit)			Damarjian, Craner, Black, and Waxman (2004)
	SCN2A	$Na_v2.1$ (α-subunit)	It is responsible for the generation and propagation of action potentials in neurons as well as muscle tissues	BFIE and EIEE11	Yoshitomi et al. (2015)

Continued

Table 1 Ion Channels and Their Associated Neurological Diseases—cont'd

Channels	Involved Genes	Associated Proteins/ Subunits	Functions	Neurological Diseases	References
	SCN9A	$Na_v1.7$ (α-subunit)	It plays a crucial role in the generation and conduction of action potentials and therefore, important for electrical signaling by most excitable cells	CIP and small fiber neuropathy	Emery et al. (2015)
	SCN4A	$Na_v1.4$ (α-subunit)	It is responsible for the generation and propagation of action potentials in neurons as well as muscle tissues	CMS, HYPP, PMC, and PAM	Basali and Prayson (2015) and Wang et al. (2015)
Cl^- channels	CLIC1	–	Functional in activated microglia during oxidative stress	AD	Skaper, Facci, and Giusti (2013)
Cyclic nucleotide-gated channel	CNGA1	α-1 Subunit	It plays a critical role in phototransduction	Autosomal recessive retinitis pigmentosa	Winkler et al. (2013)
	CNGA3	α-3 Subunit	It is responsible for normal vision and olfactory signal transduction	CRD	Shaikh et al. (2015)
	CNGB1	β-1 Subunit	It regulates the ion influx into the rod photoreceptor outer segment in response to light-induced alteration in intracellular cGMP concentrations	Autosomal recessive retinitis pigmentosa	Michalakis et al. (2014)

	CNGB3	β-3 Subunit	It regulates the ion influx into the rod photoreceptor outer segment in response to light-induced alteration in intracellular cGMP concentrations	ACHM3 and JMD	Nishiguchi, Sandberg, Gorji, Berson, and Dryja (2005)
Glycine receptors	GLRA1	α-1 Subunit	Directs inhibitory role in brainstem and spinal cord	FH and hereditary hyperekplexia	Chau, Roitfarb, Carabuena, and Camann (2015)
	GLRB	β-Subunit	It has an inhibitory role against the glycine receptor that mediates postsynaptic inhibition in the spinal cord and other regions of CNS	Hereditary hyperekplexia	Chau et al. (2015)
Nicotinic Ach receptors	CHRNB2	β-2 Subunit	Involves in modulation of neurotransmitters release	ADNFLE and NFLE	Boillot and Baulac (2015) and Conti et al. (2015)
	CHRNA2	α-2 Subunit	Involves in modulation of neurotransmitters release	ADNFLE and NFLE	Boillot and Baulac (2015) and Conti et al. (2015)
	CHRNA4	α-4 Subunit	It is involved in modulation of neurotransmitters release	ADNFLE	Boillot and Baulac (2015)
	CHRNA1	α-1 Subunit	It plays a crucial role in ACh binding/channel gating and neuromuscular organogenesis. Further, it leads to opening of an ion-conducting channel across the plasma membrane	Multiple pterygium syndrome	Chen (2012)
	CHRND	δ-Subunit			
	CHRNG	γ-Subunit			

Continued

Table 1 Ion Channels and Their Associated Neurological Diseases—cont'd

Channels	Involved Genes	Associated Proteins/ Subunits	Functions	Neurological Diseases	References
	CHRNB1	β-1 Subunit	It is involved in opening of an ion-conducting channel across the plasma membrane	CMS	Mihaylova et al. (2010)
	CHRNE	ε-Subunit			
GABA receptors	GABRG2	γ-2 Subunit	It directs inhibitory role against GABAergic neurons	GEFS + and Dravet syndrome	Helbig (2015) and Ishii et al. (2014)
	GABRA1	α-1 Subunit	It directs inhibitory role against GABAergic neurons in the brain that act as a ligand-gated Cl^- channels	GEFS +, Dravet syndrome, CAE, and JME	Helbig (2015), Hirose (2014), and Ishii et al. (2014)
	GABRA6	α-6 Subunit			
	GABRB3	β-3 Subunit			

PD, Parkinson's disease; CAE, childhood absence epilepsy; JME, juvenile myoclonic epilepsy; EA5, episodic ataxia type 5; FHM, familial hemiplegic migraine; SCA6, spinocerebellar ataxia type 6; EA2, episodic ataxia type 2; AD, Alzheimer's disease; HD, Huntington disease; MHS, malignant hyperthermia susceptibility; CCD, central core disease; FEPS, familial episodic pain syndrome; CSNB1C, congenital stationary night blindness type 1C; congenital dSMA, congenital distal spinal muscular atrophy; CMT2C, Charcot–Marie–Tooth neuropathy type 2C; SCA13, spinocerebellar ataxia type 13; EIEE7, early infantile epileptic encephalopathy type 7; DFNB2A, deafness, autosomal dominant, type 2A; EAST syndrome (epilepsy, ataxia, sensorineural deafness, and tubulopathy); GEPD, generalized epilepsy with paroxysmal dyskinesia; NFLE5, nocturnal frontal lobe epilepsy type 5; EIEE14, early infantile epileptic encephalopathy type 14; FHM3, familial hemiplegic migraine type 3; EAE, experimental autoimmune encephalomyelitis; BFIE, benign familial infantile epilepsy; EIEE11, early infantile epileptic encephalopathy type 11; CIP, congenital indifference to pain; CMS, congenital myasthenic syndrome; HYPP, hyperkalemic periodic paralysis; PMC, paramyotonia congenital; PAM, potassium-aggravated myotonia; EA1, episodic ataxia type 1; MS, multiple sclerosis; BFNS, benign familial neonatal seizures; IGE, idiopathic generalized epilepsy; GEFS +, generalized epilepsy with febrile seizures plus; FH, familial hyperekplexia; ADNFLE, autosomal dominant nocturnal frontal lobe epilepsy; CRD, juvenile cone–rod dystrophy; ACHM3, achromatopsia type 3; JMD, juvenile macular degeneration; NFLE, nocturnal frontal lobe epilepsy.

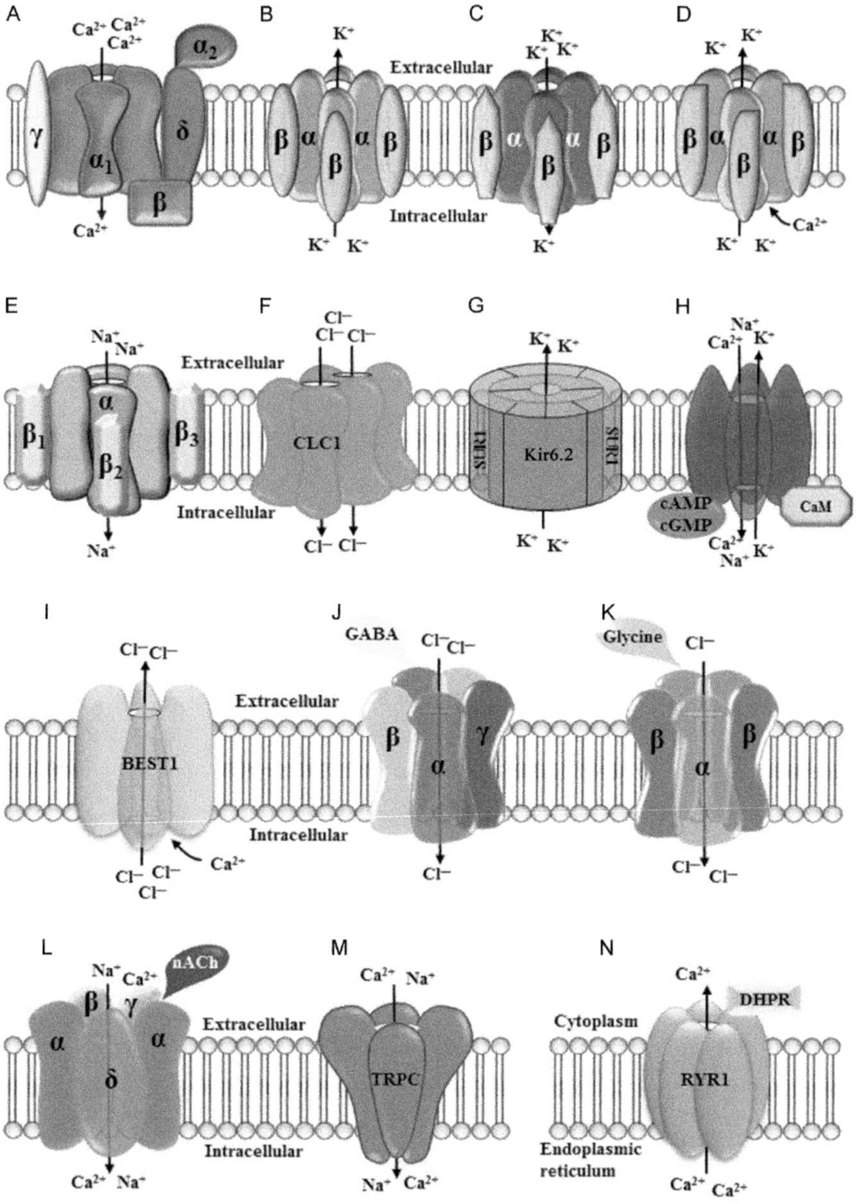

Figure 1 Architecture of ion channels involved in neurological disorders (A) voltage-gated calcium channel, (B) voltage-gated potassium channel, (C) inward rectifier potassium channel, (D) calcium-activated potassium channel, (E) voltage-gated sodium channel, (F) voltage-gated chloride channel, (G) ATP-sensitive potassium channel, (H) cyclic nucleotide-gated ion channel, (I) calcium-activated chloride ion channel; BEST1-Bestrophin-1, (J) GABA (γ-amino butyric acid) receptor, (K) glycine receptor, (L) nACH (nicotinic acetylcholine) receptor, (M) TRP (transient receptor potential) channel, and (N) RYR1 (ryanodine) receptor; DHPR, dihydropyridine receptor.

synaptic plasticity, neuronal growth, and survival. However, in case of AD, Ca^{2+} flux gets perturbed in response to increased oxidative stress and disturbed energy metabolism that alter the normal functioning of glutamate receptor, glucose transporters, and ion-motive ATPases (Itkin et al., 2011). For instance, accumulated Aβ has been found to elevate the cellular Ca^{2+} ion level by plasma membrane L-type Ca^{2+} channels and Na^+/K^+-ATPase activity in the hippocampus and cortex region in the brain thereby causing excessive excitatory responses, i.e., glutamate excitotoxicity and ultimately neuronal death (Camandola & Mattson, 2011). Interestingly, catalytic subunit of γ-secretase, i.e., presenilin-1 is also found to be responsible for leaking Ca^{2+} ion from endoplasmic reticulum to the cytoplasm through Ca^{2+} leak channels and, thus increases the cellular load of Ca^{2+} ion in AD brain (Tu et al., 2006). Moreover, recent studies reported the Ca^{2+} homeostasis disrupting role of transient receptor potential (TRP) channels in AD. Thus, increased intracellular Ca^{2+} ion also modulates amyloid-β precursor protein (AβPP) processing and affects multiple downstream cascades, including tau metabolism, suppression of housekeeping genes, and loss of autophagic function, thereby exacerbating the symptoms of AD (Yamamoto, Wajima, Hara, Nishida, & Mori, 2007). Furthermore, K^+ channel aberrations have also been reported in AD patients. Since the potassium channel plays a crucial role in action potential generation and in maintaining the resting potential, therefore, any hindrance in K^+ channel causes impaired neurotransmission leading to neuronal damage. Further, accumulated Aβ has shown to inhibit voltage-dependent fast-inactivating K^+ currents in hippocampal neurons (Poulopoulou et al., 2010). Moreover, voltage-gated K^+ channels ($K_v1.3$, $K_v1.5$) and calcium-activated K^+ channel (KCNN4/KCa3.1) have been reported to cause neurodegeneration in response to neuroinflammation caused by toxic Aβ via microglial activation (Kaushal, Koeberle, Wang, & Schlichter, 2007). Likewise, the K_v3 subfamilies of K^+ channel subunits, which possess the ability of fast repolarization of action potential, are also found to be compromised and downregulated in AD (Bhullar & Rupasinghe, 2013). In another case, 113-pSTEA-sensitive K^+ channel was absent in fibroblasts from AD patients compared with normal human fibroblasts in peripheral tissues. Thus, affected K^+ channel activity also causes intracellular Ca^{2+} overload, resulting in altered neuronal excitability that may eventually lead to neuronal death (Mocali et al., 2014). Recently, a new intracellular chloride channel 1 (CLIC1) has been identified on plasma membrane of activated microglial cells in the hippocampus of mild AD patients. Upon Aβ stimulation of microglia, CLIC1 channels get highly expressed and are responsible for the change in membrane anion permeability of the neuron thereby leading to neuronal death (Novarino et al., 2004). In addition to these channels, nAChR also acts as a major culprit of AD brain because cholinergic depletion may increase the production of Aβ and aggravate its neurotoxicity by altering signal transduction events coupled with cholinergic neurotransmission (Buckingham, Jones, Brown, & Sattelle, 2009). Further, the expressions of nAChR subtypes, i.e., a7 and a4b2 are reported to be highly expressed in AD-affected brain regions, thereby suggesting a role of these receptors in the AD etiopathology. The degree of reduction in cortical acetylcholine correlates well with the severity of symptoms in AD (Nery et al., 2013).

2.2 Parkinson's Disease

PD is the second most common brain illness affecting 1% of the old age (60–65 years) population, primarily characterized by postural instability, bradykinesia, tremor at rest, and rigidity. There are diverse factors responsible for PD pathogenesis, including activity related to cellular Ca^{2+} overload, mitochondrial dysfunction, oxidative or metabolic stress, and particularly, few neurotoxins that make neuronal cells more prone to cell death (Massano & Bhatia, 2012). However, ion channels also play a decisive role in regulating various metabolic processes in the brain; therefore, any perturbations in channels have been reported to cause malfunctioning of neurons (Fig. 2B). For instance, ATP-sensitive potassium channel Kir6.2 along with Sur1 protein gets highly expressed in dopaminergic (DA) neurons in the substantia nigra pars compacta (SNc) and causes excitotoxicity, thus have been implicated in disease progression (Gong et al., 2014). Likewise, mutations in Kir3.2 channel (G protein-gated inwardly rectifying K1 channel) make it nonselective that leads to the increase in conduction of Na^+ ions as a substitute for highly selective K^+ ions thereby causing loss of cells in cerebellum and DA neurons in SNc (Shieh, Coghlan, Sullivan, & Gopalakrishnan, 2000). Moreover, Kir3.1 together with Kir3.2 form the G protein-gated inwardly rectifying potassium-1/2 (GIRK-1/GIRK-2) channels in neurons and direct degeneration of DA neurons in PD brain (Shieh et al., 2000). Similarly, voltage-gated T-type Ca^{2+} channels (TTCCs) and Ca^{2+}-sensitive voltage-gated A-type K^+ channels together with voltage-gated LTCCs (L-type Ca^{2+} channels) and ATP-sensitive K^+ (K-ATP) channels contribute toward basal ganglia dysfunction in SNc DA neurons thereby leading to progressive loss of neuronal firing, thus causing PD (Dragicevic, Schiemann, & Liss, 2015). Further, LTCCs act as an autonomous pacemaker of the neuron that is responsible for the maintenance of basal DA tone in target, such as the striatum. For instance, $Ca_v1.2$ and $Ca_v1.3$ LTCCs are active during pacemaking activity of DA neurons in SNc and cause a continuous increase in intracellular Ca^{2+} levels that result in increased oxidative and mitochondrial stress, which provoke selective vulnerability of neuronal cell toward death (Dragicevic et al., 2015). Besides in a subpopulation of medial SNc DA neurons, K-ATP channel activity also facilitated the switching from tonic firing to NMDAR (glutamate receptor)-mediated bursting *in vivo* that lead to phasic DA release. When glutamate binds to the receptor, they elicit opening of the NMDAR channel with the subsequent influx of Ca^{2+} into the cell. Consequently, any alterations in glutamate transmission cause dyskinesias in PD (Schiemann et al., 2012). Recently, a new ion channel, i.e., Hv1 proton channel, has been shown to be expressed in human brain microglia, immune tissues, which are required for NADPH oxidase generation of superoxide during the respiratory burst in phagocytic leukocytes and lead to neurodegeneration like PD (Wu et al., 2012).

2.3 Huntington's Disease

HD is a hereditary NDD characterized by cognitive loss, emotional imbalance, and uncoordinated movements. It is caused by an autosomal dominant mutation in Huntingtin (Htt) gene responsible for the expansion of CAG trinucleotide repeat >36 that leads to the synthesis of polyglutamine tract, thus mutated HTT (mHTT) protein is prone to aggregation and found to form intracellular accumulations in different cell types (Labbadia & Morimoto, 2013). Tong et al. investigated the functional implication of ion channels in different cell types for identifying the etiopathology of HD using mouse models. Altered Kir4.1 channel activity has been reported in mHTT expressing striatal astrocytes that disrupted the extracellular K^+ homeostasis thereby causing hyperexcitability, i.e., HD motor symptoms in striatal neurons. However, normal Kir4.1 channel is one of the predominant astrocytal K^+ channels that are crucial for maintaining resting membrane potential of the cells and also for maintaining extracellular K^+ buffering in the brain (Tong et al., 2014). Moreover, mHTT in HD is reported to alter the function of high-voltage-activated (HVA) Ca^{2+} channels (Miller & Bezprozvanny, 2010). Interestingly, apart from Ca^{2+} channel dysfunction, various other ion channels (Na^+, K^+, Cl^-) have also shown their reduced expression in numerous studies on HD mouse models;

for instance, in a study, auxiliary subunit (β4) of sodium channel is found to be downregulated in the striatum of mouse models and patients, while induced expression of β4 leads to the neurite outgrowth in Neuro2a cells in hippocampus signifying their role in neurite degeneration in transgenic mice and HD patients (Oyama et al., 2006). Similarly, other groups detected decreased expression of K^+ channel subunits (Kir2.1/Kir2.3) in striatal neurons of HD transgenic mouse models (Ariano et al., 2005). Moreover, expression of muscular ClC-1 chloride channel is also significantly reduced in R6/2 HD mouse model (Waters, Varuzhanyan, Talmadge, & Voss, 2013); therefore, functional alteration of these channels disturbs the ion homeostasis in cortical pyramidal neurons, thereby affecting the neurotransmitter release, synaptic integration, and genetic expression, which play a contributing role in cortical dysfunction in HD (Fig. 2C).

2.4 Multiple Sclerosis

MS is an immune-mediated chronic degenerative disorder of the central nervous system, where gradual demyelination takes place in patches throughout the brain and spinal cord. The typical symptoms of this disease include loss of coordination, muscular weakness, visual and lingual disturbances, which are most frequent among young individuals in industrialized societies. It has been characterized with the inflammatory neuronal damage caused by the abundance of macrophages, T lymphocytes, microglial, and dendritic cells (Fitzner & Simons, 2010). Here, neurons are primarily injured by the infiltrating lymphocytes and macrophages either by the direct cell contact or by toxicity mediators, i.e., glutamate or nitric oxide and indirectly by the loss of oligodendrocytes and myelin sheath. Apart from inflammatory mediators, redistribution of certain voltage/ligand-gated ion channels and transporters has been reported to be associated with altered electrical activity, intracellular calcium overload, compromised mitochondrial activity, and subsequently neuronal death (Fig. 2D; Meuth, Melzer, Kleinschnitz, Budde, & Wiendl, 2009). Further, alterations in the expression pattern of specific voltage-gated Na^+ channel isoforms ($Na_v1.2$, $Na_v1.5$, $Na_v1.6$, and $Na_v1.8$) have been reported in MS, and their overloading is involved in axonal degeneration followed by cerebellar dysfunction (Craner et al., 2004; Waxman, 2006). Moreover, Na_v channels trigger the Na^+ influx into axons that ultimately raise the level of intra-axonal Ca^{2+} ions and interfere with the myelination of axons thereby leading to the pathogenicity of MS. Furthermore, calcium channel ($Ca_v1.2$, $Ca_v1.3$, $Ca_v1.4$, and N-type; $Ca_v2.2$) and potassium channel isoforms ($K_v1.1$, $K_v1.2$, $K_v1.3$, and

K_v3.1b subtype) are also found to be upregulated that interfered with the conduction in demyelinating axons (Judge, Lee, Bever, & Hoffman, 2006; Kornek et al., 2001). Thus, elevated calcium levels are responsible for triggering apoptotic signals and neuronal degeneration in MS. Recently, a new cation channel, i.e., transient receptor potential melastatin-4 (TRPM4) channel has been implicated in MS that gets expressed in response to inflammatory CNS lesions (Schattling et al., 2012). However, there is no direct evidence for chloride channel dysfunction in MS, but few studies have reported autosomal recessive ClC-2 chloride channel deficiency in patients affected with leukoencephalopathy (Depienne et al., 2013).

2.5 Amyotrophic Lateral Sclerosis

ALS is a chronic motor neurodegenerative disease with invariable fatality and is characterized by substantial loss of motor neurons in motor cortex, brain stem, and spinal cord. The patients experience gradual muscular weakness, fasciculation, atrophy that ultimately leads to impairment in voluntary movement (Guatteo et al., 2007). However, exact mechanism for ALS is still illusive, but studies are going on animal models to investigate the plausible cause. Previous reports suggested that spontaneous activation of voltage-gated Na^+ channel (Na_v1.5) is responsible for the contraction of mammalian denervated muscle fibers (Kallen et al., 1990; Zona, Pieri, & Carunchio, 2006). Moreover, substantial reduction of potassium channel (K_v1.2) expression has also been observed in human sporadic ALS (Shibuya et al., 2011). The persistent Na^+ ion conduction followed by sudden drop in K^+ ion conductance is responsible for axonal hyperexcitability thereby leading to the symptoms of ALS (Kanai et al., 2006). Furthermore, the motor neurons that innervate tongue muscles are also vulnerable to degeneration in ALS which is found to be linked with differential expression of VGCCs. Other studies have also shown immunoreactivity with various calcium ion channels (L-type, N-type, P/Q-type, and T-type) in ALS patients and animal models (Gonzalez et al., 2011; Kimura et al., 1994; Lennon & Lambert, 1989). Interestingly, implications of mitochondrial channelopathy in the progression of ALS have been investigated by Israelson et al. who found that mutant superoxide dismutase 1 (SOD1) inhibits the mitochondrial voltage-dependent anion channel-1 (VDAC1/porin-1) and causes mitochondrial-dependent apoptosis, i.e., fatal paralysis in ALS. However, the plausible mechanism associated with ion channel dysfunction in ALS has been illustrated in Fig. 2E, but further researches are going

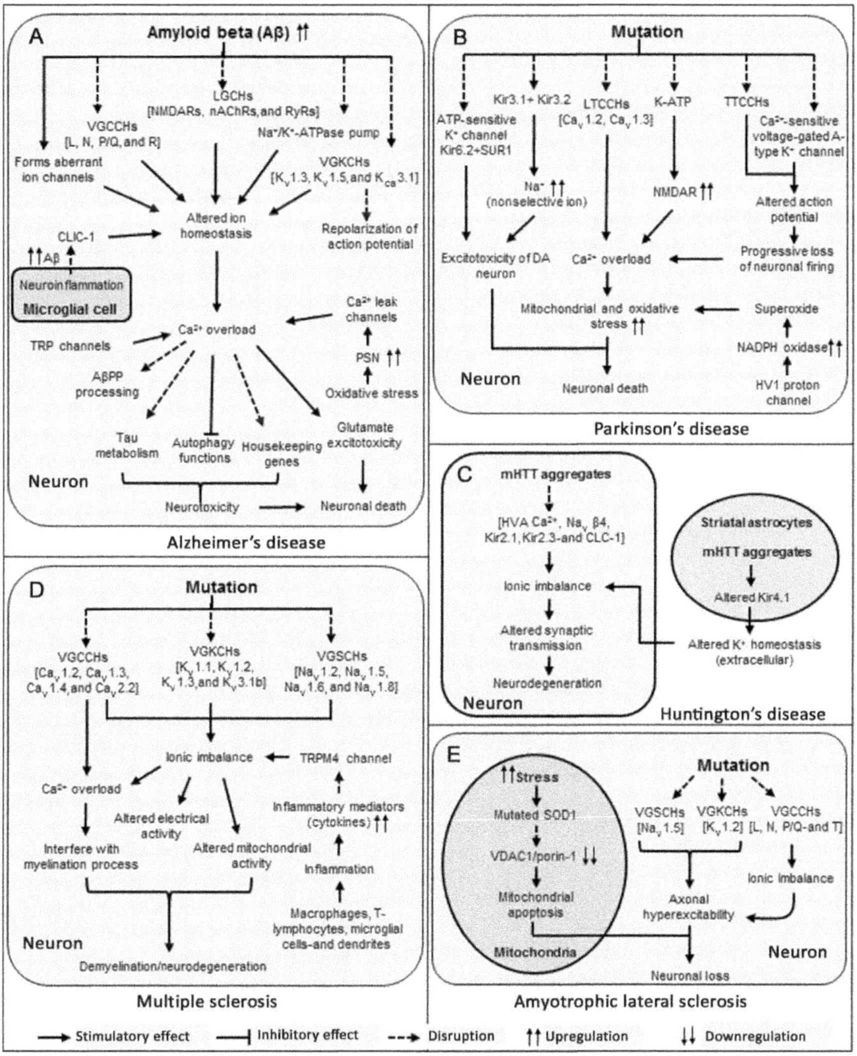

Figure 2 Molecular mechanism of ion channel-mediated pathogenesis of neurodegenerative disorders: (A) Alzheimer's disease, (B) Parkinson's disease, (C) Huntington's disease, (D) multiple sclerosis, and (E) amyotrophic lateral sclerosis. VGCCHs, voltage-gated calcium channels; LGCHs, ligand-gated channels; VGKCHs, voltage-gated potassium channels; NMDARs, *N*-methyl-D-aspartate receptors; nAChRs, nicotinic acetylcholine receptors; RyRs, ryanodine receptors; CLIC-1, chloride intracellular channel 1; TRP, transient receptor potential channel; AβPP, amyloid-β protein precursor; PSN, presenilins; LTCCHs, L-type calcium channels; TTCCHs, T-type calcium channels; VGSCHs, voltage-gated sodium channels; mHTT, mutant Huntingtin protein; VDAC1, voltage-dependent anion channel 1.

on to investigate the exact mechanism involved in its pathogenesis (Israelson et al., 2010).

Moreover, the pathological mechanism associated with these aberrant ion channels in NDDs has been summarized in Table 2.

Table 2 Defective Channels and Their Associated Mechanisms in NDDs

Neurodegenerative Disorders	Defective Channels	Associated Mechanisms	References
Alzheimer's disease	L-, N-, P/Q-, R-type Ca^{2+} channel, Ca^{2+} leak channel, $K_v1.3$, $K_v1.5$, $K_{Ca}3.1$ channel, CLIC1 channel, Aβ channel, TRP channel, Na^+/K^+ ATPase pump, ligand-gated NMDARs, nAChRs, RyRs channel	In response to Aβ burden and oxidative stress, K_v^+, Ca_v^{2+}, K_{ca}^+, Cl^-, Ca^{2+} leaky channel, and ligand-gated channels disrupt ionic homeostasis thereby leading to calcium overload, which alters the neuronal excitability and membrane ion permeability causing neurodegeneration	(Camandola and Mattson (2011), Danysz and Parsons (2012), Itkin et al. (2011), Neelands et al. (2000), Novarino et al. (2004), and Yamamoto et al. (2007)
Parkinson's disease	$Ca_v1.2$, Ca_v 1.3, T-type Ca^{2+} channel, KCa_v^+ A-type channel, K_{ATP}^+ channel, Kir3.1, Kir3.2, Kir6.2 + SUR1 channel, Hv1 proton channel, Ligand-gated NMDARs channel	Mutational defects in Ca_v^{2+}, KCa_v^+, K_{ATP}^+, Kir, Hv1, and ligand-gated channels create sodium and calcium overburden causing excitotoxicity in DA neurons that lead to progressive loss of neuronal firing	Dragicevic et al. (2015), Gong et al. (2014), Schiemann et al. (2012), Shieh et al. (2000), and Wu et al. (2012)
Huntington's disease	HVA Ca^{2+} channel, Kir2.1, Kir2.3, Kir4.1 channel, $Na_v\beta4$ channel, CLIC1 channel	Mutant HTT protein affects the function of HVA Ca^{2+}, Navβ4, Kir, and Cl^- channel thereby disrupting K^+/Cl^- homeostasis and causes hyperexcitability in striatal neurons	(Ariano et al. (2005), Labbadia and Morimoto (2013), Miller and Bezprozvanny (2010), Oyama et al. (2006), and Waters et al. (2013)

Table 2 Defective Channels and Their Associated Mechanisms in NDDs—cont'd

Neurodegenerative Disorders	Defective Channels	Associated Mechanisms	References
Multiple sclerosis	$Na_v1.2$, $Na_v1.5$, $Na_v1.6$, $Na_v1.8$ channel, $K_v1.1$, $K_v1.2$, $K_v1.3$, $K_v3.1b$ channel, $Ca_v1.2$, $Ca_v1.3$, $Ca_v1.4$, $Ca_v2.2$ channel, TRPM-4 channel	Altered Na_v^+, K_v^+, Ca_v^{2+} channels along with TRPM-4 cation channel are responsible for ionic imbalance and Ca^{2+} overburden that interfere with the myelination process leading to cerebellar dysfunction	Craner et al. (2004), Judge et al. (2006), Kornek et al. (2001), Meuth et al. (2009), Schattling et al. (2012), and Waxman (2006)
Amyotrophic lateral sclerosis	$Na_v1.5$ channel, $K_v1.2$ channel, L-, N-, P/Q-, T-type Ca_v^{2+} channels, VDAC1/porin1 channel	Stress-affected defective Na_v^+, K_v^+, Ca_v^{2+}, VDAC1/porin1 channels are responsible for ionic imbalance, axonal hyperexcitability, and mitochondrial apoptosis thereby leading to neurodegeneration	Gonzalez et al. (2011), Israelson et al. (2010), Kanai et al. (2006), Shibuya et al. (2011), and Zona et al. (2006)

3. THERAPEUTICS APPROACH TO CORRECT ALTERED CHANNEL FUNCTION IN NDDs

The malfunctioning of ion channels associated with neurological disorders is gradually being identified and makes it a remarkable area of neuroscience research. It has been spreading in a wide range of neurological outcomes, including memory, movement, and neuromuscular disorders. Most of these diseases develop in response to the genetic aberrations in the channel coding proteins that disturb the ionic balance in the brain and have been classified as neurological channelopathies. In different cases, these channels either get overly expressed or get repressed thereby contributing to the symptoms of the diseases. Therefore, it is crucial to control the ionic imbalance caused by the defective ion channels with the help of their blockers or stimulators in order to regain the ionic homeostasis. For this purpose, nowadays, several biological compounds are being implicated to target defective ion channels and their associated receptors, which reduce cell toxicity. For instance, antiepileptic drug, levetiracetam, has been used to counteract the depletion of $K_v4.2$ channel in AD in order to maintain the normal dendritic excitability and synaptic plasticity of neurons (Hall et al., 2015). Likewise, zonisamide has benefitted the PD patients by inhibiting T-type Ca^{2+} currents in the subthalamic nucleus, and this drug is now in clinical trials (Yang, Tai, Pan, & Kuo, 2014). Similarly, pyrimidine-2,4,6-triones has been used as an antagonist for $Ca_v1.3$ (L-type) channel against cell loss in PD without any side effects (Kang et al., 2012). Moreover, L-type calcium channels are blocked with the orally deliverable dihydropyridines (DHPs: Adalat, Bayer; Norvasc, Pfizer; Plendil, AstraZeneca) in early-onset PD (Surmeier, Guzman, Sanchez-Padilla, & Schumacker, 2011). However, DHPs along with memantine are also used nowadays for the treatment of late-onset AD. They reduce the toxic effect of calcium ions by blocking Ca^{2+} entry into the neurons (Nimmrich & Eckert, 2013). Furthermore, in case of HD, benzamil (Ben), a chemical agent is used to block the function of acid-sensing ion channel (ASIC) in order to maintain the ion permeability in the brain (Wong et al., 2008). Similarly, many other biological compounds have been implicated and identified so far to target the defective ion channels in the brain that are summarized in Table 3. These biological compounds have their own specific targets, where they bind to act and reduce the severity of neurological outcomes. Additionally, an alternative approach for targeting ion channels could be modulation

Table 3 Ion Channel Modulators in Neurodegenerative Disorders

Biomolecules	Specific Targets	Principle Phenotypes	References
Waixenicin A	TRPM7 calcium-permeable divalent cation channels	NDDs (neuronal cell death under ischemic stresses)	Turlova et al. (2014)
Memantine (MN)	NMDA receptor	HD, AD, PD, and SCA1	(Anitha, Nandhu, Anju, Jes, and Paulose (2011), Iizuka, Nakamura, and Hirai (2015), Tronci et al. (2014), and Tu, Okamoto, Lipton, and Xu (2014)

Table 3 Ion Channel Modulators in Neurodegenerative Disorders—cont'd

Biomolecules	Specific Targets	Principle Phenotypes	References
Tetrahydrohyperforin	TRPC channels	AD	Montecinos-Oliva, Schuller, Parodi, Melo, and Inestrosa (2014)
CFM6104	VGSCs	MS	Al-Izki et al. (2014)
Oxcarbazepine	VGSCs	MS	Lidster et al. (2013)
Zinc (Zn^{2+})	Amyloid channels	AD	Di Scala et al. (2014)
HC-067047, apamin, and charybdotoxin	TRPV4 channels	Aβ-induced oxidative stress	Zhang, Papadopoulos, and Hamel (2013)
Dalfampridine	VGKCs and ASIC1a	MS	Boiko, Kucher, Eaton, and Stockand (2013)
Mifepristone	Glucocorticoid receptors, VGCCs, and NMDA receptors	Hypobaric hypoxia (HH)-induced NDDs	Baitharu, Deep, Jain, Prasad, and Ilavazhagan (2013)
Nimodipine	VGCCs and NMDA receptors	Iron-induced neurotoxicity in brain	Lockman et al. (2012)
Methyllycaconitine, α-bungarotoxin, and mecamylamine	nAChRs	AD	Ni, Marutle, and Nordberg (2013)
Tetrodotoxin (TTX)	VGSCs	Axonal degeneration	Persson et al. (2013)
Black tea extract and rosmarinic acid	Mitochondrial ion channels	AD and PD	Camilleri et al. (2013)
Epigallocatechin gallate (EGCG)	nAChRs	AD	Zhang et al. (2014)
Dihydropyridine and benzothiazepine	VDCCs and mitochondrial Na (+)/Ca(2+) exchanger (MNCX)	AD	Fernández-Morales et al. (2012)
Scutellarin	nAChRs	Dementia	Guo, Wang, and Huang (2011)
Baicalein	NMDA receptors	NDDs	Wang et al. (2011)

Continued

Table 3 Ion Channel Modulators in Neurodegenerative Disorders—cont'd

Biomolecules	Specific Targets	Principle Phenotypes	References
Icariin	VGCCs	AD	Li, Tsai, Li, and Wang (2010)
Total flavonoids of Jiawei Wuzi Yanzong prescription	VGCCs	AD	Li, Cai, Li, and Wang (2009)
Minocycline	Mitochondrial ion channels	HD and ALS	Antonenko, Rokitskaya, Cooper, and Krasnikov (2010)
Nicotine	nAChRs	PD	Quik, Perez, and Bordia (2012)
Iptakalim	Mitochondrial ATP-sensitive potassium channels	PD	Hu et al. (2005)
4-Hydroxy-2,3-nonenal (4HN)	NMDA receptors	Oxidative stress-associated NDDs	Lu, Chan, Haughey, Lee, and Mattson (2001)
PcTx1 venom	ASIC1a	Neuronal injury following cerebral ischemia (stroke)	McCarthy, Rash, Chassagnon, King, and Widdop (2015)
GS967	VGSCs	Epilepsy	Anderson et al. (2014)
Flunarizine (FLN)	VGSCs	Migraine prophylaxis	Ye et al. (2011)
Nifedipine and Nimodipine	VGCCs	PD	Ritz et al. (2010)
Gabapentin	VGCCs	Epilepsy	Omori et al. (2009)
Amantadine	NMDA receptors	HD and PD	Verhagen et al. (2002) and Crosby, Deane, and Clarke (2003)
Dimebon	NMDA receptors	AD, HD, and schizophrenia	Egea et al. (2014)
BL-1020	$GABA_A$	Schizophrenia	Geffen et al. (2009)

of their accessory proteins, as they are responsible for controlling the channel kinetics. Moreover, neurological ion channel function may also be controlled by changes in voltage, chemical interaction, or by mechanical perturbation. These strategies may also be beneficial for therapeutic purposes to prevent neurological diseases caused by defective ion channels.

4. CONCLUSION

In this chapter, we described the potential role of ion channels in controlling the membrane physiology and maintaining brain homeostasis. Moreover, their impact on BBB and gap junctions has been demonstrated for better understanding of neurological diseases that are caused by ion channel dysfunctioning. Numerous ion channels have been catalogued that are associated with the onset of various chronic neurological disorders, and their functional roles have been elucidated. The experimental evidence reports that NDDs are supplemented by inflammations, neurotoxic protein accumulations, physiological stress, and mitochondrial dysfunctions. These pathological changes are responsible for disturbance in the normal physiological process and brain homeostasis that lead to the disease progression. We also described the architecture of ion channels associated with neurological disorders to demonstrate the mechanism of ion conductance in neurons that could be beneficial for designing novel drugs at specific target for therapeutic intervention. Further disease pathology of AD, PD, HD, MS, and ALS has been elucidated with respect to defective ion channels. These channels were found to be a causative factor for neurodegeneration by various mechanisms that include calcium overload, neuronal hyperexcitability, and altered membrane ion permeability, progressive loss of neuronal firing, mitochondrial apoptosis and interference with the myelination process. Additionally, channel modulators have been identified that play a crucial role in the reversing the chronic effects of aberrant ion channels. Moreover, unraveling their regulatory mechanism in neurodegeneration could shed light on emerging better therapeutic strategies.

ACKNOWLEDGMENTS

We would like to thanks the senior management of Delhi Technological University for constant support and encouragement. There is no conflict or competing interest declared by authors.

GLOSSARY

Voltage-gated ion channels A class of transmembrane pores forming proteins, which are establishing the resting potential, action potential, electrical firing and regulating the efflux and influx of ions across the membrane by opening and closing of gates.

Blood–Brain Barrier (BBB) A dynamic semipermeable filter that allows only the essential and rich nutrients to cross the central nervous system and interplay a complex function between different types of cell, for instance, endothelial cells, astrocytes, and pericytes. It inhibits the entry of toxic chemicals and pathogens that harm the central nervous system. This endothelial membrane is very important to regulate and maintain the brain homeostasis.

Gap junction A bunch of intracellular channels which allow direct cell–cell transfer of ions and small molecules.

Channelopathy Heterogeneous group of disorders caused due to improper function of ion channels that are located in the membrane or in various cell organelles.

REFERENCES

Abbott, N. J., Ronnback, L., & Hansson, E. (2006). Astrocyte–endothelial interactions at the blood–brain barrier. *Nature Reviews. Neuroscience*, *7*, 41–53.

Abbracchio, M. P., Burnstock, G., Verkhratsky, A., & Zimmermann, H. (2009). Purinergic signaling in the nervous system: An overview. *Trends in Neurosciences*, *32*, 19–29.

Al-Izki, S., Pryce, G., Hankey, D. J., Lidster, K., von Kutzleben, S. M., Browne, L., et al. (2014). Lesional-targeting of neuroprotection to the inflammatory penumbra in experimental multiple sclerosis. *Brain*, *137*, 92–108.

Ambrosino, P., Alaimo, A., Bartollino, S., Manocchio, L., De Maria, M., Mosca, I., et al. (2015). Epilepsy-causing mutations in Kv7.2 C-terminus affect binding and functional modulation by calmodulin. *Biochimica et Biophysica Acta*, *1852*, 1856–1866.

Anderson, L. L., Thompson, C. H., Hawkins, N. A., Nath, R. D., Petersohn, A. A., & Rajamani, S. (2014). Antiepileptic activity of preferential inhibitors of persistent sodium current. *Epilepsia*, *55*, 1274–1283.

Anitha, M., Nandhu, M. S., Anju, T. R., Jes, P., & Paulose, C. S. (2011). Targeting glutamate mediated excitotoxicity in Huntington's disease: Neural progenitors and partial glutamate antagonist—Memantine. *Medical Hypotheses*, *76*, 138–140.

Antonenko, Y. N., Rokitskaya, T. I., Cooper, A. J., & Krasnikov, B. F. (2010). Minocycline chelates Ca2+, binds to membranes, and depolarizes mitochondria by formation of Ca2+-dependent ion channels. *Journal of Bioenergetics and Biomembranes*, *42*, 151–163.

Ariano, M. A., Cepeda, C., Calvert, C. R., Flores-Hernández, J., Hernández-Echeagaray, E., Klapstein, G. J., et al. (2005). Striatal potassium channel dysfunction in Huntington's disease transgenic mice. *Journal of Neurophysiology*, *93*, 2565–2574.

Ariano, P., Erriquez, J., Gilardino, A., Ferraro, M., Lovisolo, D., & Distasi, C. (2006). Calcium signals and the in vitro migration of chick ciliary ganglion cells. *Cell Calcium*, *40*, 63–71.

Baitharu, I., Deep, S. N., Jain, V., Prasad, D., & Ilavazhagan, G. (2013). Inhibition of glucocorticoid receptors ameliorates hypobaric hypoxia induced memory impairment in rat. *Behavioural Brain Research*, *240*, 76–86.

Barnett, M. W., & Larkman, P. M. (2007). The action potential. *Practical Neurology*, *7*, 192–197.

Basali, D., & Prayson, R. A. (2015). Episodic weakness and vacuolar myopathy in hypokalemic periodic paralysis. *Journal of Clinical Neuroscience*, *22*, 1846–1847.

Bellone, R. R., Holl, H., Setaluri, V., Devi, S., Maddodi, N., Archer, S., et al. (2013). Evidence for a retroviral insertion in TRPM1 as the cause of congenital stationary night blindness and leopard complex spotting in the horse. *PLoS One*, *8*, e78280.

Bhullar, K. S., & Rupasinghe, H. P. (2013). Polyphenols: Multipotent therapeutic agents in neurodegenerative diseases. *Oxidative Medicine and Cellular Longevity*, *2013*, 891748.

Black, J. A., Newcombe, J., Trapp, B. D., & Waxman, S. G. (2007). Sodium channel expression within chronic multiple sclerosis plaques. *Journal of Neuropathology & Experimental Neurology*, *66*, 828–837.

Black, J. A., Newcombe, J., & Waxman, S. G. (2010). Astrocytes within multiple sclerosis lesions upregulate sodium channel *Nav1.5*. *Brain*, *133*, 835–846.

Boiko, N., Kucher, V., Eaton, B. A., & Stockand, J. D. (2013). Inhibition of neuronal degenerin/epithelial Na+ channels by the multiple sclerosis drug 4-aminopyridine. *The Journal of Biological Chemistry*, *288*, 9418–9427.

Boillot, M., & Baulac, S. (2015). Genetic models of focal epilepsies. *Journal of Neuroscience Methods*, *S0165–0270*, 00220–00224.

Boycott, K. M., Pearce, W. G., & Bech-Hansen, N. T. (2000). Clinical variability among patients with incomplete X-linked congenital stationary night blindness and a founder mutation in CACNA1F. *Canadian Journal of Ophthalmology*, *35*, 204–213.

Bruun, M., Hjermind, L. E., Thomsen, C., Danielsen, E., Thomsen, L. L., Pinborg, L. H., et al. (2015). Familial hemiplegic migraine type 1 associated with parkinsonism: A case report. *Case Reports in Neurology*, *14*, 84–89.

Buckingham, S. D., Jones, A. K., Brown, L. A., & Sattelle, D. B. (2009). Nicotinic acetylcholine receptor signalling: Roles in Alzheimer's disease and amyloid neuroprotection. *Pharmacological Reviews*, *61*, 39–61.

Camandola, S., & Mattson, M. P. (2011). Aberrant subcellular neuronal calcium regulation in aging and Alzheimer's disease. *Biochimica et Biophysica Acta*, *1813*, 965–973.

Camilleri, A., Zarb, C., Caruana, M., Ostermeier, U., Ghio, S., & Högen, T. (2013). Mitochondrial membrane permeabilisation by amyloid aggregates and protection by polyphenols. *Biochimica et Biophysica Acta*, *1828*, 2532–2543.

Cepeda, C., Wu, N., André, V. M., Cummings, D. M., & Levine, M. S. (2006). The corticostriatal pathway in Huntington's disease. *Progress in Neurobiology*, *81*, 253–271.

Chau, A., Roitfarb, M., Carabuena, J. M., & Camann, W. (2015). Anesthetic management of a parturient with hyperekplexia. *A & A Case Reports*, *4*, 103–106.

Chen, C. P. (2012). Prenatal diagnosis and genetic analysis of fetal akinesia deformation sequence and multiple pterygium syndrome associated with neuromuscular junction disorders: A review. *Taiwanese Journal of Obstetrics & Gynecology*, *51*, 12–17.

Chung, S., Lee, J., Joe, E. H., & Uhm, D. Y. (2001). Beta-amyloid peptide induces the expression of voltage dependent outward rectifying K+ channels in rat microglia. *Neuroscience Letters*, *300*, 67–70.

Conti, V., Aracri, P., Chiti, L., Brusco, S., Mari, F., & Marini, C. (2015). Nocturnal frontal lobe epilepsy with paroxysmal arousals due to CHRNA2 loss of function. *Neurology*, *14*, 1520–1528.

Cooper, E. C., & Jan, L. Y. (1999). Ion channel genes and human neurological disease: Recent progress, prospects, and challenges. *Proceedings of the National Academy of Sciences of the United States of America*, *96*, 4759–4766.

Craner, M. J., Newcombe, J., Black, J. A., Hartle, C., Cuzner, M. L., & Waxman, S. G. (2004). Molecular changes in neurons in multiple sclerosis: Altered axonal expression of Nav1.2 and Nav1.6 sodium channels and Na+/Ca2+ exchanger. *Proceedings of the National Academy of Sciences of the United States of America*, *101*, 8168–8173.

Crosby, N. J., Deane, K. H., & Clarke, C. E. (2003). Amantadine for dyskinesia in Parkinson's disease. *Cochrane Database of Systematic Reviews*, *2*, CD003467.

Cross, J. H., Arora, R., Heckemann, R. A., Gunny, R., Chong, K., Carr, L., et al. (2013). Neurological features of epilepsy, ataxia, sensorineural deafness, tubulopathy syndrome. *Developmental Medicine and Child Neurology*, *55*, 846–856.

Damarjian, T. G., Craner, M. J., Black, J. A., & Waxman, S. G. (2004). Upregulation and colocalization of p75 and Nav1.8 in Purkinje neurons in experimental autoimmune encephalomyelitis. *Neuroscience Letters*, *369*, 186–190.

Danysz, W., & Parsons, C. G. (2012). Alzheimer's disease, β-amyloid, glutamate, NMDA receptors and memantine—Searching for the connections. *British Journal of Pharmacology*, *167*, 324–352.

Delgado-Escueta, A. V., Koeleman, B. P., Bailey, J. N., Medina, M. T., & Durón, R. M. (2013). The quest for juvenile myoclonic epilepsy genes. *Epilepsy & Behavior*, *1*, 52–57.

Depienne, C., Bugiani, M., Dupuits, C., Galanaud, D., Touitou, V., Postma, N., et al. (2013). Brain white matter oedema due to ClC-2 chloride channel deficiency: An observational analytical study. *Lancet Neurology*, *12*, 659–668.

Di Scala, C., Troadec, J. D., Lelièvre, C., Garmy, N., Fantini, J., & Chahinian, H. (2014). Mechanism of cholesterol-assisted oligomeric channel formation by a short Alzheimer β-amyloid peptide. *Journal of Neurochemistry*, *128*, 186–195.

Dragicevic, E., Poetschke, C., Duda, J., Schlaudraff, F., Lammel, S., Schiemann, J., et al. (2014). Cav1.3 channels control D2-autoreceptor responses via NCS-1 in substantia nigra dopamine neurons. *Brain*, *137*, 2287–2302.

Dragicevic, E., Schiemann, J., & Liss, B. (2015). Dopamine midbrain neurons in health and Parkinson's disease: Emerging roles of voltage-gated calcium channels and ATP-sensitive potassium channels. *Neuroscience*, *284*, 798–814.

Eckle, V. S., Shcheglovitov, A., Vitko, I., Dey, D., Yap, C. C., Winckler, B., et al. (2014). Mechanisms by which a CACNA1H mutation in epilepsy patients increases seizure susceptibility. *The Journal of Physiology*, *592*, 795–809.

Egea, J., Romero, A., Parada, E., León, R., Dal-Cim, T., & López, M. G. (2014). Neuroprotective effect of dimebon against ischemic neuronal damage. *Neuroscience*, *267*, 11–21.

Eijkelkamp, N., Linley, J. E., Baker, M. D., Minett, M. S., Cregg, R., Werdehausen, R., et al. (2012). Neurological perspectives on voltage-gated sodium channels. *Brain*, *135*, 2585–2612.

Emery, E. C., Habib, A. M., Cox, J. J., Nicholas, A. K., Gribble, F. M., Woods, C. G., et al. (2015). Novel SCN9A mutations underlying extreme pain phenotypes: Unexpected electrophysiological and clinical phenotype correlations. *The Journal of Neuroscience*, *35*, 7674–7681.

Etemad, S., Campiglio, M., Obermair, G. J., & Flucher, B. E. (2014). The juvenile myoclonic epilepsy mutant of the calcium channel β(4) subunit displays normal nuclear targeting in nerve and muscle cells. *Channels (Austin, Tex.)*, *8*, 334–343.

Evangelista, T., Bansagi, B., Pyle, A., Griffin, H., Douroudis, K., Polvikoski, T., et al. (2015). Phenotypic variability of TRPV4 related neuropathies. *Neuromuscular Disorders*, *6*, 516–521.

Fernández-Morales, J. C., Arranz-Tagarro, J. A., Calvo-Gallardo, E., Maroto, M., Padín, J. F., & García, A. G. (2012). Stabilizers of neuronal and mitochondrial calcium cycling as a strategy for developing a medicine for Alzheimer's disease. *ACS Chemical Neuroscience*, *3*, 873–883.

Figueroa, V. A., Retamal, M. A., Cea, L. A., Salas, J. D., Vargas, A. A., Verdugo, C. A., et al. (2014). Extracellular gentamicin reduces the activity of connexion hemichannels and interferes with purinergic Ca(2+) signaling in HeLa cells. *Frontiers in Cellular Neuroscience*, *4*(8), 265.

Fitzner, D., & Simons, M. (2010). Chronic progressive multiple sclerosis—Pathogenesis of neurodegeneration and therapeutic strategies. *Current Neuropharmacology*, *8*, 305–315.

Friedman, A., Kaufer, D., & Heinemann, U. (2009). Blood–brain barrier break-down-inducing astrocytic transformation: Novel targets for the prevention of epilepsy. *Epilepsy Research*, *85*, 142–149.

Gao, Y., Yechikov, S., Vázquez, A. E., Chen, D., & Nie, L. (2013). Impaired surface expression and conductance of the KCNQ4 channel lead to sensorineural hearing loss. *Journal of Cellular and Molecular Medicine*, *17*, 889–900.

Geffen, Y., Nudelman, A., Gil-Ad, I., Rephaeli, A., Huang, M., Savitsky, K., et al. (2009). BL-1020: A novel antipsychotic drug with GABAergic activity and low catalepsy, is efficacious in a rat model of schizophrenia. *European Neuropsychopharmacology*, *1*, 1–13.

George, A. L., Jr. (2004). Inherited channelopathies associated with epilepsy. *Epilepsy Currents—American Epilepsy Society*, *4*, 65–70.

Gong, X. G., Sun, H. M., Zhang, Y., Zhang, S. J., Gao, Y. S., Feng, J., et al. (2014). Da-bu-yin-wan and qian-zheng-san to neuroprotect the mouse model of Parkinson's disease. *Evidence-Based Complementary and Alternative Medicine*, *2014*, 729195.

Gonzalez, L. E., Kotler, M. L., Vattino, L. G., Conti, E., Reisin, R. C., Mulatz, K. J., et al. (2011). Amyotrophic lateral sclerosis-immunoglobulins selectively interact with neuromuscular junctions expressing P/Q-type calcium channels. *Journal of Neurochemistry*, *119*, 826–838.

Goodenough, D. A., Paul, D. L., & Jesaitis, L. (1988). Topological distribution of two connexin32 antigenic sites in intact and split rodent hepatocyte gap junctions. *The Journal of Cell Biology*, *107*, 1817–1824.

Gouaux, E., & Mackinnon, R. (2005). Principles of selective ion transport in channels and pumps. *Science*, *310*, 1461–1465.

Guatteo, E., Carunchio, I., Pieri, M., Albo, F., Canu, N., Mercuri, N. B., et al. (2007). Altered calcium homeostasis in motor neurons following AMPA receptor but not voltage-dependent calcium channels' activation in a genetic model of amyotrophic lateral sclerosis. *Neurobiology of Disease*, *28*, 90–100.

Guo, L. L., Wang, Y. L., & Huang, Y. (2011). Effect of scutellarin on expressions of nicotinic acetylcholine receptor protein and mRNA in the brains of dementia rats. *Zhongguo Zhong xi yi jie he za zhi Zhongguo Zhongxiyi jiehe zazhi*, *31*, 789–793.

Gürsoy, S., & Erçal, D. (2015 Aug 13). Diagnostic approach to genetic causes of early-onset epileptic encephalopathy. *Journal of Child Neurology*. pii: 0883073815599262. [Epub ahead of print] PubMed PMID:26271793.

Hall, A. M., Throesch, B. T., Buckingham, S. C., Markwardt, S. J., Peng, Y., Wang, Q., et al. (2015). Tau-dependent Kv4.2 depletion and dendritic hyperexcitability in a mouse model of Alzheimer's disease. *The Journal of Neuroscience*, *35*, 6221–6230.

Hardy, J. A., & Higgins, G. A. (1992). Alzheimer's disease—The amyloid cascade hypothesis. *Science*, *256*, 184–185.

Hartfield, E. M., Yamasaki-Mann, M., Ribeiro Fernandes, H. J., Vowles, J., James, W. S., Cowley, S. A., et al. (2014). Physiological characterisation of human iPS-derived dopaminergic neurons. *PLoS One*, *9*, e87388.

Helbig, I. (2015). Genetic causes of generalized epilepsies. *Seminars in Neurology*, *35*, 288–292.

Hirose, S. (2014). Mutant GABA(A) receptor subunits in genetic (idiopathic) epilepsy. *Progress in Brain Research*, *213*, 55–85.

Hu, L. F., Wang, S., Shi, X. R., Yao, H. H., Sun, Y. H., Ding, J. H., et al. (2005). ATP-sensitive potassium channel opener iptakalim protected against the cytotoxicity of MPP + on SH-SY5Y cells by decreasing extracellular glutamate level. *Journal of Neurochemistry*, *94*, 1570–1579.

Iizuka, A., Nakamura, K., & Hirai, H. (2015). Long-term oral administration of the NMDA receptor antagonist memantine extends life span in spinocerebellar ataxia type 1 knock-in mice. *Neuroscience Letters*, *592*, 37–41.

Irie, T., Matsuzaki, Y., Sekino, Y., & Hirai, H. (2014). Kv3.3 channels harbouring a mutation of spinocerebellar ataxia type 13 alter excitability and induce cell death in cultured cerebellar Purkinje cells. *The Journal of Physiology*, *592*, 229–247.

Ishii, A., Kanaumi, T., Sohda, M., Misumi, Y., Zhang, B., Kakinuma, N., et al. (2014). Association of nonsense mutation in GABRG2 with abnormal trafficking of GABAA receptors in severe epilepsy. *Epilepsy Research*, *108*, 420–432.

Israelson, A., Arbel, N., Da Cruz, S., Ilieva, H., Yamanaka, K., Shoshan-Barmatz, V., et al. (2010). Misfolded mutant SOD1 directly inhibits VDAC1 conductance in a mouse model of inherited ALS. *Neuron*, *67*, 575–587.

Itkin, A., Dupres, V., Dufrêne, Y. F., Bechinger, B., Ruysschaert, J. M., & Raussens, V. (2011). Calcium ions promote formation of amyloid β-peptide (1-40) oligomers causally implicated in neuronal toxicity of Alzheimer's disease. *PLoS One*, *6*, e18250.

Judge, S. I., Lee, J. M., Bever, C. T., Jr., & Hoffman, P. M. (2006). Voltage-gated potassium channels in multiple sclerosis: Overview and new implications for treatment of central nervous system inflammation and degeneration. *Journal of Rehabilitation Research and Development*, *43*, 111–122.

Kallen, R. G., Sheng, Z. H., Yang, J., Chen, L. Q., Rogart, R. B., & Barchi, R. L. (1990). Primary structure and expression of a sodium channel characteristic of denervated and immature rat skeletal muscle. *Neuron*, *4*, 233–242.

Kanai, K., Kuwabara, S., Misawa, S., Tamura, N., Ogawara, K., Nakata, M., et al. (2006). Altered axonal excitability properties in amyotrophic lateral sclerosis: Impaired potassium channel function related to disease stage. *Brain*, *129*, 953–962.

Kang, S., Cooper, G., Dunne, S. F., Dusel, B., Luan, C. H., Surmeier, D. J., et al. (2012). CaV1.3-selective L-type calcium channel antagonists as potential new therapeutics for Parkinson's disease. *Nature Communications*, *3*, 1146.

Kanno, Y., & Loewenstein, W. R. (1964). Low-resistance coupling between gland cells. Some observations on intercellular contact membranes and intracellular pace. *Nature*, *201*, 194–195.

Kaushal, V., Koeberle, P. D., Wang, Y., & Schlichter, L. C. (2007). The Ca2+-activated K+ channel KCNN4/KCa3.1 contributes to microglia activation and nitric oxide-dependent neurodegeneration. *The Journal of Neuroscience*, *27*, 234–244.

Kim, J. B. (2014). Channelopathies. *Korean Journal of Pediatrics*, *57*, 1–18.

Kim, S., & Rhim, H. (2011). Effects of amyloid-β peptides on voltage-gated L-type Ca(V)1.2 and Ca(V)1.3 Ca(2+) channels. *Molecules and Cells*, *32*, 289–294.

Kimura, F., Smith, R. G., Delbono, O., Nyormoi, O., Schneider, T., Nastainczyk, W., et al. (1994). Amyotrophic lateral sclerosis patient antibodies label Ca2+ channel alpha 1 subunit. *Annals of Neurology*, *35*, 164–171.

Komuro, H., & Kumada, T. (2005). Ca2+ transients control CNS neuronal migration. *Cell Calcium*, *37*, 387–393.

Koran, M. E., Hohman, T. J., & Thornton-Wells, T. A. (2014). Genetic interactions found between calcium channel genes modulate amyloid load measured by positron emission tomography. *Human Genetics*, *133*, 85–93.

Kornek, B., Storch, M. K., Bauer, J., Djamshidian, A., Weissert, R., Wallstroem, E., et al. (2001). Distribution of a calcium channel subunit in dystrophic axons in multiple sclerosis and experimental autoimmune encephalomyelitis. *Brain*, *24*, 1114–1124.

Kremeyer, B., Lopera, F., Cox, J. J., Momin, A., Rugiero, F., Marsh, S., et al. (2010). A gain-of-function mutation in TRPA1 causes familial episodic pain syndrome. *Neuron*, *66*, 671–680.

Kumar, D., Ambasta, R. K., & Kumar, P. (2014). Mutational consequences of aberrant ion channels in neurological disorders. *The Journal of Membrane Biology*, *247*, 1083–1127.

Kurachi, Y., & North, A. (2004). Ion channels: Their structure, function and control an overview. *The Journal of Physiology*, *554*, 245–247.

Labbadia, J., & Morimoto, R. I. (2013). Huntington's disease: Underlying molecular mechanisms and emerging concepts. *Trends in Biochemical Sciences*, *38*, 378–385.

Lassche, S., Lainez, S., Bloem, B. R., van de Warrenburg, B. P., Hofmeijer, J., Lemmink, H. H., et al. (2014). A novel KCNA1 mutation causing episodic ataxia type I. *Muscle & Nerve*, *2*, 289–291.

Lawrence, T. S., Beers, W. H., & Gilula, N. B. (1978). Transmission of hormonal stimulation by cell-to-cell communication. *Nature*, *272*, 501–506.

Lee, U. S., & Cui, J. (2009). β subunit-specific modulations of BK channel function by a mutation associated with epilepsy and dyskinesia. *The Journal of Physiology*, *587*, 1481–1498.

Lemoine, D., Jiang, R., Taly, A., Chataigneau, T., Specht, A., & Grutter, T. (2012). Ligand-gated ion channels: New insights into neurological disorders and ligand recognition. *Chemical Reviews*, *112*, 6285–6318.

Lennon, V. A., & Lambert, E. H. (1989). Autoantibodies bind solubilized calcium channel-omega-conotoxin complexes from small cell lung carcinoma: A diagnostic aid for Lambert-Eaton myasthenic syndrome. *Mayo Clinic Proceedings*, *64*, 1498–1504.

Li, L., Cai, H., Li, L., & Wang, X. (2009). Effect of total flavonoids of jiawei wuzi yanzong prescription on VGCCs induced by amyloid-beta25-35 peptide in CA1 pyramidal neurons of rat hippocampal slice. *Zhongguo Zhong xi yi jie he za zhi Zhongguo Zhongxiyi jiehe zazhi*, *34*, 1975–1978.

Li, L., Tsai, H. J., Li, L., & Wang, X. M. (2010). Icariin inhibits the increased inward calcium currents induced by amyloid-beta(25-35) peptide in CA1 pyramidal neurons of neonatal rat hippocampal slice. *The American Journal of Chinese Medicine*, *38*, 113–125.

Lidster, K., Jackson, S. J., Ahmed, Z., Munro, P., Coffey, P., Giovannoni, G., et al. (2013). Neuroprotection in a novel mouse model of multiple sclerosis. *PLoS One*, *8*, e79188.

Lin, Y., Gao, H., Liu, Y., Liang, X., Liu, X., Wang, Z., et al. (2015). Two novel mutations in the bestrophin-1 gene and associated clinical observations in patients with best vitelliform macular dystrophy. *Molecular Medicine Reports*, *12*, 2584–2588.

Liu, T., Kong, D., Shah, B. P., Ye, C., Koda, S., Saunders, A., et al. (2012). Fasting activation of AgRP neurons requires NMDA receptors and involves spinogenesis and increased excitatory tone. *Neuron*, *73*, 511–522.

Lo, E. H., Dalkara, T., & Moskowitz, M. A. (2003). Mechanisms, challenges and opportunities in stroke. *Nature Reviews. Neuroscience*, *4*, 399–415.

Lockman, J. A., Geldenhuys, W. J., Jones-Higgins, M. R., Patrick, J. D., Allen, D. D., & Van der Schyf, C. J. (2012). NGP1-01, a multi-targeted polycyclic cage amine, attenuates brain endothelial cell death in iron overload conditions. *Brain Research*, *1489*, 133–139.

Lodish, H., Berk, A., Zipursky, S. L., Matsudaira, P., Baltimore, D., & Darnell, J. (2000). *Molecular cell biology* (4th ed.). New York: W.H. Freeman. Section 21.4, Chapter 15.1–15.8.

Lu, C., Chan, S. L., Haughey, N., Lee, W. T., & Mattson, M. P. (2001). Selective and biphasic effect of the membrane lipid peroxidation product 4-hydroxy-2,3-nonenal on N-methyl-D-aspartate channels. *Journal of Neurochemistry*, *78*, 577–589.

Lüscher, C., & Slesinger, P. A. (2010). Emerging roles for G protein-gated inwardly rectifying potassium (GIRK) channels in health and disease. *Nature Reviews. Neuroscience*, *5*, 301–315.

Madry, C., Haglerød, C., & Attwell, D. (2010). The role of pannexin hemichannels in the anoxic depolarization of hippocampal pyramidal cells. *Brain*, *133*, 3755–3763.

Marmorstein, A. D., Kinnick, T. R., Stanton, J. B., Johnson, A. A., Lynch, R. M., & Marmorstein, L. Y. (2015). Bestrophin-1 influences transepithelial electrical properties and Ca2+ signaling in human retinal pigment epithelium. *Molecular Vision*, *21*, 347–359.

Massano, J., & Bhatia, K. P. (2012). Clinical approach to Parkinson's disease: Features, diagnosis, and principles of management. *Cold Spring Harbor Perspectives in Medicine*, *2*, a008870.

McCarthy, C. A., Rash, L. D., Chassagnon, I. R., King, G. F., & Widdop, R. E. (2015). PcTx1 affords neuroprotection in a conscious model of stroke in hypertensive rats via selective inhibition of ASIC1a. *Neuropharmacology, 99*, 650–657.

Meisler, M. H., & Kearney, J. A. (2005). Sodium channel mutations in epilepsy and other neurological disorders. *Journal of Clinical Investigation, 115*, 2010–2017.

Meşe, G., Richard, G., & White, T. W. (2006). Gap junctions: Basic structure and function. *Journal of Investigative Dermatology, 127*, 2516–2524.

Meuth, S. G., Melzer, N., Kleinschnitz, C., Budde, T., & Wiendl, H. (2009). Multiple sclerosis a channelopathy? Targeting ion channels and transporters in inflammatory neurodegeneration. *Der Nervenarzt, 80*, 422–429.

Miceli, F., Striano, P., Soldovieri, M. V., Fontana, A., Nardello, R., & Robbiano, A. (2015). A novel KCNQ3 mutation in familial epilepsy with focal seizures and intellectual disability. *Epilepsia, 56*, e15–e20.

Michalakis, S., Koch, S., Sothilingam, V., Garcia Garrido, M., Tanimoto, N., Schulze, E., et al. (2014). Gene therapy restores vision and delays degeneration in the CNGB1(-/-) mouse model of retinitis pigmentosa. *Advances in Experimental Medicine and Biology, 801*, 733–739.

Mihaylova, V., Scola, R. H., Gervini, B., Lorenzoni, P. J., Kay, C. K., Werneck, L. C., et al. (2010). Molecular characterisation of congenital myasthenic syndromes in Southern Brazil. *Journal of Neurology, Neurosurgery & Psychiatry, 81*, 973–977.

Miller, B. R., & Bezprozvanny, I. (2010). Corticostriatal circuit dysfunction in Huntington's disease: Intersection of glutamate, dopamine and calcium. *Future Neurology, 5*, 735–756.

Mocali, A., Della Malva, N., Abete, C., Mitidieri Costanza, V. A., Bavazzano, A., & Boddi, V. (2014). Altered proteolysis in fibroblasts of Alzheimer patients with predictive implications for subjects at risk of disease. *International Journal of Alzheimer's Disease, 2014*, 520152.

Møller, R. S., Heron, S. E., Larsen, L. H., Lim, C. X., Ricos, M. G., Bayly, M. A., et al. (2015). Mutations in KCNT1 cause a spectrum of focal epilepsies. *Epilepsia, 56*, e114–e120.

Montecinos-Oliva, C., Schuller, A., Parodi, J., Melo, F., & Inestrosa, N. C. (2014). Effects of tetrahydrohyperforin in mouse hippocampal slices: Neuroprotection, long-term potentiation and TRPC channels. *Current Medicinal Chemistry, 21*, 3494–3506.

Murayama, T., Kurebayashi, N., Yamazawa, T., Oyamada, H., Suzuki, J., Kanemaru, K., et al. (2015). Divergent activity profiles of type 1 ryanodine receptor channels carrying malignant hyperthermia and central core disease mutations in the amino-terminal region. *PLoS One, 10*, e0130606.

Mylvaganam, S., Ramani, M., Krawczyk, M., & Carlen, P. L. (2014). Roles of gap junctions, connections, and pannexins in epilepsy. *Frontiers in Physiology, 5*, 172.

Nakayama, J. (2009). Progress in searching for the febrile seizure susceptibility genes. *Brain & Development, 5*, 359–365.

Neelands, T. R., King, A. P., & Macdonaldm, R. L. (2000). Functional expression of L-, N-, P/Q-, and R-type calcium channels in the human NT2-N cell line. *Journal of Neurophysiology, 84*, 2933–2944.

Nery, A. A., Magdesian, M. H., Trujillo, C. A., Sathler, L. B., Juliano, M. A., Juliano, L., et al. (2013). Rescue of amyloid-beta-induced inhibition of nicotinic acetylcholine receptors by a peptide homologous to the nicotine binding domain of the alpha 7 subtype. *PLoS One, 8*, e67194.

Ni, R., Marutle, A., & Nordberg, A. (2013). Modulation of α7 nicotinic acetylcholine receptor and fibrillar amyloid-β interactions in Alzheimer's disease brain. *Journal of Alzheimer's Disease, 33*, 841–851.

Nimmrich, V., & Eckert, A. (2013). Calcium channel blockers and dementia. *British Journal of Pharmacology, 169*, 1203–1210.

Nimmrich, V., & Gross, G. (2012). P/Q-type calcium channel modulators. *British Journal of Pharmacology*, *167*, 741–759.

Nishiguchi, K. M., Sandberg, M. A., Gorji, N., Berson, E. L., & Dryja, T. P. (2005). Cone cGMP-gated channel mutations and clinical findings in patients with achromatopsia, macular degeneration, and other hereditary cone diseases. *Human Mutation*, *25*, 248–258.

Novarino, G., Fabrizi, C., Tonini, R., Denti, M. A., Malchiodi-Albedi, F., Lauro, G. M., et al. (2004). Involvement of the intracellular ion channel CLIC1 in microglia-mediated beta-amyloid-induced neurotoxicity. *The Journal of Neuroscience*, *24*, 5322–5330.

Ohba, C., Kato, M., Takahashi, N., Osaka, H., Shiihara, T., Tohyama, J., et al. (2015). De novo KCNT1 mutations in early-onset epileptic encephalopathy. *Epilepsia*, *56*, e121–e128.

Omidi, Y., & Gumbleton, M. (2005). Biological membranes and barriers. In R. Mahato (Ed.), *Biomaterials for delivery and targeting of proteins and nucleic acids* (pp. 220–263). New York: CRC Press.

Omori, Y., Kagaya, K., Enomoto, R., Sasaki, A., Andoh, T., Nojima, H., et al. (2009). A mouse model of sural nerve injury-induced neuropathy: gabapentin inhibits pain-related behaviors and the hyperactivity of wide-dynamic range neurons in the dorsal horn. *Journal of Pharmaceutical Sciences*, *109*, 532–539.

Ördög, B., Hategan, L., Kovács, M., Seprényi, G., Kohajda, Z., Nagy, I., et al. (2015). Identification and functional characterisation of a novel KCNJ2 mutation, Val302del, causing Andersen-Tawil syndrome. *Canadian Journal of Physiology and Pharmacology*, *17*, 1–7.

Orellana, J. A., Sáez, P. J., Shoji, K. F., Schalper, K. A., Palacios-Prado, N., Velarde, V., et al. (2009). Modulation of brain hemichannels and gap junction channels by pro-inflammatory agents and their possible role in neurodegeneration. *Antioxidants & Redox Signaling*, *2*, 369–399.

Oulès, B., Del Prete, D., Greco, B., Zhang, X., Lauritzen, I., Sevalle, J., et al. (2012). Ryanodine receptor blockade reduces amyloid-β load and memory impairments in Tg2576 mouse model of Alzheimer disease. *The Journal of Neuroscience*, *32*, 11820–11834.

Oyama, F., Miyazaki, H., Sakamoto, N., Becquet, C., Machida, Y., Kaneko, K., et al. (2006). Sodium channel beta4 subunit: Down-regulation and possible involvement in neuritic degeneration in Huntington's disease transgenic mice. *Journal of Neurochemistry*, *98*, 518–529.

Parri, R. H., & Dineley, T. K. (2010). Nicotinic acetylcholine receptor interaction with beta-amyloid: Molecular, cellular, and physiological consequences. *Current Alzheimer Research*, *7*, 27–39.

Parton, L. E., Ye, C. P., Coppari, R., Enriori, P. J., Choi, B., Zhang, C. Y., et al. (2007). Glucose sensing by POMC neurons regulates glucose homeostasis and is impaired in obesity. *Nature*, *449*, 228–232.

Persson, A. K., Kim, I., Zhao, P., Estacion, M., Black, J. A., & Waxman, S. G. (2013). Sodium channels contribute to degeneration of dorsal root ganglion neurites induced by mitochondrial dysfunction in an in vitro model of axonal injury. *The Journal of Neuroscience*, *3319*, 250–261.

Pietrobon, D. (2002). Calcium channels and channelopathies of the central nervous system. *Molecular Neurobiology*, *25*, 31–50.

Poulopoulou, C., Markakis, I., Davaki, P., Tsaltas, E., Rombos, A., Hatzimanolis, A., et al. (2010). Aberrant modulation of a delayed rectifier potassium channel by glutamate in Alzheimer's disease. *Neurobiology of Disease*, *37*, 339–348.

Qin, L. Y., Wu, X. F., Block, M. L., Liu, Y. X., Bresse, G. R., Hong, J. S., et al. (2007). Systemic LPS causes chronic neuroinflammation and progressive neurodegeneration. *Glia*, *55*, 453–462.

Quik, M., Perez, X. A., & Bordia, T. (2012). Nicotine as a potential neuroprotective agent for Parkinson's disease. *Movement Disorders*, *8*, 947–957.

Quintanilla, R. A., Orellana, J. A., & Von Bernhardi, R. (2012). Understanding risk factors for Alzheimer's disease: Interplay of neuroinflammation, connexin-based communication and oxidative stress. *Archives of Medical Research*, *43*, 632–644.

Ritz, B., Rhodes, S. L., Qian, L., Schernhammer, E., Olsen, J. H., & Friis, S. (2010). L-type calcium channel blockers and Parkinson disease in Denmark. *Annals of Neurology*, *67*, 600–606.

Rubin, L. L., & Staddon, J. M. (1999). The cell biology of the blood-brain barrier. *Annual Review of Neuroscience*, *22*, 11–28.

Rus, H., Pardo, C. A., Hu, L., Darrah, E., Cudrici, C., Niculescu, T., et al. (2005). The voltage-gated potassium channel Kv1.3 is highly expressed on inflammatory infiltrates in multiple sclerosis brain. *Proceedings of the National Academy of Sciences of the United States of America*, *102*, 11094–11099.

Schattling, B., Steinbach, K., Thies, E., Kruse, M., Menigoz, A., Ufer, F., et al. (2012). TRPM4 cation channel mediates axonal and neuronal degeneration in experimental autoimmune encephalomyelitis and multiple sclerosis. *Nature Medicine*, *18*, 1805–1811.

Schiemann, J., Schlaudraff, F., Klose, V., Bingmer, M., Seino, S., Magill, P. J., et al. (2012). K-ATP channels in dopamine substantia nigra neurons control bursting and novelty-induced exploration. *Nature Neuroscience*, *15*, 1272–1280.

Shaikh, R. S., Reuter, P., Sisk, R. A., Kausar, T., Shahzad, M., Maqsood, M. I., et al. (2015). Homozygous missense variant in the human CNGA3 channel causes cone-rod dystrophy. *European Journal of Human Genetics*, *23*, 473–480.

Shibuya, K., Misawa, S., Arai, K., Nakata, M., Kanai, K., Yoshiyama, Y., et al. (2011). Markedly reduced axonal potassium channel expression in human sporadic amyotrophic lateral sclerosis: An immunohistochemical study. *Experimental Neurology*, *232*, 149–153.

Shieh, C. C., Coghlan, M., Sullivan, J. P., & Gopalakrishnan, M. (2000). Potassium channels: Molecular defects, diseases, and therapeutic opportunities. *Pharmacological Reviews*, *52*, 557–594.

Skaper, S. D., Facci, L., & Giusti, P. (2013). Intracellular ion channel CLIC1: Involvement in microglia-mediated β-amyloid peptide(1-42) neurotoxicity. *Neurochemical Research*, *38*, 1801–1808.

Sohn, J. W. (2013). Ion channels in the central regulation of energy and glucose homeostasis. *Frontiers in Neuroscience*, *7*, 85.

Strange, K. (2004). Cellular volume homeostasis. *Advances in Physiology Education*, *28*, 155–159.

Surmeier, D. J., Guzman, J. N., Sanchez-Padilla, J., & Schumacker, P. T. (2011). The role of calcium and mitochondrial oxidant stress in the loss of substantia nigra pars compacta dopaminergic neurons in Parkinson's disease. *Neuroscience*, *198*, 221–231.

Takahashi, S., Glatt, S. J., Uchiyama, M., Faraone, S. V., & Tsuang, M. T. (2015). Meta-analysis of data from the Psychiatric Genomics Consortium and additional samples supports association of CACNA1C with risk for schizophrenia. *Schizophrenia Research*, *168*, 429–433.

Tong, X., Ao, Y., Faas, G. C., Nwaobi, S. E., Xu, J., Haustein, M. D., et al. (2014). Astrocyte Kir4.1 ion channel deficits contribute to neuronal dysfunction in Huntington's disease model mice. *Nature Neuroscience*, *17*, 694–703.

Tronci, E., Fidalgo, C., Zianni, E., Collu, M., Stancampiano, R., Morelli, M., et al. (2014). Effect of memantine on L-DOPA-induced dyskinesia in the 6-OHDA-lesioned rat model of Parkinson's disease. *Neuroscience*, *265*, 245–252.

Tu, H., Nelson, O., Bezprozvanny, A., Wang, Z., Lee, S. F., Hao, Y. H., et al. (2006). Presenilins form ER Ca2+ leak channels, a function disrupted by familial Alzheimer's disease-linked mutations. *Cell*, *126*, 981–993.

Tu, S., Okamoto, S., Lipton, S. A., & Xu, H. (2014). Oligomeric Aβ-induced synaptic dysfunction in Alzheimer's disease. *Molecular Neurodegeneration*, *9*, 48.

Turlova, E., Bae, C. Y., Deurloo, M., Chen, W., Barszczyk, A., Horgen, F. D., et al. (2014 Dec 11). TRPM7 regulates axonal outgrowth and maturation of primary hippocampal neurons. *Molecular Neurobiology*. [Epub ahead of print] PubMed PMID:25502295.

Vega, J. L., Subiabre, M., Figueroa, F., Schalper, K. A., Osorio, L., González, J., et al. (2013). Role of gap junctions and hemichannels in parasitic infections. *BioMed Research International, 2013*, 589130.

Verhagen, M. L., Morris, M. J., Farmer, C., Gillespie, M., Mosby, K., Wuu, J., et al. (2002). Huntington's disease: A randomized, controlled trial using the NMDA-antagonist amantadine. *Neurology, 59*, 694–699.

Vinken, M., Vanhaecke, T., Papeleu, P., Snykers, S., Henkens, T., & Rogiers, V. (2006). Connexins and their channels in cell growth and cell death. *Cellular Signalling, 18*, 592–600.

Wacker, S. J., Jurkowski, W., Simmons, K. J., Fishwick, C. W., Johnson, A. P., Madge, D., et al. (2012). Identification of selective inhibitors of the potassium channel Kv1.1-1.2(3) by high-throughput virtual screening and automated patch clamp. *ChemMedChem, 7*, 1775–1783.

Wang, X. Y., Ren, B. W., Yong, Z. H., Xu, H. Y., Fu, Q. X., & Yao, H. B. (2015). Mutation analysis of CACNA1S and SCN4A in patients with hypokalemic periodic paralysis. *Molecular Medicine Reports, 12*, 6267–6274.

Wang, W., Wang, F., Yang, Y. J., Hu, Z. L., Long, L. H., Fu, H., et al. (2011). The flavonoid baicalein promotes NMDA receptor-dependent long-term potentiation and enhances memory. *British Journal of Pharmacology, 162*, 1364–1379.

Waters, C. W., Varuzhanyan, G., Talmadge, R. J., & Voss, A. A. (2013). Huntington disease, skeletal muscle is hyperexcitable owing to chloride and potassium channel dysfunction. *Proceedings of the National Academy of Sciences of the United States of America, 110*, 9160–9165.

Waxman, S. G. (2006). Axonal conduction and injury in multiple sclerosis: The role of sodium channels. *Nature Reviews. Neuroscience, 7*, 932–941.

Weilinger, N. L., Maslieieva, V., Bialecki, J., Sridharan, S. S., Tang, P. L., & Thompson, R. J. (2013). Ionotropic receptors and ion channels in ischemic neuronal death and dysfunction. *Acta Pharmacologica Sinica, 34*, 39–48.

White, T. W., & Paul, D. L. (1999). Genetic diseases and gene knockouts reveal diverse connexin functions. *Annual Review of Physiology, 61*, 283–310.

Winkler, P. A., Ekenstedt, K. J., Occelli, L. M., Frattaroli, A. V., Bartoe, J. T., Venta, P. J., et al. (2013). A large animal model for CNGB1 autosomal recessive retinitis pigmentosa. *PLoS One, 8*, e72229.

Wolburg, H., & Lippoldt, A. (2002). Tight junctions of the blood–brain barrier: Development, composition and regulation. *Vascular Pharmacology, 38*, 323–337.

Wong, H. K., Bauer, P. O., Kurosawa, M., Goswami, A., Washizu, C., Machida, Y., et al. (2008). Blocking acid-sensing ion channel alleviates Huntington's disease pathology via an ubiquitin–proteasome system-dependent mechanism. *Human Molecular Genetics, 17*, 3223–3235.

Wu, L. J., Wu, G., Akhavan Sharif, M. R., Baker, A., Jia, Y., Fahey, F. H., et al. (2012). The voltage-gated proton channel Hv1 enhances brain damage from ischemic stroke. *Nature Neuroscience, 15*, 565–573.

Yamamoto, S., Wajima, T., Hara, Y., Nishida, M., & Mori, Y. (2007). Transient receptor potential channels in Alzheimer's disease. *Biochimica et Biophysica Acta, 1772*, 958–967.

Yang, Y. C., Tai, C. H., Pan, M. K., & Kuo, C. C. (2014). The T-type calcium channel as a new therapeutic target for Parkinson's disease. *Pflügers Archiv—European Journal of Physiology, 466*, 747–755.

Ye, Q., Yan, L. Y., Xue, L. J., Wang, Q., Zhou, Z. K., Xiao, H., et al. (2011). Flunarizine blocks voltage-gated Na(+) and Ca(2+) currents in cultured rat cortical neurons: A possible locus of action in the prevention of migraine. *Neuroscience Letters, 487*, 394–399.

Yoshitomi, S., Takahashi, Y., Ishizuka, M., Yamaguchi, T., Watanabe, A., Nasu, H., et al. (2015). Three patients manifesting early infantile epileptic spasms associated with 2q24.3 microduplications. *Brain & Development*, *37*, 874–879.

Zhang, L., Papadopoulos, P., & Hamel, E. (2013). Endothelial TRPV4 channels mediate dilation of cerebral arteries: Impairment and recovery in cerebrovascular pathologies related to Alzheimer's disease. *British Journal of Pharmacology*, *170*, 661–670.

Zhang, X., Wu, M., Lu, F., Luo, N., He, Z. P., & Yang, H. (2014). Involvement of α7 nAChR signaling cascade in epigallocatechin gallate suppression of β-amyloid-induced apoptotic cortical neuronal insults. *Molecular Neurobiology*, *49*, 66–77.

Zhao, Y., & Zhao, B. (2013). Oxidative stress and the pathogenesis of Alzheimer's disease. *Oxidative Medicine and Cellular Longevity*, *2013*, 316523.

Zlokovic, B. V. (2008). The blood–brain barrier in health and chronic neurodegenerative disorders. *Neuron*, *57*, 178–201.

Zona, C., Pieri, M., & Carunchio, I. (2006). Voltage-dependent sodium channels in spinal cord motor neurons display rapid recovery from fast inactivation in a mouse model of amyotrophic lateral sclerosis. *Journal of Neurophysiology*, *96*, 3314–3322.

CHAPTER FOUR

Acid-Sensing Ion Channels as Potential Pharmacological Targets in Peripheral and Central Nervous System Diseases☆

Beatrice Mihaela Radu*^,†^, Adela Banciu†, Daniel Dumitru Banciu†, Mihai Radu*^,‡,1^

*Department of Neurological and Movement Sciences, Section of Anatomy and Histology, University of Verona, Verona, Italy
†Department of Anatomy, Animal Physiology and Biophysics, Faculty of Biology, University of Bucharest, Bucharest, Romania
‡Department of Life and Environmental Physics, 'Horia Hulubei' National Institute for Physics and Nuclear Engineering, Magurele, Romania
[1]Corresponding author: e-mail address: mihai.radu@univr.it

Contents

☆Thematic volume: Ion channels as therapeutic targets thematic volume.

Advances in Protein Chemistry and Structural Biology, Volume 103
ISSN 1876-1623
http://dx.doi.org/10.1016/bs.apcsb.2015.10.002

Abstract

Acid-sensing ion channels (ASICs) are widely expressed in the body and represent good sensors for detecting protons. The pH drop in the nervous system is equivalent to ischemia and acidosis, and ASICs are very good detectors in discriminating slight changes in acidity. ASICs are important pharmacological targets being involved in a variety of pathophysiological processes affecting both the peripheral nervous system (e.g., peripheral pain, diabetic neuropathy) and the central nervous system (e.g., stroke, epilepsy, migraine, anxiety, fear, depression, neurodegenerative diseases, etc.). This review discusses the role played by ASICs in different pathologies and the pharmacological agents acting on ASICs that might represent promising drugs. As the majority of abovementioned pathologies involve not only neuronal dysfunctions but also microvascular alterations, in the next future, ASICs may be also considered as potential pharmacological targets at the vasculature level. Perspectives and limitations in the use of ASICs antagonists and modulators as pharmaceutical agents are also discussed.

1. ACID-SENSING ION CHANNELS: GENERAL VIEW

1.1 Biophysical Properties: Pharmacological Relevance

1.1.1 Structural and Conformational Considerations on ASICs

Acid-sensing ion channels (ASICs) have six subunits ASIC1a, ASIC1b, ASIC2a, ASIC2b, ASIC3, ASIC4 that assemble in functional homomeric and heteromeric channels (for review, see Gründer & Pusch, 2015; Kellenberger & Schild, 2015) being functionally expressed in the peripheral nervous system (PNS) and the central nervous system (CNS). ASICs are steeply activated and slowly desensitized, and these biophysical characteristics dictate their role as sensor for protons (Gründer & Pusch, 2015) in different physiological and pathological conditions.

ASIC1a undergoes conformational changes upon proton binding, and the amino acidic residues Asp-238, Glu-239, and Asp-260, which line the finger domains, are essential for this conformational modification (Ramaswamy, MacLean, Gorfe, & Jayaraman, 2013) that precedes desensitization (Bonifacio, Lelli, & Kellenberger, 2014). Transitions between ASIC1a functional states are controlled by extracellular subunit interactions, where the open state of the channel is set by the finger-knuckle disulfide bond E235C/K393C, while the transition to the nonconducting state is set by E235C/Y389C (Gwiazda, Bonifacio, Vullo, & Kellenberger, 2015).

Mutations of the PDZ-binding motif of ASIC1a increased its surface expression and current density, while mutations of the RRGK or KEAKR motif reduced the channel surface expression and its current density

(Jing, Chu, & Zha, 2013). The PDZ-binding motif of ASIC1 and ASIC2 was proved to be essential for their protein kinase C (PKC)-dependent PICK1-regulatory trafficking (Baron et al., 2002; Joch et al., 2007), and PICK1 has multiple protein partners implicated in various neurological diseases (Focant & Hermans, 2013). Therefore, the structural and conformational changes underwent by ASICs should be considered in an extended pharmacological context, where protein–protein interactions generate a network event cascade.

1.1.2 Protons Act as Neurotransmitters

ASIC1a and ASIC2a play an important role in synaptic plasticity, learning, and memory (Wemmie et al., 2002). Recent data indicate that knocking down ASIC2a expression in primary cortical neurons interferes with the development of dendrites and dendritic spines (Wu et al., 2015). Recently, the idea that protons are acting as neurotransmitters and that ASICs represent their synaptic receptors has emerged, and this proton/ASIC pair is critical for amygdala-dependent learning and memory (Du et al., 2014).

1.1.3 ASICs Couple with a Multiple Receptors, Ion Channels, Transporters, and Pumps

An essential biophysical property of ASICs is their ability to couple with different receptors (Fig. 1) and to create functional signaling tandems that play important roles in the neuronal physiology and pathology. To date, ASIC1a is functionally coupled by the activation of Ca^{2+}/calmodulin-dependent protein kinase II (CaMKII) cascade with the glutamate receptor NMDA-R and has implications in acidotoxicity during ischemia (Gao et al., 2005). α-Methyl-5-HT, a specific agonist for 5-HT_2 receptor, potentiates ASICs by activating 5-HT_2 receptors and the corresponding PKC signaling pathway in rat primary sensory neurons and contributing to acid-mediated pain in tissue injury and inflammation (Qiu et al., 2012). G-protein-coupled receptors, such as muscarinic (M1) or somatostatin (ST1), regulate ASIC1a channel, and their agonists oxotremorine M and somatostatin, respectively, are able to block ASIC1a currents (Staruschenko & Stockand, 2007). Moreover, ASIC1a has a functional coupling with the light-activated proton pump archaerhodopsin-3 and with the voltage-gated proton channel Hv1 (Zeng et al., 2015). ASICs are reversibly inhibited by activation of GABA(A) and glycine receptors that is intermediated by the chloride channels opening (Chen, Whissell, Orser, & MacDonald, 2011).

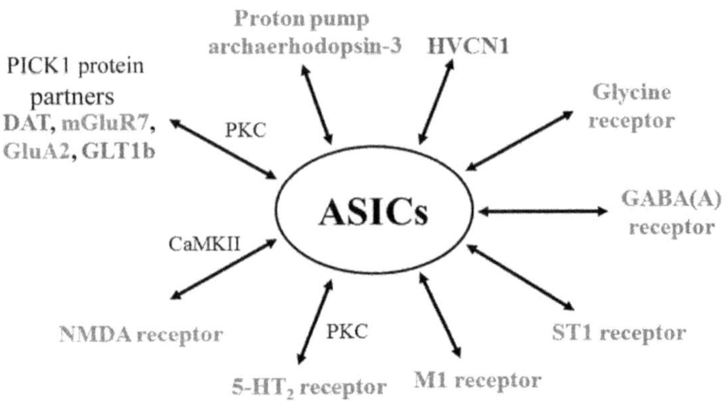

Figure 1 ASIC channels couple to different receptors, ion channels, pumps, and transporters: *N*-methyl-D-aspartate (NMDA) receptor, serotonin (5-HT_2) receptor, muscarinic acetylcholine (M1) receptor, somatostatin (ST1) receptor, gamma-aminobutyric acid (GABA(A)) receptor, glycine receptor, hydrogen voltage-gated channel 1 (HVCN1), proton pump archaerhodopsin-3, protein interacting with C kinase 1 (PICK1) protein partners—dopamine transporter (DAT), metabotropic glutamate receptor 7 (mGluR7), glutamate AMPA receptor (GluA2), glutamate transporter (GLT1b). The following color correspondences were attributed, red (dark gray in the print version): G-coupled protein receptors or metabotropic receptors; green (dark gray in the print version): ionotropic receptors; blue (dark gray in the print version): transporters; magenta (light gray in the print version): pump; violet (gray in the print version): voltage-gated ion channel.

It was also evidenced that PICK1 has multiple protein partners (ASICs, DAT, mGluR7, GluA2, GLT1b, etc.) and is targeted to the internal side of the plasma membrane contributing to the anchoring of these partners and influencing their synaptic localization and their implication in various neurological diseases (Focant & Hermans, 2013) and therefore, PICK1 might represent an integrative signaling platform.

1.2 Agonists, Antagonists, and Modulators: Pharmacological Relevance

Common drugs (see Table 1) have been reported to act on ASICs pointing out the relevance of these channels as pharmacological targets. Nonsteroid anti-inflammatory drugs (NSAIDs; flurbiprofen, ibuprofen, salicylic acid, aspirin, diclofenac, CHF5074), diuretics (amiloride), local anesthetics (tetracaine, lidocaine, propofol), antibiotics (streptomycin, neomycin), anticoagulants (nafamostat mesilate), antiparasitics (diarylamidines), and antimalaria (chloroquine) drugs have been characterized as ASICs inhibitors (Li et al., 2014; Mango et al., 2014; Voilley, 2004; Voilley et al., 2001).

Table 1 Agonists, Antagonists, and Modulators Acting on ASIC Channels

Pharmacology	ASIC1a	ASIC2	ASIC3
Agonists	pH_{50} 6.8 (Gunthorpe, Smith, Davis, & Randall, 2001; Marin, Prica, Amuzescu, Neaga, & Flonta, 2008; Neaga et al., 2005)	ASIC2a: pH_{50} 4–5 (Baron et al., 2002; Harding, Kusama, Hattori, Gautam, & Benson, 2014; Lingueglia, 2007) ASIC1a/2a: pH_{50} 4.8–5.4 (Gründer & Pusch, 2015; Harding et al., 2014) ASIC1a/2b: pH_{50} 6.4 (Gründer & Pusch, 2015)	hASIC3: pH_{50} 3.66 (fast), 3.82 (slow) (Babinski, Lê, & Séguéla, 1999) DRASIC: pH_{50} 6.5–6.7 (Sutherland, Benson, Adelman, & McCleskey, 2001) ASIC2a/ASIC3 pH_{50} 6.5 ASIC2b/ASIC3 pH_{50} 5.1 (Hesselager, Timmermann, & Ahring, 2004)
Modulators/ activators	α-Actinins (O'Bryant, Vann, & Xiong, 2014) AKAP79/150 (O'Bryant et al., 2014) Amitriptyline (Marin et al., 2008) Arachidonic acid (Smith, Cadiou, & McNaughton, 2007) Ba^{2+} (Babini, Paukert, Geisler, & Grunder, 2002) Ca^{2+}, Mg^{2+} (Babini et al., 2002; Baron, Schaefer, Lingueglia, Champigny, & Lazdunski, 2001) Calcineurin (O'Bryant et al., 2014) Dynorphins (Sherwood & Askwith, 2009) FMRF-amide (Askwith et al., 2000; Marin et al., 2008) Kallikrein (O'Bryant et al., 2014) Lactate (Immke & McCleskey, 2001a) MitTx (Bohlen et al., 2011) Neuropeptide FF (Askwith et al., 2000) Nitric oxide (NO; Cadiou et al., 2007) p11 (O'Bryant et al., 2014) Serine proteases (Marin et al., 2008; Neaga et al., 2005) Spermine (Babini et al., 2002) Sulfhydryl compounds (DTT; Cho & Askwith, 2007)	AKAP79/150 (O'Bryant et al., 2014) Calcineurin (O'Bryant et al., 2014) MitTx (Bohlen et al., 2011) Zn^{2+} (Lingueglia, 2007)	Arcaine (Li, Yu, Zhang, Cao, & Xu, 2010) Agmatine (Li et al., 2010) GMQ (Yu et al., 2010) Prolactin (Liu et al., 2015) Neuropeptide SF (Deval et al., 2003)

Continued

Table 1 Agonists, Antagonists, and Modulators Acting on ASIC Channels—cont'd

Pharmacology	ASIC1a	ASIC2	ASIC3
Antagonists	Amiloride (Waldmann, Bassilana, et al., 1997; Waldmann, Champigny, Bassilana, Heurteaux, Lazdunski, 1997) 4-Aminopyridine (4-AP; Boiko, Kucher, Eaton, & Stockand, 2013) Ca^{2+} (Zhang, Sigworth, & Canessa, 2006) CaMKII (Gao et al., 2005) CHF5074 (Mango et al., 2014) Diminazene (Chen et al., 2010) DTNB (Chu, Close, Saugstad, & Xiong, 2006) Flurbiprofen (Voilley, de Weille, Mamet, & Lazdunski, 2001) Lidocaine (Lin et al., 2011) Mambalgins (Diochot et al., 2012) Nafamostat (Ugawa et al., 2007) Ni^{2+} (Staruschenko, Dorofeeva, Bolshakov, & Stokand, 2007) Pb^{2+} (Wang, Duan, Xu, Xu, & Xu, 2006) Piroxicam (Mazumder & Borah, 2015) Psalmotoxin (PcTx1; Escoubas et al., 2000; Radu, Banciu, Banciu, & Radu, 2014; Radu, Dumitrescu, et al., 2014) Tetracaine (Leng, Lin, Cottrel, & Xiong, 2013) Trypsin (Vukicevic, Weder, Boillat, Boesch, & Kellenberger, 2006) Zinc (Zn^{2+}; Chu et al., 2004; Hey, Chu, Seeds, Simon, & Xiong, 2007)	4-Aminopyridine (4-AP; Boiko et al., 2013) Amiloride, A-317567 (Lingueglia, 2007) Nafamostat (Ugawa et al., 2007)	A-317567 (Leng & Xiong, 2013) APETx2 (Diochot, Salinas, Baron, Escoubas, & Lazdunski, 2007; Radu, Banciu, et al., 2014; Radu, Dumitrescu, et al., 2014) Aspirin (Voilley et al., 2001) Diclofenac (Voilley et al., 2001) Gd^{3+} (Babinski, Catarsi, Biagini, & Séguéla, 2000) Nafamostat (Ugawa et al., 2007) Salicylate (Voilley et al., 2001) Zn^{2+} (Jiang, Papasian, Wang, Xiong, & Chu, 2010)

Beside the pharmacological agents already used in clinical practice, there are a multitude of ASIC antagonists and modulators (Table 1), some of them being endogenous mammalian products (e.g., divalent cations, prolactin, serine proteases, spermine, etc.) while others are synthetic drugs.

2. PNS PATHOLOGIES: ASICs AS A PHARMACOLOGICAL TARGET

2.1 Peripheral Pain

2.1.1 Description

Hypoxia-induced pain is a common clinical manifestation, and acidosis induced by hypoxia contributes to the activation of ASICs (Yingjun & Xun, 2013). Indeed, acidosis is present in many painful inflammatory and ischaemic pathologies (Waldmann, Bassilana, et al., 1997; Waldmann, Champigny, et al., 1997) that occur in clinical conditions, such as hematomas, fractures, lesions, or postoperative states (Deval et al., 2011; Yingjun & Xun, 2013). Moreover, chronic hypoxia determines acidosis and activates ASICs that are involved in chemotransmission between oxygen-sensitive type I cells and chemoafferent nerve terminals (Liu, He, Dinger, & Fidone, 2011). Acid-activated channels are expressed in nociceptors of dorsal root ganglia (Deval et al., 2008, 2011; Mamet, Baron, Lazdunski, & Voilley, 2002; Poirot, Berta, Decosterd, & Kellenberger, 2006) and contribute in pain perception and hyperalgesia (Fig. 2).

2.1.2 Pharmacology

A-317567 inhibits the sustained ASIC3 current component and has a greater efficacy compared to amiloride in diminishing acid-induced pain (Dube et al., 2005). On the other hand, A-317567 has an inhibitory effect of hypoxic responses only under chronic hypoxia exposure (Liu et al., 2011). Extracellular Zn^{2+} also inhibits ASIC3 in a pH- and Ca(2+)-independent manner (Jiang et al., 2010) that might play an important role in pain perception. Prolactin potentiates ASIC activity and plays a role in hyperalgesia (Liu et al., 2015).

2.2 Diabetic Neuropathy

2.2.1 Description

Impairment of peripheral neurons in diabetes is associated with painful or painless symptoms. Most frequently can be observed a decrease in pain sensitivity which leads to ignoring chronic skin lesions or even the occurrence of acute myocardial infarction without pain. The causes of the increase in pain threshold are the *vasa nervorum* damage and nonenzymatic glycosylation

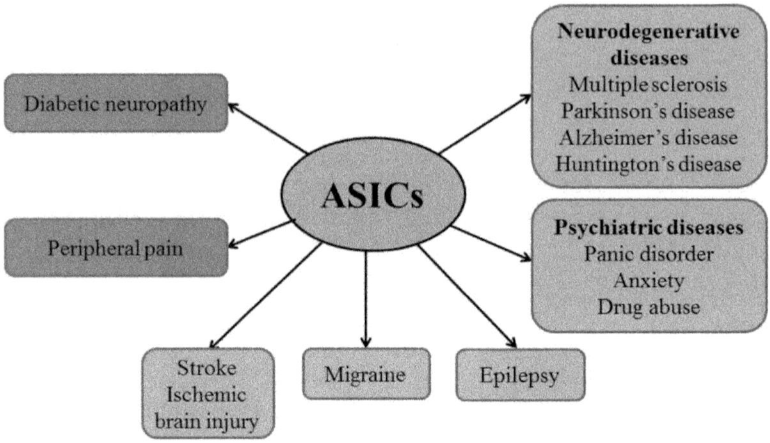

Figure 2 ASICs as pharmaceutical targets. ASIC channels play an essential role in the PNS (diabetic neuropathy, peripheral pain) and the CNS (stroke, migraine, epilepsy, psychiatric diseases, neurodegenerative diseases) pathologies being both as proton sensors and as therapeutical sites of action for different pharmacological agents. Blue (dark gray in the print version) squares are attributed to PNS pathologies, while green (light gray in the print version) squares are attributed to CNS pathologies.

of surface proteins that modify their structure and function (Cameron, Eaton, Cotter, & Tesfaye, 2001).

In a transgenic mice model for type 1 diabetes was shown that ASIC1 current inactivates faster, while ASIC2 inactivates slower in dorsal root ganglia neurons compared to the normoglycemic animals (Radu, Banciu, et al., 2014; Radu, Dumitrescu, et al., 2014). Insulin was demonstrated to regulate the ASIC1a surface trafficking and its presence in serum maintains a low level of ASIC1a on the plasma membrane (Chai, Li, Branigan, Xiong, & Simon, 2010). ASIC3$^{-/-}$ mice are protected against age-dependent glucose intolerance (Huang, Yang, Lin, Wang, & Chen, 2008) and have reduced anxiety levels and enhanced insulin sensitivity (Wu, Cheng, Sun, Wong, & Chen, 2012). Moreover, ASIC3$^{-/-}$ mice have high baseline levels of Akt phosphorylation, the signaling pathway of insulin receptors, and high number of mitochondrial DNA in adipose tissue (Huang et al., 2008).

Although extensive literature was dedicated to changes in expression of various ion channels (including ASICs) described in dorsal root ganglia sensory neurons in relationship with experimental or clinical diabetic neuropathy, yet clinical practice is limited by the unknown contribution of channelopathies versus vasculitis in diabetic neuropathy (Radu, Banciu, et al., 2014).

2.2.2 Pharmacology

In transgenic diabetic mice, PcTx1 partly reverts the ASIC1 and ASIC2 inactivation kinetics, while APETx2 maintains the ASIC3 current amplitude unaltered but slows the ASIC3 inactivation (Radu, Banciu, et al., 2014; Radu, Dumitrescu, et al., 2014). APETx2 was proved to improve the glucose control and to increase the insulin sensitivity in older ASIC3$^{+/+}$ mice (Huang et al., 2008).

Several agents for treating painful diabetic neuropathy were included and tested in clinical trials (Kaur, Hota, et al., 2011; Kaur, Pandhi, & Dutta, 2011). Among these, antidepressants like amitriptyline, imipramine, duloxetine proved to be efficient. Amitriptyline, a tricyclic antidepressant, acts on peripheral and CNS and has an analgesic effect. *In vitro*, it has been shown that amitriptyline has a dual effect on ASIC1a transients, activation at 1–10 μM, and inhibition at 100–500 μM, possibly by a direct interaction with the extracellular loop of the channel, as indicated by removal of activation by trypsin (Marin et al., 2008). Since topical application of amitriptyline (Kaur, Hota, et al., 2011; Kaur, Pandhi, et al., 2011) has been proved efficient in reducing pain, it can be assumed that ASIC1a is involved in diabetic neuropathic pain.

Several clinical trials were focused on finding an efficient therapies for painful diabetic neuropathy such as decreasing reactive oxygen species formation, preventing vascular damage, decreasing flux via the polyol pathway, preventing hyperglycaemia through activation of PKC, increasing tissue oxygenation (Javed, Alam, & Malik, 2015).

2.3 Perspectives

Acidosis was described to induce painful and painless neuropathies in both patients and animal models, and the most important pathological feature is represented by the peripheral blood flow alteration (Groeneweg, Huygen, Coderre, & Zijlstra, 2009; Nukada, 2014). Patients affected by complex regional pain syndrome have impaired microcirculation in the affected limb that is correlated with increased vasoconstriction, tissue hypoxia, and metabolic tissue acidosis (Groeneweg et al., 2009), and pain might be neuropathic, inflammatory, nociceptive, with functional nature or of mixed origin.

Since ASICs have multiple advanced functions, including in the CNS, it is useful that BBB to be impermeable for pharmacological products targeting PNS (Fig. 3). ASIC1a antagonists (Bohlen et al., 2011) and ASIC1b antagonists (Diochot et al., 2012) could be effective in the treatment of peripheral pain. This therapy should not inhibit ASIC3 (Benson, Eckert, &

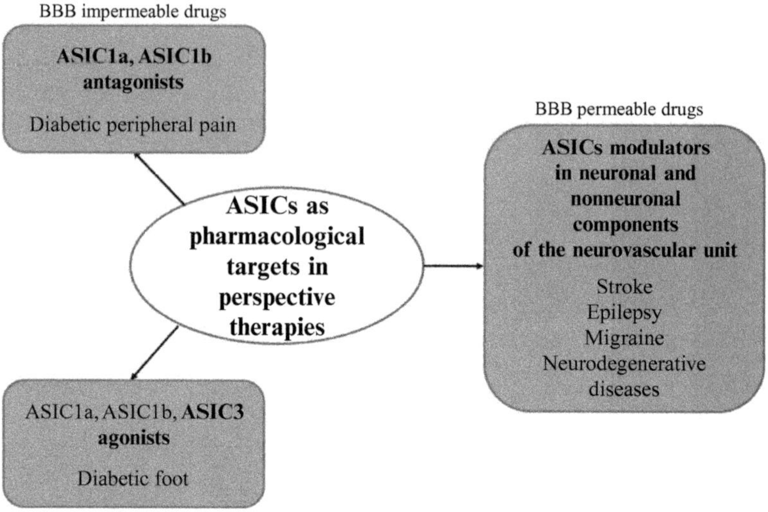

Figure 3 Foreseen ASICs targeted drugs with corresponding therapies. Perspectives in considering ASIC modulators in the therapy of the PNS diseases and in targeting ASICs expressed in hyperactive brain areas in the therapy of the CNS diseases with focal activity.

McCleskey, 1999; Immke & McCleskey, 2001b; Sutherland et al., 2001) or heteromeric ASIC3/2A (Hattori et al., 2009), in order to prevent stimuli deprivation in the case of acute myocardial infarction.

Due to chronic diabetic foot lesions ignored by patients and that are insufficiently treated, it is necessary to have a therapy that decreases peripheral pain threshold. ASIC1a agonists (Bohlen et al., 2011) and ASIC1b agonists (Diochot et al., 2012) could be effective in adapting the pain threshold of patients with diabetic foot. In addition, reduced expression of ASIC3 (Deval et al., 2008, 2011; Walder, Gautam, Wilson, Benson, & Sluka, 2011), with an effect similar to nonenzymatic glycosylation damage by altering the structure and function, suggests that ASIC3 agonist can be effective in treating diabetic alteration of pain threshold.

ASIC3 agonists may be more effective than ASIC1a and ASIC1b agonists in treating diabetic alterations of pain threshold because ASIC3 current is responsible for nonadapting pain caused by tissue acidosis (Waldmann, Bassilana, et al., 1997; Waldmann, Champigny, et al., 1997). In addition, there is a high threshold for cardiac pain, with the possibility of developing myocardial infarction with minimum symptoms or even without pain, and that creates the opportunity of using ASIC3 agonists (Benson et al., 1999; Immke & McCleskey, 2001b; Sutherland et al., 2001) and heteromeric ASIC3/2A agonists (Hattori et al., 2009) as heart pain modulators.

3. CNS PATHOLOGIES: ASICs AS A PHARMACOLOGICAL TARGET

3.1 Stroke: Ischemic Brain Injury

3.1.1 Description

The main mechanism of ischemic brain injuries is represented by vascular thrombosis, and three major therapeutic approaches have been developed, including endovascular thrombectomy, mechanical clot disruption, and augmented fibrinolysis (Nesbit, Luh, Tien, & Barnwell, 2004). There have been various attempts to delay the negative effects of ischemia, due to difficulties of thrombolytic treatment (Wardlaw et al., 2012) as soon as the occurrence of thrombosis.

Lack of vascular perfusion in affected areas leads to oxygen deprivation associated with ATP deficiency, the transition from aerobic metabolism to anaerobic production of lactic acid and protons, and acidification in the affected area (Rehncrona, 1985). ATP deficiency leads to impaired function of the Na^+-K^+-ATPase and membrane depolarization (Lipton, 1999). An increased duration of this depolarization leads to long-lasting activation of voltage-gated calcium channels and excessive glutamate neurotransmitter release (Benveniste, Drejer, Schousboe, & Diemer, 1984; Nicholls & Attwell, 1990). During cerebral ischemia, the pH decreases to 6.5–6.0 in normal glycemic conditions and below 6.0 during severe ischemia or under hyperglycemic conditions (Nedergaard, Kraig, Tanabe, & Pulsinelli, 1991; Rehncrona, 1985; Siesjo, Katsura, & Kristian, 1996). Activation of ASIC1a is involved in ischemic stroke by means of acidosis-mediated brain injury mechanisms (Hey et al., 2007; O'Bryant et al., 2014; Xiong et al., 2004). Other proton-sensitive channels or receptors, e.g., transient receptor potential melastatin 7 (TRPM7) and Na^+/H^+ exchanger isoform 1 (NHE1), have been also described to contribute in ischemic brain injury (Leng, Shi, Xiong, & Sun, 2014).

Studies on the proton dynamics (Fig. 4) in focal ischemia indicated a more accentuated drop of the extracellular pH (pH_{ext}) than of the intracellular pH (pH_{int}) in ischemic brain compared to the normal brain (Nedergaard et al., 1991). Indeed, low pH_{ext} results in lowering pH_{int}, and there are physiological mechanisms that contribute to the rapid recovery of pH_{int}. $ASIC^{-/-}$-cultured neurons have a alteration in restoring the pH_{int} compared to the $ASIC^{+/+}$ neurons (Feldhaus, Grunder, & Schnizler, 2011). On the other hand, pH_{int} regulates the properties of ASICs, such as activation, inactivation, and recovery from desensitization (Wang, Chu, et al., 2006).

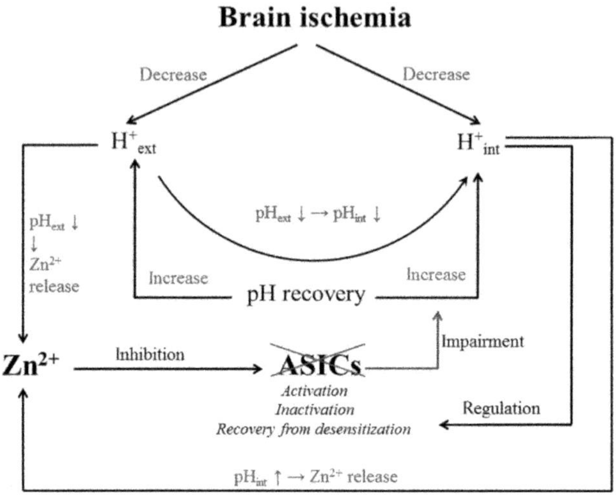

Figure 4 Role of ASICs in brain ischemia/stroke. Brain ischemia alters proton homeostasis decreasing both pH_{ext} and pH_{int}, and the drop of pH_{ext} also generates the drop of pH_{int}. There is a physiological mechanism for the pH recovery, but the deletion of ASICs impairs it. In turn, pH_{int} regulates ASICs' biophysical properties (activation, inactivation, recovery from desensitization). Zn^{2+} inhibits ASICs and exerts a protective effect in brain ischemia. On the other, Zn^{2+} release from the acidic organelles is stimulated by the pH_{int} increase and inhibited by the pH_{ext} decrease. "Brain ischemia," "ASICs," and "Zn^{2+}" marked in bold are the three main players involved in the mechanism of stroke.

3.1.2 Pharmacology

ASIC1a channel activity is modulated by brain ischemia-related signals, such as lactate, arachidonic acid, spermine, blood-derived proteases (thrombin, plasmin, and tissue plasminogen activator), CaMKII, nitric oxide (NO), dynorphins (Leng et al., 2014). ASIC1a blockade with amiloride or PcTx1 has a neuroprotective role in different animal models of stroke (Xiong et al., 2004). However, PcTx1 may be inappropriate for stroke therapy due to low blood–brain barrier permeability and long-term stability (Mishra, Verma, & Raghubir, 2010; Pignataro, Simon, & Xiong, 2007). Additionally, small molecule inhibitors, such as A-317567, NSAIDs, diamidines, and local anesthetics, are inhibiting ASICs and might play a role in the stroke intervention (Leng & Xiong, 2013; O'Bryant et al., 2014).

The alkaloid compound sophocarpine, isolated from *Sophora pachycarpa*, used on rats with transient cerebral ischemia, suggested a neuronal anti-apoptotic effect with a reduction of affected volume, probably by down-regulation of ASIC1a (Yifeng et al., 2011). ASIC1a gene knockout and

ASIC2a upregulation have a neuroprotective effect in neuronal ischemia (Johnson, Jin, Minami, Chen, & Simon, 2001; Miao, Zhang, Lin, Lu, & Qiu, 2010; Pignataro et al., 2007). Combined blocking of ASIC1a and NMDA receptors seems to have greater effect than singular therapy (Pignataro et al., 2007).

It was observed on animal models that flurbiprofen, a nonsteroid anti-inflammatory drug that inhibits ASIC1a, has a neuroprotective effect both in acute phase and late administration after stroke onset (Mishra et al., 2010). A recent study highlights that Homer1a, a scaffolding protein located at postsynaptic density, is overexpressed at long-term ASIC activation, and Homer1a knockdown exacerbates *in vitro* neuronal lesions induced by acidosis (Su et al., 2014). Moreover, Homer1a protects against traumatic brain injury via modulation of mGluR1 and mGluR5 (Luo et al., 2014).

ASICs can be modulated by several divalent ions like Zn^{2+}, Ca^{2+}, Mg^{2+}, Cu^{2+}, Ba^{2+}, Pb^{2+}, Cd^{2+}, Ni^{2+} (Babini et al., 2002; Baron et al., 2001; Staruschenko et al., 2007; Wang, Duan, et al., 2006). Extracellular Zn^{2+} (Fig. 4) inhibits the activity of ASIC1a channels and exerts an important protective effect against acidosis-induced injury of cells (Hey et al., 2007). Interestingly, the pH_{ext} drop (from 6.6 to 6.1) in cultured cortical and hippocampal neurons generates the release in micromolar amounts of Zn^{2+} from the cytosolic zinc–cysteine complexes, but not from the mitochondria or acidic organelles (Kiedrowski, 2014). On the other hand, Zn^{2+} release is determined by a Na^{+}-induced pH_{int} increase (Kiedrowski, 2011), and Zn^{2+} homeostasis correlated with extracellular/intracellular proton balance seems to play an important role in brain recovery from ischemia.

3.2 Epilepsy

3.2.1 Description

Seizures determine brain acidosis by dropping the pH from 7.35 to 6.8 (Somjen, 1984; Wang & Sonnenschein, 1955), in particular *status epilepticus* (SE) being a condition characterized by neuronal hyperexcitability leading to acidosis. Reversely, brain acidosis stops seizure activity both in brain slices (Velisek, Dreier, Stanton, Heinemann, & Moshe, 1994) and in patients' brain (Lennox, 1929). ASIC1a was proved to be the link between seizure interruption mechanisms and acidosis (Ziemann et al., 2008). It seems that ASIC1a plays the role of a sensor that "is" or "is not" triggering seizure-like activity depending on its up- or downregulation (Ziemann et al., 2008). Recent clinical data reinforce the potential role of ASICs as pharmaceutical targets in epilepsy and screening of the temporal lobe epilepsy (TLE) patients from the Han Chinese population for the six tag single-nucleotide polymorphisms (SNP) of the gene encoding ASIC1a revealed a significant allelic frequency for the rs844347-A SNP, but not of the rs590460, rs404751, rs653576, rs450423, rs706793 SNPs (Lv et al., 2011).

ASIC2 is downregulated in all hippocampal areas after SE, while ASIC1a is downregulated only in CA1–2 area rats (Biagini, Babinski, Avoli, Marcinklewicz, & Séguéla, 2011), and this downregulation persists even after diazepam or pentobarbital treatment (Biagini et al., 2011).

ASIC1a and ASIC2a expressed in piriform cortex (PC) might play an important role in seizure-like activity as this brain region is highly susceptible to the pH_{ext} changes and several pieces of evidence support this idea (Wu et al., 2015): (i) in second and third layers of PC in TLE patients and epileptic pilocarpine-induced rats, the ASIC1a positive neurons presented abnormal morphology (swollen cells with dendritic elongation) and significant diminishment of their number; (ii) in the acute phase of SE (24, 48, and 72 h), the mRNA levels in the rat PC encoding for ASIC1a have a twofold significant reduction and ASIC2a present a twofold significant increase, before gradually returning to control levels in latent phase (2 weeks); (iii) immediately after SE (1 and 2 h), fast changes in ASIC1a and ASIC2a protein levels occurred with the same pattern as their mRNA levels but preceding by far the gene expression modification; (iv) changes in ASIC subunit composition occur after SE presenting a transient pattern with a significant decrease in ASIC1a/2a complex after 1 day followed by a significant increase after 2 weeks.

ASIC3 seems to play an important role in epilepsy, and its upregulation inhibits seizure-like events by activating GABAergic interneurons (Cao et al., 2014). Indeed, ASIC3 expression is upregulated in the temporal cortex of TLE patients and in the cortex and CA1/CA3 hippocampus of epileptic pilocarpine-induced rats (Cao et al., 2014).

Focal cortical dysplasia (FCD) is a common cause of intractable epilepsy and in patients with cortical lesions affected by FCD type I and type II was detected a significant decrease of mRNA ASIC1a levels and consequent ASIC1a downregulation in cortex (Guo et al., 2014).

3.3.2 Pharmacology

Amiloride, a potent and nonselective ASIC blocker, has been proven to interfere with seizure-like activity by: (i) suppressing generalized seizures (e.g., convulsions) in pentylenetetrazole (PTZ) and maximal electroshock models (Ali, Ahmad, Pillai, & Vohora, 2004, 2005), (ii) delaying the development of kindling induced by PTZ (Ali et al., 2005), (iii) delaying the onset of the first episode of limbic seizures and the occurrence of SE following pilocarpine administration (N'Gouemo, 2008), (iv) preventing the loss of ASIC1a-positive neurons in PC and increasing the survival rate in the first 12–48 h postinjection of pilocarpine (Wu et al., 2015), and (v) reducing cerebral hypoxia-induced neurodegeneration, seizures, and audiogenic myoclonic jerks upon intracisternal administration (Tai & Truong, 2013).

PcTx1, a specific inhibitor of ASIC1a, has a controversial effects in epilepsy, in a kainate-induced model of epilepsy determining the reduction of seizures (Coryell et al., 2009) but oppositely it determines the increase of seizure severity (Ziemann et al., 2008).

APETx2, a specific ASIC3 inhibitor, after *in vivo* intracisternal injection is able to shorten the latency and to prolong the duration of seizures in pilocarpine- and PTZ-induced models and increased the incidence of generalized tonic clonic seizures (Cao et al., 2014). Moreover, after *in vitro* application on hippocampal slices, it increases the amplitude and frequency of interictal-like epileptiform activity, decreases the action potential firing frequency in interneurons and pyramidal neurons, and decreases the amplitude of miniature inhibitory postsynaptic currents (mIPSCs) in CA1 seizure neurons (Cao et al., 2014).

3.3 Migraine

3.3.1 Description

ASIC subtypes are involved in the three phases of migraine: (i) hypothalamic activity involves ASIC1, ASIC2, and ASIC3 and is driving premonitory symptoms and central modulation of meningeal afferent input; (ii) cortical-spreading depression (CSD) implies ASIC1 activity and underlies events such as aura, dural mast cell degranulation, and meningeal afferent activation; (iii) meningeal afferent activity is correlated with ASIC3 and ASIC1 activation and contributes to headache, middle meningeal artery vasodilation (Dussor, 2015).

The pathophysiology of migraine seems to be intrinsically correlated with the decrease of pH_{ext} in the meninges that contributes to headache. Activation of trigeminal neurons innervating the intracranial meninges is associated to the headache experienced during a migraine attack. In fact, approximately 80% of the Sprague–Dawley rat dural afferents respond at pH 6.0 and 50% at pH 7.0 (Yan et al., 2011), with a desensitization time constant similar to that reported for ASIC3 homomeric or heteromeric channels. Meanwhile, only 64% of the Wistar rat dural afferents are activated at pH 6.1 (Schepelmann, Ebersberger, Pawlak, Oppmann, & Messlinger, 1999) and 37% of the Hartley guinea pig dural afferents respond to pH 5.0 (Bove & Moskowitz, 1997). Other ion channels (e.g., TRPA1, TRPV1, TRPV4, and P2X) have been also identified in the dural afferents to play an important role in migraine (Yan & Dussor, 2014).

3.3.2 Pharmacology

Amiloride inhibited the CSD (the experimental correlate of aura) and the trigeminal activation in *in vivo* migraine models (Holland et al., 2012). In a pilot clinical trial, amiloride reduced aura and headache symptoms in four of seven patients with intractable aura (Holland et al., 2012).

APETx2, but not capsazepine, represses the proton stimulation of the calcitonin gene-related peptide (CGRP) secretion in cultured primary trigeminal ganglia neurons, suggesting that ASIC3, but not TRPV1, might play an important role in signal transduction mechanisms associated to migraine that occur in trigeminal ganglia neurons (Durham & Masterson, 2013). Antimigraine drugs, such as rizatriptan and onabotulinumtoxin, inhibited the proton-regulated release of CGRP from trigeminal neurons (Durham & Masterson, 2013).

3.4 Neurodegenerative Diseases

3.4.1 Description

3.4.1.1 Multiple Sclerosis

ASIC1a channels, mainly expressed in CNS, have an important contribution in the neuronal degeneration described in multiple sclerosis (MS). ASIC1a is associated with axonal degeneration and detects the pH decrease during the spinal cord acidosis in experimental autoimmune encephalomyelitis (EAE) mice used as MS model (Friese et al., 2007). Recently, MRI spectroscopy evidenced that CNS acidosis had negative effects on neuronal network activity in EAE mice, diminishing the firing and bursting of neurons and the number of presynaptic terminals (de Ceglia et al., 2015). $ASIC1^{-/-}$ mice have a significant reduction in neurodegeneration comparing to $ASIC1^{+/+}$ mice (Friese et al., 2007). ASIC1 is upregulated in axons and oligodendrocytes in brain lesions from EAE mice and the colocalization of ASIC1 expression with the presence of axonal injury marker beta amyloid precursor protein (APP) was considered to be associated with the role played by ASIC1 in the neuronal degeneration (Vergo et al., 2011). Moreover,

ASIC1 is overexpressed in axons from chronic brain lesions of patients with progressive MS (Arun et al., 2013).

3.4.1.2 Parkinson's Disease

Acidosis and the impairment of oxidative energy metabolism are neurological clinical conditions in Parkinson's disease (Bowen et al., 1995) and most probably ASICs are detecting the pH changes in brain. Another important feature in Parkinson's disease is the degradation of midbrain dopaminergic neurons that was evidenced both in patients (German, Manaye, Smith, Woodward, & Saper, 1989) and in rodent models (Drui et al., 2014), and studies on murine midbrain slices have evidenced an abundant expression of ASIC1 in these neurons (Pidoplichko & Dani, 2006). Mutations of alpha-synuclein and its autophagic degradation are important mechanisms in Parkinson's disease contributing to the dopaminergic neurons degradation (Pan, Kondo, Le, & Jankovic, 2008; Sun et al., 2011), and ASICs play an important role in the autophagic degradation (Sun et al., 2011). *Parkin*, the gene responsible for the autosomal recessive juvenile form of Parkinson's disease, promotes PICK1 monoubiquitination and suppresses PICK1-mediated potentiation of ASIC2a currents (Focant & Hermans, 2013; Joch et al., 2007) and activation of the PKC stimulates potentiation of ASIC2a currents by a PICK1-mediated mechanism (Baron et al., 2002). Actually, Parkin$^{-/-}$ mice have a potentiation of native ASIC currents in hippocampal neurons that are endogenously suppressed by *parkin* in Parkin$^{+/+}$ mice (Joch et al., 2007).

3.4.1.3 Alzheimer's Disease

The occurrence of local brain ischemia is a common feature in Alzheimer's disease (Humpel, 2011). The APP was markedly upregulated in aging slices compared with developing slices and the exposure of aging slice to acidosis (pH 6.0) triggered the release and aggregation of beta-amyloid (Marksteiner & Humpel, 2008). Oppositely, a protective effect against amyloid-induced neurotoxicity was observed when coexposing cultured hippocampal cells to mild pH 7.05–7.4 and nonfibrillar beta-amyloid (Uribe-San Martín, Herrera-Molina, Olavarría, Ramírez, & von Bernhardi, 2009). The learning and memory deficit present in Alzheimer's disease seems to be associated with a reduced function of ASIC1a channels that are inhibited by nanomolar concentrations of amyloid beta peptide (Maysami, Branigan, Simon, & Xiong, 2009).

3.4.1.4 Huntington's Disease

It is presumed that acidosis and ASICs are involved in Huntington's disease (HD). NMR spectra recorded on tissue sample collected from R6/2 transgenic mouse, an HD animal model, evidenced an enhanced concentration of lactate in the brain after the 12th week of age, especially in striatum (Tsang et al., 2006; Zacharoff et al., 2012) and cortex (Zacharoff et al., 2012), compared with wild type. The lactate increase is accompanied by a NAA decrease that was considered to be an evidence for impaired mitochondrial activity in the old R6/2 mice (Tsang et al., 2006). Oppositely, magnetic resonance spectroscopy studies on HD patients show very heterogeneous responses, but the trend is to prove the presence of lactate both in premanifest and manifest status of disease (Jenkins, Koroshetz, Beal, & Rosen, 1993; Reynolds, Prost, & Mark, 2005). Insulin secretion and sensitivity were found decreased in normoglycemic HD patients (Lalić et al., 2008), suggesting a higher neuronal sensitivity to acidosis in HD patients, most probably due to ASIC upregulation in the presence of low insulin levels (Chai et al., 2010).

3.4.2 Pharmacology

Amiloride was successfully used for neuroprotection in both preclinical and clinical studies on MS. Amiloride (10 mg/kg daily for 10 days after EAE induction) impaired the neurodegeneration in wild-type EAE mice in a similar extend as in ASIC$^{-/-}$ EAE mice (Friese et al., 2007; Vergo et al., 2011). Three years treatment with amiloride (10 mg/day) in a cohort of 14 patients significantly reduced the rate of brain atrophy and the tissue diffusivity in different brain regions, such as corpus callosum, corticospinal tract, thalamus (Arun et al., 2013), and the project was further extended in a clinical trial still in progress (http://www.ms-smart.org/). Alternatively, intraventricular administration in EAE mice of diarylamidine diminazene aceturate, an ASIC channels blocker, prevented acidosis-induced reduction of synchronized firing in cortical networks, ameliorated EAE severity, and reduced axonal derangement (de Ceglia et al., 2015).

Amiloride (10 mg/kg) administered in 1-methyl-4-phenyl-1,2,3,6-tetrahydropyridine (MPTP)-treated mice, a murine model for Parkinson's disease, impairs neurodegeneration by rescuing 94% of the dopaminergic neurons in striatum and reduces dopamine level (Arias et al., 2008). Benzamil, a derivative of amiloride, reduced the level of cytoplasmic aggregates formation in a cell model of HD, and reduced cytoplasmic aggregates, the motor deficits and increased the body weight and lifetime in R6/2 mice, an animal model of HD (Wong et al., 2008) and acted by means of a mechanism involving the ubiquitin–proteasome system and ASIC1a channels.

Paeoniflorin, the active principle from the root of *Paeoniae alba*, is commonly used to treat neurodegenerative diseases, such as Parkinson's disease, and it modulates ASICs activity and protein expression being able to protect neurons from acidosis (Sun et al., 2011). Meanwhile, the PKC activator 1-oleyl-2-acetyl-*sn*-glycerol (OAG) stimulates ASIC2a only in the presence of PDZ domain-containing protein PICK1 (Baron et al., 2002) with great relevance in Parkinson's disease.

CHF5074, a nonsteroidal anti-inflammatory derivative, has a great potential in the treatment of Alzheimer's disease and was shown to reduce ASIC1a-mediated current at pH 5.5 and to protect hippocampus against *in vitro* ischemia (Mango et al., 2014).

3.5 Psychiatric Disorders

3.5.1 Description

ASIC1 has been proved to play an important role in fear conditioning animal models that mimic human anxiety disorders. Immunohistochemical analysis indicated that ASIC1a is expressed in the bed nucleus of the stria terminalis, medial amygdala, and periaqueductal gray; these brain areas being important in the generation and expression of innate fear (Coryell et al., 2007). Indeed, ASIC1a is a chemosensor in amygdala that detects carbon dioxide and acidosis in fear-related behavior (Ziemann et al., 2009).

ASIC1-null mice present a deficit (Wemmie et al., 2003), while ASIC1-overexpressing transgenic mice have an enhanced (Wemmie et al., 2004) fear conditioning. ASIC1a$^{-/-}$ mice had a reduction in the innate fear-related behavior (Coryell et al., 2007). Restoring ASIC1a in the basolateral amygdala of ASIC1a$^{-/-}$ mice rescued context-dependent fear memory, but not unconditioned fear responses (Coryell et al., 2008). ASIC1a$^{-/-}$ mice were resistant to depression, whereas restoration of the ASIC1a expression in the amygdala suppressed the antidepressant-like phenotype (Coryell et al., 2009). ASIC1a disruption in nucleus accumbens indicated an important role in cocaine addiction-related behavior (Kreple et al., 2014).

The chromosomal location of ASIC1 indicates that it might be a key player in anxiety-like disorders. Thus, *ACCN2* gene encoding for ASIC1 is located on the chromosome 12q13.13, in the same region that was previously reported to be associated with panic disorder (Smoller et al., 2001). However, the clinical genetic screening on human ASIC1 for SNPs that represented the major allelic risk for anxiety spectrum phenotypes (e.g., susceptibility to neuroticism, major depression, generalized anxiety disorder, panic disorder, agoraphobia, social phobia, etc.) did not reveal consistent replicability of the scoring and did not provide any significant association with the studied psychiatric phenotypes (Hettema et al., 2008). In a case–control study, two SNPs (rs685012 and rs10875995) at the *ACCN2* locus were associated with panic disorder and with amygdala phenotypes susceptible to anxiety (Smoller et al., 2014).

3.5.2 Pharmacology

Intracerebroventricular administration of PcTx1 reduced trimethylthiazoline-evoked freezing in ASIC1a$^{+/+}$ mice but not ASIC1a$^{-/-}$ mice in correlation with innate fear mechanisms (Coryell et al., 2007), prevented stress-induced hyperthermia underlying anxiolytic-like effects (Dwyer et al., 2009), and had antidepressant effects (Coryell et al., 2009). Intracerebroventricular injection of A-317567 exerts antidepressant effects by blocking ASIC1a channels (Coryell et al., 2009). The antidepressant action of fluoxetine (serotonergic action), desipramine (norepinephrinergic action), and bupropion (dopaminergic action) is independent of ASIC1a mechanisms (Coryell et al., 2009). Amiloride, but not PcTx1, blocks ASICs present in the lateral amygdala neurons that are generating excitatory postsynaptic currents being involved in synaptic plasticity and amygdala-dependent fear-related learning and memory (Du et al., 2014).

3.6 Perspectives

The most important commonality of the above-described CNS pathologies is the activation of the neurovascular unit (Bertini et al., 2013; Iadecola, 2005; Lee & Pienaar, 2014; Najjar, Pearlman, Devinsky, Najjar, & Zagzag, 2013; Posada-Duque, Barreto, & Cardona-Gomez, 2014; Radu et al., 2013). ASICs have been described to be expressed in different non-neuronal components of the neurovascular unit, including microglia, vascular wall, or astrocytes. To date, ASIC1 was identified to be expressed in the wall of the rat and human cerebral arteries, and its presence is correlated with the involvement in modulating the cerebrovascular tone (Lin, Jin, Nashelsky, & Talman, 2014). ASIC1, ASIC2a, and ASIC3 expression was evidenced in cultured and *in situ* rat microglia and with role of promoting the inflammation and the microglia migration (Yu et al., 2015). Moreover, 10% of the rat astrocytes present ASIC-like currents and most of the ASICs receptors are expressed in the nucleus but not in the plasma membrane or cytoplasm as occurs in neurons (Huang et al., 2010). Opposite effects of the extracellular acidosis have been reported on different components of the neurovascular unit from activation to cell damage. Extracellular acidosis activates ASIC-like channels in 46% of the smooth muscle cells cultivated from freshly isolated cerebral artery (Chung et al., 2010). On the other hand, pH values below 6.6 determine the damage of both endothelial cells and the cholinergic neurons compared to pH 7.4, 7.0, and 6.6 that has no effect on the cell survival (Pirchl, Marksteiner, & Humpel, 2006). Therefore, promising pharmacological targets in the CNS might be considered the ASICs in different nonneuronal components of the neurovascular unit.

Considering the multiple functions played by ASICs in the CNS and the potential side effects upon their pharmacological activation or inhibition, it is necessary to create pharmacological strategies to regulate ASICs depending on their localization in a pathological area (Fig. 3). In brain areas with high activity of ASIC1a and ASIC1b, like in epilepsy, cell swelling occurs (Wu et al., 2015). This creates the therapeutical opportunity to target ASICs in certain brain areas that present specific changes in function, metabolism, or cell volume.

4. CONCLUSIONS

ASICs may seem to be promising pharmacological targets in a variety of PNS and CNS pathologies. ASICs biophysical properties, their capacity as proton sensors, and their ability to form a receptor–neurotransmitter pair definitely contribute to the rising interest in pharmacological design. On the other hand, ASICs capacity to couple with a wide range of receptors, ion channels, transporters, and pumps, and their regulation by common drugs used in the treatment of various clinical conditions (e.g., malaria, diuresis, anesthesia, inflammation, coagulation, etc.) and by endogenous secreted compounds limit the design of efficient drugs targeting these channels with reduced side effects.

ACKNOWLEDGMENTS

B.M.R. is the holder of a PhD fellowship financed by the Italian Ministry of Education, University and Research (MIUR). D.D.B. is currently supported by the Sectorial Operational Programme Human Resources Development (SOPHRD), financed by the European Social Fund and the Romanian Government under the contract number POSDRU/159/1.5/S/141531.

REFERENCES

Ali, A., Ahmad, F. J., Pillai, K. K., & Vohora, D. (2004). Evidence of the antiepileptic potential of amiloride with neuropharmacological benefits in rodent models of epilepsy and behavior. *Epilepsy & Behavior*, *5*, 322–328.

Ali, A., Ahmad, F. J., Pillai, K. K., & Vohora, D. (2005). Amiloride protects against pentylenetetrazole-induced kindling in mice. *British Journal of Pharmacology*, *145*, 880–884.

Arias, R. L., Sung, M. L., Vasylyev, D., Zhang, M. Y., Albinson, K., et al. (2008). Amiloride is neuroprotective in an MPTP model of Parkinson's disease. *Neurobiology of Disease*, *31*(3), 334–341.

Arun, T., Tomassini, V., Sbardella, E., de Ruiter, M. B., Matthews, L., Leite, M. I., et al. (2013). Targeting ASIC1 in primary progressive multiple sclerosis: Evidence of neuroprotection with amiloride. *Brain*, *136*, 106–115.

Askwith, C. C., Cheng, C., Ikuma, M., Benson, C., Price, M. P., & Welsh, M. J. (2000). Neuropeptide FF and FMRFamide potentiate acid-evoked currents from sensory neurons and proton-gated DEG/ENaC channels. *Neuron*, *26*, 133–141.

Babini, E., Paukert, M., Geisler, H. S., & Grunder, S. (2002). Alternative splicing and interaction with di- and polyvalent cations control the dynamic range of acid-sensing ion channel 1 (ASIC1). *Journal of Biological Chemistry*, *277*, 41597–41603.

Babinski, K., Catarsi, S., Biagini, G., & Séguéla, P. (2000). Mammalian ASIC2a and ASIC3 subunits co-assemble into heteromeric proton-gated channels sensitive to Gd3+. *Journal of Biological Chemistry*, *275*(37), 28519–28525.

Babinski, K., Lê, K. T., & Séguéla, P. (1999). Molecular cloning and regional distribution of a human proton receptor subunit with biphasic functional properties. *Journal of Neurochemistry*, *72*(1), 51–57.

Baron, A., Deval, E., Salinas, M., Lingueglia, E., Voilley, N., & Lazdunski, M. (2002). Protein kinase C stimulates the acid-sensing ion channel ASIC2a via the PDZ domain-containing protein PICK1. *Journal of Biological Chemistry*, *277*(52), 50463–50468.

Baron, A., Schaefer, L., Lingueglia, E., Champigny, G., & Lazdunski, M. (2001). Zn2+ and H+ are coactivators of acid-sensing ion channels. *Journal of Biological Chemistry*, *276*(38), 35361–35367.

Benson, C. J., Eckert, S. P., & McCleskey, E. W. (1999). Acid-evoked currents in cardiac sensory neurons: A possible mediator of myocardial ischemic sensation. *Circulation Research*, *84*, 921–928.

Benveniste, H., Drejer, J., Schousboe, A., & Diemer, N. H. (1984). Elevation of the extracellular concentrations of glutamate and aspartate in rat hippocampus during transient cerebral ischemia monitored by intracerebral microdialysis. *Journal of Neurochemistry*, *43*, 1369–1374.

Bertini, G., Bramanti, P., Constantin, G., Pellitteri, M., Radu, B. M., Radu, M., et al. (2013). New players in the neurovascular unit: Insights from experimental and clinical epilepsy. *Neurochemistry International*, *63*(7), 652–659.

Biagini, G., Babinski, K., Avoli, M., Marcinklewicz, M., & Séguéla, P. (2011). Regional and subunit-specific downregulation of acid-sensing ion channels in the pilocarpine model of epilepsy. *Neurobiology of Disease*, *8*, 45–58.

Bohlen, C. J., Chesler, A. T., Sharif-Naeini, R., Medzihradszky, K. F., Zhou, S., King, D., et al. (2011). A heteromeric Texas coral snake toxin targets acid-sensing ion channels to produce pain. *Nature*, *479*, 410–414.

Boiko, N., Kucher, V., Eaton, B. A., & Stockand, J. D. (2013). Inhibition of neuronal degenerin/epithelial Na channels by the multiple sclerosis drug 4-aminopyridine. *The Journal of Biological Chemistry*, *288*, 9418–9427.

Bonifacio, G., Lelli, C. I., & Kellenberger, S. (2014). Protonation controls ASIC1a activity via coordinated movements in multiple domains. *Journal of General Physiology*, *143*(1), 105–118.

Bove, G. M., & Moskowitz, M. A. (1997). Primary afferent neurons innervating guinea pig dura. *Journal of Neurophysiology*, 77(1), 299–308.

Bowen, B. C., Block, R. E., Sanchez-Ramos, J., Pattany, P. M., Lampman, D. A., Murdoch, J. B., et al. (1995). Proton MR spectroscopy of the brain in 14 patients with Parkinson disease. *American Journal of Neuroradiology*, *16*, 61–68.

Cadiou, H., Studer, M., Jones, N. G., Smith, E. S., Ballard, A., McMahon, S. B., et al. (2007). Modulation of acid-sensing ion channel activity by nitric oxide. *Journal of Neuroscience*, *27*, 13251–13260.

Cameron, N., Eaton, S. E., Cotter, M. A., & Tesfaye, S. (2001). Vascular factors and metabolic interactions in the pathogenesis of diabetic neuropathy. *Diabetologia*, *44*, 1973–1988.

Cao, Q., Wang, W., Gu, J., Jiang, G., Wang, K., Xu, Z., et al. (2014). Elevated expression of acid sensing-ion channel 3 inhibits epilepsy via activation of interneurons. *Molecular Neurobiology*. http://dx.doi.org/10.1007/s12035-014-9014-0.

Chai, S., Li, M., Branigan, D., Xiong, Z. G., & Simon, R. P. (2010). Activation of acid-sensing ion channel 1a (ASIC1a) by surface trafficking. *Journal of Biological Chemistry*, *285*(17), 13002–13011.

Chen, X., Qiu, L., Li, M., Durrnagel, S., Orser, B. A., Xiong, Z. G., et al. (2010). Diarylamidines: High potency inhibitors of acid-sensing ion channels. *Neuropharmacology*, *58*, 1045–1053.

Chen, X., Whissell, P., Orser, B. A., & MacDonald, J. F. (2011). Functional modifications of acid-sensing ion channels by ligand-gated chloride channels. *PLoS One*, *6*(7), e21970.

Cho, J. H., & Askwith, C. C. (2007). Potentiation of acid-sensing ion channels by sulfhydryl compounds. *American Journal of Physiology. Cell Physiology*, *292*, C2161–C2174.

Chu, X. P., Close, N., Saugstad, J. A., & Xiong, Z. G. (2006). ASIC1a-specific modulation of acid-sensing ion channels in mouse cortical neurons by redox reagents. *Journal of Neuroscience*, *26*, 5329–5339.

Chu, X. P., Wemmie, J. A., Wang, W. Z., Zhu, X. M., Saugstad, J. A., Price, M. P., et al. (2004). Subunit-dependent high-affinity zinc inhibition of acid-sensing ion channels. *Journal of Neuroscience*, *24*, 8678–8689.

Chung, W. S., Farley, J. M., Swenson, A., Barnard, J. M., Hamilton, G., Chiposi, R., et al. (2010). Extracellular acidosis activates ASIC-like channels in freshly isolated cerebral artery smooth muscle cells. *American Journal of Physiology—Cell Physiology*, *298*(5), C1198–C1208.

Coryell, M. W., Wunsch, A. M., Haenfler, J. M., Allen, J. E., McBride, J. L., Davidson, B. L., et al. (2008). Restoring acid-sensing ion channel-1a in the amygdala of knock-out mice rescues fear memory but not unconditioned fear responses. *Journal of Neuroscience*, *28*(51), 13738–13741.

Coryell, M. W., Wunsch, A. M., Haenfler, J. M., Allen, J. E., Schnizler, M., Ziemann, A. E., et al. (2009). Acid-sensing ion channel-1a in the amygdala, a novel therapeutic target in depression-related behavior. *Journal of Neuroscience*, *29*, 5381–5388.

Coryell, M. W., Ziemann, A. E., Westmoreland, P. J., Haenfler, J. M., Kurjakovic, Z., Zha, X. M., et al. (2007). Targeting ASIC1a reduces innate fear and alters neuronal activity in the fear circuit. *Biological Psychiatry*, *62*(10), 1140–1148.

de Ceglia, R., Chaabane, L., Biffi, E., Bergamaschi, A., Ferrigno, G., Amadio, S., et al. (2015). Down-sizing of neuronal network activity and density of presynaptic terminals by pathological acidosis are efficiently prevented by diminazene aceturate. *Brain, Behavior, and Immunity*, *45*, 263–276.

Deval, E., Baron, A., Lingueglia, E., Mazarguil, H., Zajac, J. M., & Lazdunski, M. (2003). Effects of neuropeptide SF and related peptides on acid sensing ion channel 3 and sensory neuron excitability. *Neuropharmacology*, *44*(5), 662–671.

Deval, E., Noel, J., Gasull, X., Delaunay, A., Alloui, A., Friend, V., et al. (2011). Acid-sensing ion channels in postoperative pain. *Journal of Neuroscience*, *31*, 6059–6066.

Deval, E., Noel, J., Lay, N., Alloui, A., Diochot, S., Friend, V., et al. (2008). ASIC3, a sensor of acidic and primary inflammatory pain. *The EMBO Journal*, *27*, 3047–3055.

Diochot, S., Baron, A., Salinas, M., Douguet, D., Scarzello, S., Dabert-Gay, A. S., et al. (2012). Black mamba venom peptides target acid-sensing ion channels to abolish pain. *Nature*, *490*, 552–555.

Diochot, S., Salinas, M., Baron, A., Escoubas, P., & Lazdunski, M. (2007). Peptides inhibitors of acid-sensing ion channels. *Toxicon*, *49*(2), 271–284.

Drui, G., Carnicella, S., Carcenac, C., Favier, M., Bertrand, A., Boulet, S., et al. (2014). Loss of dopaminergic nigrostriatal neurons accounts for the motivational and affective deficits in Parkinson's disease. *Molecular Psychiatry*, *19*(3), 358–367.

Du, J., Reznikov, L. R., Price, M. P., Zha, X. M., Lu, Y., Moninger, T. O., et al. (2014). Protons are a neurotransmitter that regulates synaptic plasticity in the lateral amygdala. *Proceedings of the National Academy of Sciences of the United States of America*, *111*(24), 8961–8966.

Dube, G. R., Lehto, S. G., Breese, N. M., Baker, S. J., Wang, X., Matulenko, M. A., et al. (2005). Electrophysiological and in vivo characterization of A-217567, a novel blocker of acid sensing ion channels. *Pain*, *117*, 88–96.

Durham, P. L., & Masterson, C. G. (2013). Two mechanisms involved in trigeminal CGRP release: Implications for migraine treatment. *Headache*, *53*(1), 67–80.

Dussor, G. (2015). ASICs as therapeutic targets for migraine. *Neuropharmacology*, *94*, 64–71.

Dwyer, J. M., Rizzo, S. J., Neal, S. J., Lin, Q., Jow, F., Arias, R. L., et al. (2009). Acid sensing ion channel (ASIC) inhibitors exhibit anxiolytic-like activity in preclinical pharmacological models. *Psychopharmacology (Berlin)*, *203*(1), 41–52.

Escoubas, P., de Weille, J. R., Lecoq, A., Diochot, S., Waldmann, R., Champigny, G., et al. (2000). Isolation of a tarantula toxin specific for a class of proton-gated Na+ channels. *Journal of Biological Chemistry*, *275*, 25116–25121.

Feldhaus, H., Grunder, S., & Schnizler, M. (2011). Acid-sensing ion channels modulate recovery of intracellular pH in mouse cortical neurons. *Acta Physiologica*, *201*(Suppl. 682), P281.

Focant, M. C., & Hermans, E. (2013). Protein interacting with C kinase and neurological disorders. *Synapse*, *67*, 532–540.

Friese, M. A., Craner, M. J., Etzensperger, R., Vergo, S., Wemmie, J. A., et al. (2007). Acid-sensing ion channel-1 contributes to axonal degeneration in autoimmune inflammation of the central nervous system. *Nature Medicine*, *13*, 1483–1489.

Gao, H., Duan, B., Wang, D. G., Deng, X. H., Zhang, G. Y., Xu, L., et al. (2005). Coupling between NMDA receptor and acid-sensing ion channel contributes to ischemic neuronal death. *Neuron*, *48*, 635–646.

German, D. C., Manaye, K., Smith, W. K., Woodward, D. J., & Saper, C. B. (1989). Midbrain dopaminergic cell loss in Parkinson's disease: Computer visualization. *Annals of Neurology*, *26*(4), 507–514.

Groeneweg, G., Huygen, F. J., Coderre, T. J., & Zijlstra, F. J. (2009). Regulation of peripheral blood flow in complex regional pain syndrome: Clinical implication for symptomatic relief and pain management. *BMC Musculoskeletal Disorders*, *10*, 116.

Gründer, S., & Pusch, M. (2015). Biophysical properties of acid-sensing ion channels (ASICs). *Neuropharmacology*, *94*, 9–18.

Gunthorpe, M. J., Smith, G. D., Davis, J. B., & Randall, A. D. (2001). Characterisation of a human acid-sensing ion channel (hASIC1a) endogenously expressed in HEK293 cells. *Pflügers Archiv—European Journal of Physiology*, *442*(5), 668–674.

Guo, W., Chen, X., He, J. J., Wei, Y. J., Zang, Z. L., Liu, S. Y., et al. (2014). Down-regulated expression of acid-sensing ion channel 1a in cortical lesions of patients with focal cortical dysplasia. *Journal of Molecular Neuroscience, 53*, 176–182.

Gwiazda, K., Bonifacio, G., Vullo, S., & Kellenberger, S. (2015). Extracellular subunit interactions control transitions between functional states of acid-sensing ion channel 1a. *Journal of Biological Chemistry, 290*(29), 17956–17966.

Harding, A. M., Kusama, N., Hattori, T., Gautam, M., & Benson, C. J. (2014). ASIC2 subunits facilitate expression at the cell surface and confer regulation by PSD-95. *PLoS One, 9*(4), e93797.

Hattori, T., Chen, J., Harding, A. M., Price, M. P., Lu, Y., Abboud, F. M., et al. (2009). ASIC2a and ASIC3 heteromultimerize to form pH-sensitive channels in mouse cardiac dorsal root ganglia neurons. *Circulation Research, 105*, 279–286.

Hesselager, M., Timmermann, D. B., & Ahring, P. K. (2004). pH dependency and desensitization kinetics of heterologously expressed combinations of acid-sensing ion channel subunits. *The Journal of Biological Chemistry, 279*, 11006–11015.

Hettema, J. M., An, S. S., Neale, M. C., van den Oord, E. J. C. G., Kendler, K. S., & Chen, X. (2008). Lack of association between the amiloride-sensitive cation channel 2 (ACCN2) gene and anxiety spectrum disorders. *Psychiatric Genetics, 18*, 73–79.

Hey, J. G., Chu, X. P., Seeds, J., Simon, R. P., & Xiong, Z. G. (2007). Extracellular zinc protects against acidosis-induced injury of cells expressing Ca^{2+}-permeable acid-sensing ion channels. *Stroke, 38*, 670–673.

Holland, P. R., Akerman, S., Andreou, A. P., Karsan, N., Wemmie, J. A., & Goadsby, P. J. (2012). Acid-sensing ion channel 1: A novel therapeutic target for migraine with aura. *Annals of Neurology, 72*(4), 559–563.

Huang, C., Hu, Z. L., Wu, W. N., Yu, D. F., Xiong, Q. J., Song, J. R., et al. (2010). Existence and distinction of acid-evoked currents in rat astrocytes. *Glia, 58*(12), 1415–1424.

Huang, S. J., Yang, W. S., Lin, Y. W., Wang, H. C., & Chen, C. C. (2008). Increase of insulin sensitivity and reversal of age-dependent glucose intolerance with inhibition of ASIC3. *Biochemical and Biophysical Research Communications, 371*(4), 729–734.

Humpel, C. (2011). Chronic mild cerebrovascular dysfunction as a cause for Alzheimer's disease? *Experimental Gerontology, 46*, 225–232.

Iadecola, C. (2005). Neurovascular regulation in the normal brain and in Alzheimer's disease. *Nature Reviews. Neuroscience, 5*, 347–360.

Immke, D. C., & McCleskey, E. W. (2001a). Lactate enhances the acid-sensing Na^{+} channel on ischemia-sensing neurons. *Nature Neuroscience, 4*, 869–870.

Immke, D. C., & McCleskey, E. W. (2001b). ASIC3: A lactic acid sensor for cardiac pain. *Scientific World Journal, 1*, 510–512.

Javed, S., Alam, U., & Malik, R. A. (2015). Burning through the pain: Treatments for diabetic neuropathy. *Diabetes, Obesity & Metabolism, 17*(12), 1115–1125.

Jenkins, B. G., Koroshetz, W. J., Beal, M. F., & Rosen, B. R. (1993). Evidence for impairment of energy metabolism in vivo in Huntington's disease using localized 1H NMR spectroscopy. *Neurology, 43*, 2689–2695.

Jiang, Q., Papasian, C. J., Wang, J. Q., Xiong, Z. G., & Chu, X. P. (2010). Inhibitory regulation of acid-sensing ion channel 3 by zinc. *Neuroscience, 169*(2), 574–583.

Jing, L., Chu, X. P., & Zha, X. M. (2013). Three distinct motifs within the C-terminus of acid-sensing ion channel 1a regulate its surface trafficking. *Neuroscience, 247*, 321–327.

Joch, M., Ase, A. R., Chen, C. X., MacDonald, P. A., Kontogiannea, M., Corera, A. T., et al. (2007). Parkin-mediated monoubiquitination of the PDZ protein PICK1 regulates the activity of acid-sensing ion channels. *Molecular Biology of the Cell, 18*, 3105–3118.

Johnson, M., Jin, K., Minami, M., Chen, D., & Simon, R. (2001). Global ischemia induces expression of acid-sensing ion channel 2a in rat brain. *Journal of Cerebral Blood Flow & Metabolism*, *21*, 734–740.

Kaur, H., Hota, D., Bhansali, A., Dutta, P., Bansal, D., & Chakrabarti, A. (2011). A comparative evaluation of amitriptyline and duloxetine in painful diabetic neuropathy: A randomized, double-blind, cross-over clinical trial. *Diabetes Care*, *34*(4), 818–822.

Kaur, S., Pandhi, P., & Dutta, P. (2011). Painful diabetic neuropathy: An update. *Annals of Neurosciences*, *18*(4), 168–175.

Kellenberger, S., & Schild, L. (2015). International Union of Basic and Clinical Pharmacology. XCI. Structure, function, and pharmacology of acid-sensing ion channels and the epithelial Na+ channel. *Pharmacological Reviews*, *67*(1), 1–35.

Kiedrowski, L. (2011). Cytosolic zinc release and clearance in hippocampal neurons exposed to glutamate—The role of pH and sodium. *Journal of Neurochemistry*, *117*(2), 231–243.

Kiedrowski, L. (2014). Proton-dependent zinc release from intracellular ligands. *Journal of Neurochemistry*, *130*(1), 87–96.

Kreple, C. J., Lu, Y., Taugher, R. J., Schwager-Gutman, A. L., Du, J., Stump, M., et al. (2014). Acid-sensing ion channels contribute to synaptic transmission and inhibit cocaine-evoked plasticity. *Nature Neuroscience*, *17*(8), 1083–1091.

Lalić, N. M., Marić, J., Svetel, M., Jotić, A., Stefanova, E., Lalić, K., et al. (2008). Glucose homeostasis in Huntington disease: Abnormalities in insulin sensitivity and early-phase insulin secretion. *Archives of Neurology*, *65*, 476–480.

Lee, H., & Pienaar, I. S. (2014). Disruption of the blood–brain barrier in Parkinson's disease: Curse or route to a cure? *Frontiers in Bioscience (Landmark Edition)*, *19*, 272–280.

Leng, T., Lin, J., Cottrel, J. E., & Xiong, Z. G. (2013). Subunit and frequency-dependent inhibition of acid sensing ion channels by local anesthetic tetracaine. *Molecular Pain*, *9*, 27.

Leng, T., Shi, Y., Xiong, Z. G., & Sun, D. (2014). Proton-sensitive cation channels and ion exchangers in ischemic brain injury: New therapeutic targets for stroke? *Progress in Neurobiology*, *115*, 189–209.

Leng, T. D., & Xiong, Z. G. (2013). The pharmacology and therapeutic potential of small molecule inhibitors of acid-sensing ion channels in stroke intervention. *Acta Pharmacologica Sinica*, *34*, 33–38.

Lennox, W. G. (1929). The effect on epileptic seizures of varying the composition of the respired air. *Journal of Clinical Investigation*, *VI*, 23–24.

Li, X., Fei, J., Lei, Z., Liu, K., Wu, J., Meng, T., et al. (2014). Chloroquine impairs visual transduction via modulation of acid sensing ion channel 1a. *Toxicology Letters*, *228*(3), 200–206.

Li, W. G., Yu, Y., Zhang, Z. D., Cao, H., & Xu, T. L. (2010). ASIC3 channels integrate agmatine and multiple inflammatory signals through the nonproton ligand sensing domain. *Molecular Pain*, *6*, 88.

Lin, J., Chu, X., Maysami, S., Li, M., Si, H., Cottrell, J. E., et al. (2011). Inhibition of acid-sensing ion channel currents by lidocaine in cultured mouse cortical neurons. *Anesthesia and Analgesia*, *112*, 977–981.

Lin, L. H., Jin, J., Nashelsky, M. B., & Talman, W. T. (2014). Acid-sensing ion channel 1 and nitric oxide synthase are in adjacent layers in the wall of rat and human cerebral arteries. *Journal of Chemical Neuroanatomy*, *61–62*, 161–168.

Lingueglia, E. (2007). Acid-sensing ion channels in sensory perception. *Journal of Biological Chemistry*, *282*(24), 17325–17329.

Lipton, P. (1999). Ischemic cell death in brain neurons. *Physiological Reviews*, *79*, 1431–1568.

Liu, X., He, L., Dinger, B., & Fidone, S. J. (2011). Chronic hypoxia-induced acid-sensitive ion channel expression in chemoafferent neurons contributes to chemoreceptor hypersensitivity. *American Journal of Physiology. Lung Cellular and Molecular Physiology, 301*(6), L985–L992.

Liu, T. T., Qu, Z. W., Ren, C., Gan, X., Qiu, C. Y., & Hu, W. P. (2015). Prolactin potentiates the activity of acid-sensing ion channels in female rat primary sensory neurons. *Neuropharmacology*. http://dx.doi.org/10.1016/j.neuropharm.2015.07.016.

Luo, P., Chen, T., Zhao, Y., Zhang, L., Yang, Y., Liu, W., et al. (2014). Postsynaptic scaffold protein Homer 1a protects against traumatic brain injury via regulating group I metabotropic glutamate receptors. *Cell death & disease, 5*, e1174.

Lv, R. J., He, J. S., Fu, Y. H., Zhang, Y. Q., Shao, X. Q., Wu, L. W., et al. (2011). *ASIC1a* polymorphism is associated with temporal lobe epilepsy. *Epilepsy Research, 96*, 74–80.

Mamet, J., Baron, A., Lazdunski, M., & Voilley, N. (2002). Proinflammatory mediators, stimulators of sensory neuron excitability via the expression of acid-sensing ion channels. *The Journal of Neuroscience, 22*, 10662–10670.

Mango, D., Barbato, G., Piccirilli, S., Panico, M. B., Feligioni, M., Schepisi, C., et al. (2014). Electrophysiological and metabolic effects of CHF5074 in the hippocampus: Protection against in vitro ischemia. *Pharmacological Research, 81*, 83–90.

Marin, A. N., Prica, C. M., Amuzescu, B. P., Neaga, E. I., & Flonta, M. L. (2008). ASIC1a activation by amitriptyline and FMRF-amide is removed by serine proteases. *Channels (Austin, Tex.), 2*(6), 419–428.

Marksteiner, J., & Humpel, C. (2008). Beta-amyloid expression, release and extracellular deposition in aged rat brain slices. *Molecular Psychiatry, 13*, 939–952.

Maysami, S., Branigan, D., Simon, R. P., & Xiong, Z. G. (2009). Amyloid beta peptide modulates the activity of acid-sensing ion channels in neurons. In *Society for neuroscience annual meeting*. 237.8.

Mazumder, M. K., & Borah, A. (2015). Piroxicam confer neuroprotection in cerebral ischemia by inhibiting cyclooxygenases, acid- sensing ion channel-1a and aquaporin-4: An *in silico* comparison with aspirin and nimesulide. *Bioinformation, 11*(4), 217–222.

Miao, Y., Zhang, W., Lin, Y., Lu, X., & Qiu, Y. (2010). Neuroprotective effects of ischemic preconditioning on global brain ischemia through up-regulation of acid-sensing ion channel 2a. *International Journal of Molecular Sciences, 11*, 140–153.

Mishra, V., Verma, R., & Raghubir, R. (2010). Neuroprotective effect of flurbiprofen in focal cerebral ischemia: The possible role of ASIC1a. *Neuropharmacology, 59*, 582–588.

N'Gouemo, P. (2008). Amiloride delays the onset of pilocarpine-induced seizures in rats. *Brain Research, 1222*, 230–232.

Najjar, S., Pearlman, D. M., Devinsky, O., Najjar, A., & Zagzag, D. (2013). Neurovascular unit dysfunction with blood–brain barrier hyperpermeability contributes to major depressive disorder: A review of clinical and experimental evidence. *Journal of Neuroinflammation, 10*, 142.

Neaga, E., Amuzescu, B., Dinu, C., Macri, B., Pena, F., & Flonta, M. L. (2005). Extracellular trypsin increases ASIC1a selectivity for monovalent versus divalent cations. *Journal of Neuroscience Methods, 144*(2), 241–248.

Nedergaard, M., Kraig, R. P., Tanabe, J., & Pulsinelli, W. A. (1991). Dynamics of interstitial and intracellular pH in evolving brain infarct. *American Journal of Physiology, 260*(3 Pt. 2), R581–R588.

Nesbit, G. M., Luh, G., Tien, R., & Barnwell, S. L. (2004). New and future endovascular treatment strategies for acute ischemic stroke. *Journal of Vascular and Interventional Radiology, 15*, S103–S110.

Nicholls, D., & Attwell, D. (1990). The release and uptake of excitatory amino acids. *Trends in Pharmacological Sciences, 11*, 462–468.

Nukada, H. (2014). Ischemia and diabetic neuropathy. *Handbook of Clinical Neurology, 126*, 469–487.

O'Bryant, Z., Vann, K. T., & Xiong, Z. G. (2014). Translational strategies for neuroprotection in ischemic stroke—Focusing on acid-sensing ion channel 1a. *Translational Stroke Research*, *5*(1), 59–68.

Pan, T., Kondo, S., Le, W., & Jankovic, J. (2008). The role of autophagy-lysosome pathway in neurodegeneration associated with Parkinson's disease. *Brain*, *131*, 1969–1978.

Pidoplichko, V. I., & Dani, J. A. (2006). Acid-sensitive ionic channels in midbrain dopamine neurons are sensitive to ammonium, which may contribute to hyperammonemia damage. *Proceedings of the National Academy of Sciences of the United States of America*, *103*, 11376–11380.

Pignataro, G., Simon, R. P., & Xiong, Z. G. (2007). Prolonged activation of ASIC1a and the time window for neuroprotection in cerebral ischaemia. *Brain*, *130*, 151–158.

Pirchl, M., Marksteiner, J., & Humpel, C. (2006). Effects of acidosis on brain capillary endothelial cells and cholinergic neurons: Relevance to vascular dementia and Alzheimer's disease. *Neurological Research*, *28*, 657–664.

Poirot, O., Berta, T., Decosterd, I., & Kellenberger, S. (2006). Distinct ASIC currents are expressed in rat putative nociceptors and are modulated by nerve injury. *The Journal of Physiology*, *576*, 215e234.

Posada-Duque, R. A., Barreto, G. E., & Cardona-Gomez, G. P. (2014). Protection after stroke: Cellular effectors of neurovascular unit integrity. *Frontiers in Cellular Neuroscience*, *8*, 231.

Qiu, F., Qiu, C. Y., Liu, Y. Q., Wu, D., Li, J. D., & Hu, W. P. (2012). Potentiation of acid-sensing ion channel activity by the activation of 5-HT2 receptors in rat dorsal root ganglion neurons. *Neuropharmacology*, *63*, 494–500.

Radu, B. M., Banciu, D. D., Banciu, A., & Radu, M. (2014). Diabetic neuropathy: Promises and disappointments from benchside to bedside. *Journal of Neurology & Stroke*, *1*(3), 00015.

Radu, B. M., Bramanti, P., Osculati, F., Flonta, M. L., Radu, M., Bertini, G., et al. (2013). Neurovascular unit in chronic pain. *Mediators of Inflammation*, *2013*, 648268.

Radu, B. M., Dumitrescu, D. I., Marin, A., Banciu, D. D., Iancu, A. D., Selescu, T., et al. (2014). Advanced type 1 diabetes is associated with ASIC alterations in mouse lower thoracic dorsal root ganglia neurons. *Cell Biochemistry and Biophysics*, *68*(1), 9–23.

Ramaswamy, S. S., MacLean, D. M., Gorfe, A. A., & Jayaraman, V. (2013). Proton-mediated conformational changes in an acid-sensing ion channel. *Journal of Biological Chemistry*, *288*(50), 35896–35903.

Rehncrona, S. (1985). Brain acidosis. *Annals of Emergency Medicine*, *14*, 770–776.

Reynolds, N. C., Jr., Prost, R. W., & Mark, L. P. (2005). Heterogeneity in 1H-MRS profiles of presymptomatic and early manifest Huntington's disease. *Brain Research*, *1031*, 82–89.

Schepelmann, K., Ebersberger, A., Pawlak, M., Oppmann, M., & Messlinger, K. (1999). Response properties of trigeminal brain stem neurons with input from dura mater encephali in the rat. *Neuroscience*, *90*(2), 543–554.

Sherwood, T. W., & Askwith, C. C. (2009). Dynorphin opioid peptides enhance acid-sensing ion channel 1a activity and acidosis-induced neuronal death. *Journal of Neuroscience*, *29*, 14371–14380.

Siesjo, B. K., Katsura, K., & Kristian, T. (1996). Acidosis-related damage. *Advances in Neurology*, *71*, 209–233. discussion 234–206.

Smith, E. S., Cadiou, H., & McNaughton, P. A. (2007). Arachidonic acid potentiates acid-sensing ion channels in rat sensory neurons by a direct action. *Neuroscience*, *145*, 686–698.

Smoller, J. W., Acierno, J. S., Jr., Rosenbaum, J. F., Biederman, J., Pollack, M. H., Meminger, S., et al. (2001). Targeted genome screen of panic disorder and anxiety disorder proneness using homology to murine QTL regions. *American Journal of Medical Genetics*, *105*, 195–206.

Smoller, J. W., Gallagher, P. J., Duncan, L. E., McGrath, L. M., Haddad, S. A., Holmes, A. J., et al. (2014). The human ortholog of acid-sensing ion channel gene ASIC1a is associated with panic disorder and amygdala structure and function. *Biological Psychiatry*, *76*(11), 902–910.

Somjen, G. G. (1984). Acidification of interstitial fluid in hippocampal formation caused by seizures and by spreading depression. *Brain Research*, *311*, 186–188.

Staruschenko, A., Dorofeeva, N. A., Bolshakov, K. V., & Stokand, J. D. (2007). Subunit-dependent cadmium and nickel inhibition of acid-sensing ion channels. *Developmental Neurobiology*, *67*, 97–107.

Staruschenko, A., & Stockand, J. D. (2007). G-protein-coupled receptor (GPCR) regulation of acid-sensing ion channel 1a. *The FASEB Journal*, *21*, 968.1.

Su, J. J., Pan, H., Zhou, H. G., Tang, Y. P., Dong, Q., & Liu, J. R. (2014). Acid-sensing ion channels activation and hypoxia upregulate Homer1a expression. *CNS Neuroscience & Therapeutics*, *20*(3), 264–274.

Sun, X., Cao, Y. B., Hu, L. F., Yang, Y. P., Li, J., Wang, F., et al. (2011). ASICs mediate the modulatory effect by paeoniflorin on α-synuclein autophagic degradation. *Brain Research*, *1396*, 77–87.

Sutherland, S. P., Benson, C. J., Adelman, J. P., & McCleskey, E. W. (2001). Acid-sensing ion channel 3 matches the acid-gated current in cardiac ischemia-sensing neurons. *Proceedings of the National Academy of Sciences of the United States of America*, *98*, 711–716.

Tai, K. K., & Truong, D. D. (2013). Amiloride but not memantine reduces neurodegeneration, seizures and myoclonic jerks in rats with cardiac arrest-induced global cerebral hypoxia and reperfusion. *PLoS One*, *8*(4), e60309.

Tsang, T. M., Woodman, B., McLoughlin, G. A., Griffin, J. L., Tabrizi, S. J., Bates, G. P., et al. (2006). Metabolic characterization of the R6/2 transgenic mouse model of Huntington's disease by high-resolution MAS 1H NMR spectroscopy. *Journal of Proteome Research*, *5*, 483–492.

Ugawa, S., Ishida, Y., Ueda, T., Inoue, K., Nagao, M., & Shimada, S. (2007). Nafamostat mesilate reversibly blocks acid-sensing ion channel currents. *Biochemical and Biophysical Research Communications*, *363*(1), 203–208.

Uribe-San Martín, R., Herrera-Molina, R., Olavarría, L., Ramírez, G., & von Bernhardi, R. (2009). Reduction of beta-amyloid-induced neurotoxicity on hippocampal cell cultures by moderate acidosis is mediated by transforming growth factor beta. *Neuroscience*, *158*, 1338–1347.

Velisek, L., Dreier, J. P., Stanton, P. K., Heinemann, U., & Moshe, S. L. (1994). Lowering of extracellular pH suppresses low-Mg(2+)-induces seizures in combined entorhinal cortex-hippocampal slices. *Experimental Brain Research*, *101*, 44–52.

Vergo, S., Craner, M. J., Etzensperger, R., Attfield, K., Friese, M. A., Newcombe, J., et al. (2011). Acid-sensing ion channel 1 is involved in both axonal injury and demyelination in multiple sclerosis and its animal model. *Brain*, *134*, 571–584.

Voilley, N. (2004). Acid-sensing ion channels (ASICs): New targets for the analgesic effects of non-steroid anti-inflammatory drugs (NSAIDs). *Current Drug Targets. Inflammation and Allergy*, *3*(1), 71–79.

Voilley, N., de Weille, J., Mamet, J., & Lazdunski, M. (2001). Nonsteroid anti-inflammatory drugs inhibit both the activity and the inflammation-induced expression of acid-sensing ion channels in nociceptors. *Journal of Neuroscience*, *21*, 8026–8033.

Vukicevic, M., Weder, G., Boillat, A., Boesch, A., & Kellenberger, S. (2006). Trypsin cleaves acid-sensing ion channel 1a in a domain that is critical for channel gating. *Journal of Biological Chemistry*, *281*, 714–722.

Walder, R. Y., Gautam, M., Wilson, S. P., Benson, C. J., & Sluka, K. A. (2011). Selective targeting of ASIC3 using artificial miRNAs inhibits primary and secondary hyperalgesia after muscle inflammation. *Pain*, *152*, 2348–2356.

Waldmann, R., Bassilana, F., de Weille, J., Champigny, G., Heurteaux, C., & Lazdunski, M. (1997). Molecular cloning of a non-inactivating proton-gated Na channel specific for sensory neurons. *The Journal of Biological Chemistry*, *272*, 20975–20978.

Waldmann, R., Champigny, G., Bassilana, F., Heurteaux, C., & Lazdunski, M. (1997). A proton-gated cation channel involved in acid-sensing. *Nature*, *386*, 173–177.

Wang, W. Z., Chu, X. P., Li, M. H., Seeds, J., Simon, R. P., & Xiong, Z. G. (2006). Modulation of acid-sensing ion channel currents, acid-induced increase of intracellular Ca2+, and acidosis-mediated neuronal injury by intracellular pH. *Journal of Biological Chemistry*, *281*(39), 29369–29378.

Wang, W., Duan, B., Xu, H., Xu, L., & Xu, T. L. (2006). Calcium-permeable acid-sensing ion channel is a molecular target of the neurotoxic metal ion lead. *Journal of Biological Chemistry*, *281*, 2497–2505.

Wang, R. I., & Sonnenschein, R. R. (1955). pH of cerebral cortex during induced convulsions. *Journal of Neurophysiology*, *18*(2), 130–137.

Wardlaw, J. M., Murray, V., Berge, E., del Zoppo, G., Sandercock, P., Lindley, R. L., et al. (2012). Recombinant tissue plasminogen activator for acute ischaemic stroke: An updated systematic review and meta analysis. *Lancet*, *379*, 2364–2372.

Wemmie, J. A., Askwith, C. C., Lamani, E., Cassell, M. D., Freeman, J. H., Jr., & Welsh, M. J. (2003). Acid-sensing ion channel 1 is localized in brain regions with high synaptic density and contributes to fear conditioning. *Journal of Neuroscience*, *23*, 5496–5502.

Wemmie, J. A., Chen, J., Askwith, C. C., Hruska-Hageman, A. M., Price, M. P., Nolan, B. C., et al. (2002). The acid-activated ion channel ASIC contributes to synaptic plasticity, learning and memory. *Neuron*, *34*(3), 463–477.

Wemmie, J. A., Coryell, M. W., Askwith, C. C., Lamani, E., Leonard, A. S., Sigmund, C. D., et al. (2004). Overexpression of acid-sensing ion channel 1a in transgenic mice increases acquired fear-related behavior. *Proceedings of the National Academy of Sciences of the United States of America*, *101*, 3621–3626.

Wong, H. K., Bauer, P. O., Kurosawa, M., Goswami, A., Washizu, C., et al. (2008). Blocking acid-sensing ion channel 1 alleviates Huntington's disease pathology via an ubiquitin-proteasome system-dependent mechanism. *Human Molecular Genetics*, *17*(20), 3223–3235.

Wu, W. L., Cheng, C. F., Sun, W. H., Wong, C. W., & Chen, C. C. (2012). Targeting ASIC3 for pain, anxiety, and insulin resistance. *Pharmacology & Therapeutics*, *134*(2), 127–138.

Wu, H., Wang, C., Liu, B., Li, H., Zhang, Y., Dong, S., et al. (2015). Altered expression pattern of acid sensing ion channel isoforms in piriform cortex after seizures. *Molecular Neurobiology*. http://dx.doi.org/10.1007/s12035-015-9130-5.

Xiong, Z. G., Zhu, X. M., Chu, X. P., Minami, M., Hey, J., Wei, W. L., et al. (2004). Neuroprotection in ischemia: Blocking calcium-permeable acid-sensing ion channels. *Cell*, *118*(6), 687–698.

Yan, J., & Dussor, G. (2014). Ion channels and migraine. *Headache*, *54*(4), 619–639.

Yan, J., Edelmayer, R. M., Wei, X., De Felice, M., Porreca, F., & Dussor, G. (2011). Dural afferents express acid-sensing ion channels: A role for decreased meningeal pH in migraine headache. *Pain*, *152*(1), 106–113.

Yifeng, M., Bin, W., Weiqiao, Z., Yongming, Q., Bing, L., & Xiaojie, L. (2011). Neuroprotective effect of sophocarpine against transient focal cerebral ischemia via down-regulation of the acid-sensing ion channel 1 in rats. *Brain Research*, *1382*, 245–251.

Yingjun, G., & Xun, Q. (2013). Acid-sensing ion channels under hypoxia. *Channels (Austin, Tex.)*, 7(4), 231–237.

Yu, Y., Chen, Z., Li, W. G., Cao, H., Feng, E. G., Yu, F., et al. (2010). A nonproton ligand sensor in the acid-sensing ion channel. *Neuron*, *68*, 61–72.

Yu, X. W., Hu, Z. L., Ni, M., Fang, P., Zhang, P. W., Shu, Q., et al. (2015). Acid-sensing ion channels promote the inflammation and migration of cultured rat microglia. *Glia*, *63*(3), 483–496.

Zacharoff, L., Tkac, I., Song, Q., Tang, C., Bolan, P. J., Mangia, S., et al. (2012). Cortical metabolites as biomarkers in the R6/2 model of Huntington's disease. *Journal of Cerebral Blood Flow and Metabolism*, *32*, 502–514.

Zeng, W. Z., Liu, D. S., Liu, L., She, L., Wu, L. J., & Xu, T. L. (2015). Activation of acid-sensing ion channels by localized proton transient reveals their role in proton signaling. *Science Reports*, *5*, 14125.

Zhang, P., Sigworth, F. J., & Canessa, C. M. (2006). Gating of acid-sensitive ion channel-1; release of Ca2+ block vs. allosteric mechanism. *The Journal of General Physiology*, *127*, 109–117.

Ziemann, A. E., Allen, J. E., Dahdaleh, N. S., Drebot, I. I., Coryell, M. W., Wunsch, A. M., et al. (2009). The amygdala is a chemosensor that detects carbon dioxide and acidosis to elicit fear behavior. *Cell*, *139*, 1012–1021.

Ziemann, A. E., Schnizler, M. K., Albert, G. W., Severson, M. A., Howard, M. A., III, Welsh, M. J., et al. (2008). Seizure termination by acidosis depends on ASIC1a. *Nature Neuroscience*, *11*(7), 816–822.

CHAPTER FIVE

Glutamatergic NMDA Receptor as Therapeutic Target for Depression

Gislaine Z. Réus*,[1], Helena M. Abelaira*, Talita Tuon*, Stephanie E. Titus†, Zuleide M. Ignácio*, Ana Lúcia S. Rodrigues‡, João Quevedo*,†,§,¶

*Laboratory of Neurosciences, Graduate Program in Health Sciences, Health Sciences Unit, University of Southern Santa Catarina, Criciuma, Santa Catarina, Brazil
†Center for Translational Psychiatry, Department of Psychiatry and Behavioral Sciences, Medical School, The University of Texas Health Science Center at Houston, Houston, Texas, USA
‡Laboratory of Neurobiology of Depression, Department of Biochemistry, Center of Biological Sciences, Federal University of Santa Catarina, Florianópolis, Santa Catarina, Brazil
§Center of Excellence on Mood Disorders, Department of Psychiatry and Behavioral Sciences, Medical School, The University of Texas Health Science Center at Houston, Houston, Texas, USA
¶Neuroscience Graduate Program, Graduate School of Biomedical Sciences, The University of Texas Health Science Center at Houston, Houston, Texas, USA
[1]Corresponding author: e-mail address: gislainezilli@hotmail.com

Contents

Abstract

Major depressive disorder (MDD) affects approximately 121 million individuals globally and poses a significant burden to the healthcare system. Around 50–60% of patients with MDD respond adequately to existing treatments that are primarily based on a monoaminergic system. However, the neurobiology of MDD has not been fully elucidated; therefore, it is possible that other biochemical alterations are involved. The glutamatergic system and its associated receptors have been implicated in the pathophysiology of MDD. In fact, the *N*-methyl-D-aspartate (NMDA) receptor, a glutamate receptor, is a binding or modulation site for both classical antidepressants and new fast-acting antidepressants. Thus, this review aims to present evidence describing the

Advances in Protein Chemistry and Structural Biology, Volume 103
ISSN 1876-1623
http://dx.doi.org/10.1016/bs.apcsb.2015.10.003

effect of antidepressants that modulate NMDA receptors and the mechanisms that contribute to the antidepressant response.

1. INTRODUCTION

Major depressive disorder (MDD) is a serious, chronic condition and has been linked to diminished quality of life, medical morbidity and mortality, and excess economic burden (Kessler et al., 2005). The prevalence of MDD in the United States is 17% and it is estimated that 121 million people are affected worldwide (Sattler & Rothstein, 2007). Almost 1 million lives are lost every year due to suicide, which translates to 3000 suicide deaths per day (WHO, 2012).

Although the treatment of MDD is generally safe and effective, it is far from optimal. In fact, standard antidepressants usually require approximately 1 month or more for antidepressant effects to manifest (Nemeroff & Owens, 2002). The lag in therapeutic effect suggests that structural and functional adaptations in the nervous system regions are mandatory for the antidepressants to take effect. There is a risk of suicide and other deliberate acts of self-harm during the first month of treatment. Thus, it is essential to investigate new treatment options for MDD with greater efficacy and faster onset.

During the second half of the twentieth century, researchers in the psychiatric field were influenced by the discovery of drugs that relieved depressive symptoms through the alteration of monoamine metabolism, particularly serotonin (5-hydroxytryptamine), norepinephrine, and dopamine. The monoamine deficiency hypothesis postulates that depressive symptoms arise from insufficient levels of monoamine neurotransmitters (Berton & Nestler, 2006). Currently, depressive patients are treated with tricyclic agents (i.e., imipramine), selective serotonin reuptake inhibitors (i.e. fluoxetine (FLX)), selective serotonin and norepinephrine reuptake inhibitors (i.e., venlafaxine), selective norepinephrine reuptake inhibitors (i.e., reboxetine), dual inhibitors of norepinephrine and dopamine reuptake (bupropion), and monoamine oxidase inhibitors (i.e., tranylcypromine). However, an open study examined the efficacy of antidepressant treatments for MDD and found that less than one-third of patients achieved remission using a single standard antidepressant after 14 weeks of treatment (Trivedi, 2006). For depressive patients who fail to respond to classical antidepressant treatments, several strategies have been proposed, such as (a) increasing the dose of the antidepressant, (b) augmentation therapy, (c) combining two antidepressants or combining antidepressants with different pharmacological profiles (combination pharmacotherapy), and (d) substituting with a new class of antidepressants (Anderson, 2000).

In the last two decades, a considerable number of basic and clinical studies have highlighted the role of the glutamatergic system in mood disorders, including bipolar disorder (BD) and MDD (Berman et al., 2000; Garcia et al., 2008b, 2008a, 2009; Zarate, Singh, Carlson, et al., 2006; Zarate, Singh, Quiroz, et al., 2006). Elevated levels of glutamate may be associated with the pathophysiology of MDD, while reduced levels of this neurotransmitter and its associated biochemical cascades may be related to the antidepressant response to classical antidepressants and modulators of the glutamatergic system. Glutamatergic modulators primarily linked to *N*-methyl-D-aspartate (NMDA) receptors have been proposed as potentially revolutionary and rapid-acting antidepressant agents for treatment-resistant mood disorders. Many of these studies are trying to unravel the mechanisms by which antidepressants exert their action as well as to find new glutamatergic NMDA receptor modulators. This chapter will focus on the evidence related to the antidepressant effects of NMDA receptor modulators and the mechanism of action underlying the antidepressant response.

2. GLUTAMATERGIC SYSTEM

Glutamate (Glu) is the major excitatory neurotransmitter in the central nervous system (CNS) (Dingledine, Borges, Bowie, & Traynelis, 1999) and plays a central role in synaptic plasticity, learning, cognition, and memory (Collingridge & Lester, 1989; Danbolt, 2001; Izquierdo, 1994). Excessive excitation of glutamate receptors can cause neuronal damage or death; this process is called excitotoxicity (Olney & de Gubareff, 1978; Olney, Labruyere, & de Gubareff, 1980). Excitotoxicity is associated with ischemic neuronal injury (Choi & Rothman, 1990; Olney, 1990), acute neurodegenerative disorders such as epilepsy (Dingledine, McBain, & McNamara, 1990; Price, 1999), chronic neurodegenerative diseases such as Alzheimer's disease and Huntington's disease, hepatic encephalopathy (Beal et al., 1986; Lipton & Rosenberg, 1994; Price, 1999), genetic diseases (Dingledine et al., 1990; Meldrum & Garthwaite, 1990), and mood disorders (Ghasemi et al., 2014).

Relatively, the brain possesses a lot of glutamate (approximately 5–15 nmol/kg, depending on the region) and only a small fraction of this total accumulates in the extracellular space (Danbolt, 2001). High densities of glutamate-reactive neurons are found in the cortex, as interneurons and pyramidal neurons of layers IV and V (Paul & Skolnick, 2003; Rajkowska, 2000). Glutamate-containing neurons are also found in subcortical structures, such as the hippocampus, caudate nucleus, thalamic nuclei, and the cerebellum (Rajkowska, 2000).

Glutamate can be synthesized from glucose through the Krebs/tricarboxylic acid cycle (Musazzi, Treccani, Mallei, & Popoli, 2013) or through recycling of glutamate by the astrocyte-neuronal, glutamate–glutamine cycle. In the glutamate–glutamine cycle, glutamate is synaptically released and taken up by surrounding astrocytes, where it is converted to glutamine, a nonneuroexcitatory amino acid, and transferred back to neurons for conversion to glutamate (Serafini et al., 2013). Glutamate is packaged in calcium-dependent synaptic vesicles by the vesicular glutamate transporters (VGLUTs) (Meloni et al., 1993), and soon after, the SNARE complex proteins (soluble *N*-ethylmaleimide-sensitive factor—attachment protein receptor) facilitate exocytosis of glutamate into the synaptic cleft (Lesch & Schmitt, 2002) where it interacts with ionotropic (iGluRs) and metabotropic (mGluRs) receptors. The iGluRs are divided into three subtypes: NMDA, AMPA (α-amino-3-hydroxy-5-methyl-4-isoxazole propionic acid), and kainate. These iGluRs are ion channels that selectively allow an influx of Ca^{2+} and Na^{+}, promoting depolarization of the neuron (Dingledine et al., 1999). Both AMPA and kainate receptors are involved in fast synaptic transmission; binding of glutamate results in a conformational change and subsequent sodium influx (Chittajallu et al., 1996; Tang, Dichter, & Morad, 1989).

In contrast, NMDA receptors do not participate in fast synaptic transmission. At a resting membrane potential, the NMDA receptor channel is blocked in a voltage-dependent manner by a magnesium ion. Removal of magnesium requires depolarization of the postsynaptic neuron, which typically occurs after glutamate binds AMPA or kainate receptors (Laube, Hirai, Sturgess, Betz, & Kuhse, 1997; Lester, Clements, Westbrook, & Jahr, 1990). The ion channel associated with the NMDA receptors requires the binding of glutamate and a co-agonist, glycine or D-serine (Henneberger, Papouin, Oliet, & Rusakov, 2010; Panatier et al., 2006). If both ligands (glutamate and a co-agonist) bind while the postsynaptic neuron is in a depolarized state, NMDA channels will open, permitting calcium to enter the cell (Jahr & Stevens, 1993). NMDA receptors have a higher permeability to calcium than AMPA or kainate receptors (Dingledine, 1983;

MacDermott, Mayer, Westbrook, Smith, & Barker, 1986). The increased influx of calcium triggers secondary messenger systems leading to the establishment of long-term potentiation, a process believed to underlie learning and memory (Bliss & Collingridge, 1993). However, an overactivation of NMDA receptors leads to an excess of intracellular calcium, which initiates a series of events that can result in cell death (Choi, Koh, & Peters, 1988).

Unlike the fast synaptic transmission of iGluRs, mGluRs are involved in slow synaptic transmission and are subdivided into three groups: Group I (mGluR1 and mGluR5), Group II (mGluR2 and mGluR3), and Group III (mGluR4, mGluR6, mGluR7, and mGluR8) on the basis of signal transduction pathways and pharmacological profiles (Masu, Tanabe, Tsuchida, Shigemoto, & Nakanishi, 1991). Group I mGluRs are positively coupled to phospholipase C (Abe et al., 1992), whereas Groups II and III mGluRs are negatively coupled to adenylyl cyclase (Nakajima et al., 1993; Okamoto et al., 1994). All three groups of mGLuRs, with the exception of mGluR6, play a role in regulating hippocampal function (Rudy, Hunsberger, Weitzner, & Reed, 2015).

Unlike many other neurotransmitters, glutamate is not significantly degraded extracellularly. Rather, it is primarily cleared through transported into glia by excitatory amino acid transporters (EAATs) (Lapidus, Soleimani, & Murrough, 2013). The glial cell, together with the presynaptic and postsynaptic neurons, makes up the "tripartite synapse" which functions to exert tight regulatory control over levels of synaptic and extrasynaptic glutamate (Parpura et al., 1994). The extensive molecular machinery regulating glutamate signaling provides multiple targets for glutamatergic drug development.

NMDA receptors are glutamate-gated cation channels that are highly permeable to Ca^{2+} and blocked by Mg^{2+} in a voltage-dependent manner. These receptors are very relevant to CNS development and maintenance, including the processes involved in brain plasticity, learning, and memory (Yuan & Bellone, 2013). On the other hand, dysfunctional NMDA receptor activity is associated with several psychiatric and neurological disorders, including schizophrenia, MDD, and stroke, due in part to extrasynaptic NMDA receptor activation. Extrasynaptic NMDA receptors are activated by enhanced glutamatergic activity and signal a cascade involved in cell death (Parsons & Raymond, 2014).

NMDA receptors are heteromeric complexes which include three different subtypes of subunits: NR1, NR2, and NR3 (GluN1, GluN2, and

GluN3). GluN1 forms the channel and GluN2 and GluN3 influence the functional properties (Lau & Zukin, 2007). GluN1 is phosphorylated by protein kinases A and C (PKA and PKC). GluN2A is phosphorylated by PKA, PKC, and cyclin-dependent kinase-5. The substrate for GluN2B is CaMKII, and GluN2B is phosphorylated by PKC and casein kinase (CK2). GluN2A and GluN2B are both phosphorylated by tyrosine (Wang, Guo, et al., 2014; Wang, Huang, et al., 2014). Changes in the phosphorylation of any of these proteins may be associated with the pathogenesis of mood disorders and neuropsychiatric illnesses.

3. NMDA RECEPTOR BINDING AND MODULATORS WITH ANTIDEPRESSANT PROPERTIES

3.1 NMDA Receptor Antagonists

3.1.1 Ketamine as an Antidepressant

Ketamine was synthesized in 1962 by Calvin Stevens and was originally used as a general venous anesthetic with a fast onset. In 1970, the Food Drugs Administration (FDA) approved the use of ketamine in humans and it became very popular as an anesthetic on the battlefield in the Vietnam War (Jansen, 2000). After that, it was introduced into clinical practice to act as a monoanesthetic drug.

There are two optically active isomers of ketamine, *S* (+) and *R* (−), which have different pharmacological properties (Luft & Mendes, 2005). In fact, studies have shown that ketamine *S* (+) has a greater affinity for the phencyclidine site in the NMDA receptor than ketamine *R* (−) (Kohrs & Durieux, 1998; Pfenninger, Durieux, & Himmelseher, 2002). Moreover, ketamine is opioid agonist that produces psychomimetic effects due to interaction with the opioid kappa receptor (Hustveit, Maurset, & Oye, 1995; Raeder & Stenseth, 2000). With the availability of the isomer *S* (+) and the potential to cause less psychomimetic effects, ketamine has been widely used as anesthetic.

Berman and collaborators first described the role of ketamine as an antidepressant agent in 2000. In that study, the authors showed that a single injection of ketamine administrated intravenously at a dose of 0.5 mg/kg was effective in improving depressive symptoms within 72 h in patients with treatment-resistant depression (Berman et al., 2000). With a similar design, Zarate, Singh, Carlson, et al. (2006), Zarate, Singh, Quiroz, et al. (2006), and Liebrenz, Stohler, and Borgeat (2009) also reported that ketamine revealed antidepressant effects in treatment-resistant MDD patients. The antidepressant effects were present at 110 min, 24 h, and 7 days after a single injection (Liebrenz et al., 2009; Zarate, Singh, Carlson, et al., 2006; Zarate, Singh, Quiroz, et al., 2006). In addition, two intravenous infusions of 0.5 mg/kg ketamine over the course of 6 weeks in a single patient with treatment-resistant major depression showed that both doses were effective but the depression score returned to baseline 7 days after the infusions (aan het Rot et al., 2010).

The antidepressant effect of ketamine has been showed previously in preclinical animal studies. We demonstrated that ketamine administered as a single dose (10 and 15 mg/kg) (Garcia et al., 2008b) or chronically (5, 10, and 15 mg/kg) (Garcia et al., 2008a) was able to decrease the immobility time of rats in the forced swimming test (FST) when compared to the classic antidepressant imipramine. The FST is one of the tests most commonly used by researchers to investigate new antidepressant drugs (Abelaira, Réus, & Quevedo, 2013). Ketamine also reduced immobility time in the FST similar to the antidepressant FLX (Owolabi, Akanmu, & Adeyemi, 2014). Interestingly, acute but not chronic treatment with ketamine increased brain-derived neurotrophic factor (BDNF) levels in the hippocampus (Garcia et al., 2008b, 2008a). BDNF is a neurotrophin that is important in MDD and antidepressant responses. Ketamine administration was also able to reverse anhedonic and depressive-like behavior induced in rodents subjected to classic animal models of depression, such as maternal care deprivation (MCD) and chronic mild stress (CMS) (Garcia et al., 2009; Réus, Carlessi et al., 2015; Réus et al., 2013, 2015a, 2015b). In addition, high levels of corticosterone and decreased BDNF levels produced by CMS were reversed by ketamine administration (Garcia et al., 2009).

Ketamine's antidepressant effect seems to go far beyond its action on the NMDA receptor. Modulation of AMPA receptors by ketamine could effect glutamatergic facilitation. Indeed, a single injection of ketamine at the dose of 10 mg/kg combined with FLX and with the antipsychotic olanzapine enhanced AMPA- and NMDA-induced currents, but the effects were more evident on AMPA receptors (Björkholm, Jardemark, Schilström, & Svensson, 2015). In addition, the inhibition of AMPA receptors by 2,3-dioxo-6-nitro-1,2,3,4-tetrahydrobenzo[*f*]quinoxaline-7-sulfonamide was able to abolish the antidepressant effects of ketamine in the FST (Koike & Chaki, 2014), suggesting that AMPA receptors are important, if not required. Inhibition of AMPA receptors also inhibited the release of serotonin induced by ketamine in the medial prefrontal cortex (Nishitani et al., 2014). A study performed by Akinfiresoye and Tizabi (2013) demonstrated that chronic treatment with AMPA in a dose-dependent manner produced antidepressant effects in the FST and the sucrose preference test. Furthermore, AMPA treatment in doses that were ineffective on their own produced an antidepressant effect when combined with ketamine (Akinfiresoye & Tizabi, 2013). The behavior observed studying the FST and sucrose preference test was associated with an increase in BDNF, synapsin, and mammalian target of rapamycin (mTOR) levels in the rodent hippocampus.

Recently, a large number of studies have highlighted the role of mTOR in both the pathophysiology of MDD and the antidepressant properties of ketamine (Abelaira, Réus, Neotti, & Quevedo, 2014; Hoeffer et al., 2008; Li et al., 2010; Park et al., 2014). In fact, ketamine administration stimulated mTOR signaling in the prefrontal cortex (Li et al., 2010). Inhibition of mTOR using rapamycin abolished ketamine's antidepressant effects (Li et al., 2010). Ketamine also induces synaptogenesis and glutamate transmission, leading to an activation of AKT and extracellular signal-regulated kinase signaling, elevated cAMP response element-binding protein (CREB), and increased levels of BDNF and its receptor tropomyosin receptor kinase B (TrkB) (Abelaira, Réus, Neotti, et al., 2014; Réus et al., 2014). Réus et al. (2011) demonstrated that the combination of a classic antidepressant imipramine plus ketamine had a more pronounced antidepressant activity in rats and increased CREB, BDNF, and protein kinases A and C. These are protein targets for NMDA and AMPA receptor phosphorylation or DNA transcription and translation.

mTOR may also be regulated by the inhibition of glycogen synthase kinase-3 (GSK-3) (Beurel, Song, & Jope, 2011). Some studies have demonstrated that ketamine exerts antidepressant effects via GSK-3. Indeed, GSK-3 was required to induce synaptogenesis and antidepressant effects after a combination treatment of ketamine and lithium in rats (Beurel et al., 2011; Liu et al., 2013), a mood stabilizer that targets GSK-3. Yang, Zhou, Gao, Shi, and Yang (2013) ran a clinical study in which ketamine (0.5 mg/kg) improved the depressive symptoms of three MDD patients; mTOR, GSK-3, and eukaryotic elongation factor 2 (eEF2) phosphorylation were acutely elevated in the plasma after ketamine treatment. However, in rats subjected to CMS, ketamine produced a rapid antidepressant effects by reversing depressive and anhedonic behavior, while SB216763 did not (Ma et al., 2013), suggesting that GSK-3 in this animal model is not required for ketamine's antidepressant effects. Thus, more studies are needed to investigate the role of the GSK-3 pathway in ketamine's mechanism of action as an antidepressant.

Oxidative stress and inflammation have also been implicated in the pathophysiology of MDD (Abelaira, Réus, Petronilho, Barichello, & Quevedo, 2014; Guan, Lin, & Tang, 2015; Réus, Fries et al., 2015; Réus, Jansen et al., 2015). Excitotoxicity mediated by NMDA receptors can produce free radicals, increase inflammatory cytokines, and lead to mitochondrial dysfunction. Previously, we showed that ketamine reduced some proinflammatory cytokines, such as tumor necrosis factor-α (TNF-α) and interleukin-6 (IL-6), in the serum and cerebrospinal fluid of stressed rats induced by MCD (Réus et al., 2015a). Ketamine inhibited TNF-α and IL-6 gene expression in lipopolysaccharide (LPS)-activated macrophages (Wu, Chen, Ueng, & Chen, 2008) and abrogated LPS-induced depressive-like behavior (Walker et al., 2013). Moreover, in maternally deprived adult rats, oxidative stress in the brain was reduced after treatment with a single dose of ketamine *S* (+) (Réus, Carlessi et al., 2015). Therefore, it is possible that modulators of the glutamatergic system may play an important role in reducing neuroinflammation and oxidative stress.

3.1.2 Memantine as an Antidepressant

Some studies have demonstrated that memantine (1-amino-3,5-dimethyladamantane) can play an important role in the treatment of MDD. Memantine acts as noncompetitive antagonist of voltage-dependent NMDA receptors and blocks the pathological effects caused by enhanced glutamate levels. It is approved to treat Alzheimer's disease in the United States and Europe and is under investigation for use in other neurological conditions (Ferguson & Shingleton, 2007).

In rodent studies, memantine administered acutely in doses of 3–20 mg/kg exerted antidepressant effects in the FST (Almeida, Felisbino, López, Rodrigues, & Gabilan, 2006; Almeida, Souza, et al., 2006; Réus et al., 2010; Rogóz, Skuza, Maj, & Danysz, 2002; Skuza & Rogóz, 2003). In rats subjected to CMS, which presented with anhedonic behavior and with corticosterone and BDNF levels dysfunctions, memantine treatment reversed these alterations (Réus, Abelaira, et al., 2012). The combination of memantine and classic antidepressants, such as imipramine, venlafaxine, and fluoxetine, produced synergistic antidepressant effects in rodents (Rogóz et al., 2002). Memantine's mechanism of action has been reported for many studies. In addition to its role as an antagonist of NMDA receptors, memantine increases neurotrophic factors, such as BDNF, NGF, and GDNF (Liu et al., 2014; Ranju, Sathiya, Kalaivani, Priya, & Saravana Babu, 2015; Réus, Abelaira, et al., 2012), has anti-inflammatory effects (Chen et al., 2012), and modulates energy metabolism (Réus, Stringari, et al., 2012).

Memantine has also demonstrated antidepressant properties clinically. In fact, in an open label study investigating the effect of memantine on depression, it was found that depressive symptoms improved within 1 week and reached and maintained maximal improvement from weeks 8 to 12 (Ferguson & Shingleton, 2007). Ketamine infusion followed by memantine administration was effective in treating MDD patients (Kollmar, Markovic, Thurauf, Schmitt, & Kornhuber, 2008). However, a randomized controlled trial failed to show antidepressant effects of memantine treatment in patients with MDD (Zarate, Singh, Carlson, et al., 2006; Zarate, Singh, Quiroz, et al., 2006). Although both ketamine and memantine are NMDA receptor antagonists, ketamine has produced better results in treatment-resistant MDD patients. These findings could be explained by the fact that intracellular cascades following NMDA receptor stimulation differ between ketamine and memantine. In fact, Gideons, Kavalali, and Monteggia (2014) demonstrated that, in contrast to ketamine, memantine does not inhibit the phosphorylation of eEF2 and therefore increases the expression of BDNF mediated by eEF2. The difference in the effects of memantine and ketamine could also be associated with the mTOR signaling pathway. For example, both ketamine and memantine reduced ethanol drinking in alcohol-preferring rats in a dose-dependent manner; however, rapamycin (mTOR inhibitor) blocked the effects ketamine, but not memantine, in alcohol-preferring rats, indicating that ketamine's effects are mediated by mTOR activation (Sabino, Narayan, Zeric, Steardo, & Cottone, 2013).

Memantine has been used as an augmentation agent in patients suffering from treatment-resistant BD I and II. It has antimanic and mood-stabilizing effects (Koukopoulos et al., 2010). Memantine combined with the mood stabilizer valproate (VPT) increased blood levels of low-density lipoprotein cholesterol but did not change affective symptoms in patients with BD (Lee et al., 2013). Similarly, a 12-week protocol involving treatment with VPT or memantine showed a decrease in Hamilton Depression Rating Scale (HDRS) and Young Mania Rating Scale (YMRS) scores. However, the combination of both VPT and memantine was not significant in improving depression and mania symptoms (Lee et al., 2014). Therefore, the literature findings suggest that memantine failed to increase the effectiveness of traditional drugs used to treat BD or MDD patients. However, memantine could be a good strategy in enhancing or maintaining the antidepressant effects of ketamine in depressive BD and MDD patients.

3.1.3 Amantadine as an Antidepressant

Amantadine hydrochloride is an NMDA receptor antagonist that also increases dopamine synthesis and blocks dopamine reuptake; it is used as an ancillary treatment for Parkinson's disease (Wu et al., 2008). Noradrenergic mechanisms may also play a role in the action of this compound (Moryl, Danysz, & Quack, 1993). Amantadine has been used since 1966 as an antiviral agent against influenza A viral infections and showed remarkable antidepressant efficacy in depressed patients infected with the Borna disease virus (Dietrich et al., 2000). In addition, this compound was shown to produce only mild antidepressant effects when administered alone, but was effective as an adjunctive therapy in combination with other antidepressants in both human and animal models (Rogóz et al., 2002, 2007). Kubera et al. (2006) showed that exposure to an FST produced an increase in plasma corticosterone levels in rats, which was significantly attenuated by adjunctive therapy with imipramine and amantadine.

Some reports argue that amantadine suppresses microglial activation and neuroinflammation (Peeters, Page, Maloteaux, & Hermans, 2002; Wang, Guo, et al., 2014; Wang, Huang, et al., 2014). In fact, Kubera et al. (2006) demonstrated that the treatment with amantadine alone or in combination evoked a significant increase in IL-10 production in stressed rats. This compound may modulate immunoreactivity indirectly by increasing dopaminergic neurotransmission; previous studies have described that the inhibitory effect activation of the dopaminergic system has on cell-mediated response (Basso, Gioino, Molin, & Cancela, 1999; Kouassi et al., 1987).

3.1.4 MK-801 as an Antidepressant

MK-801 ((+)-5-methyl-10,11-dihydro-5-H-dibenzo [*a*,*d*]cycloheptan-5,10-imine maleate) is a noncompetitive NMDA receptor antagonist that blocks the neurophysiological effects of the NMDA receptor complex by binding to a site in the ion channel of the receptor, thereby blocking the channel (Wong, Knight, & Woodruff, 1988). MK-801 also inhibits the reuptake of dopamine, noradrenaline, and, to a lesser extent, serotonin *in vitro* (Tan et al., 2015) and has a wide spectrum of pharmacological activities including anticonvulsant, anxiolytic, and neuroprotective properties (Clineschmidt, Martin, Bunting, & Papp, 1982; Gill, Foster, & Woodruff, 1987).

In another study, Zarrindast, Nasehi, Pournaghshband, and Yekta (2012) showed that the administration of MK-801 in rats induced anxiolytic behavior and influenced mobility, learning, and memory. Moreover, studies have shown that the use of MK-801 with FLX induced faster and more pronounced antidepressant effect when compared to treatment with antidepressants alone (Pruus, Rudissaar, Allikmets, & Harro, 2010). Furthermore, pretreatment with fluvoxamine, MK-801, and ketamine and a combination of these NMDA antagonists with fluvoxamine produced a protective effect against the depression induced by shock in mice (Chaturvedi, Bapna, & Chandra, 2001).

3.1.5 Ro25-6981 as an Antidepressant

Evidence *in vitro* demonstrates that the NR2B-containing NMDA subtype glutamate receptor antagonist Ro25-6981 prevents the behavioral stress-facilitated hippocampal long-term depression (Wang, Yang, Dong, Cao, & Xu, 2006). Li et al. (2010) showed that a single subanesthetic dose of the Ro25-6981 demonstrated rapid antidepressant effects and also counteracted depressive-like behaviors in chronically stressed rodents. In addition, studies have reported that Ro25-6981 has antidepressant actions in the FST and learned helplessness paradigms (Li et al., 2010; Maeng et al., 2008). Ro25-6981 also rapidly reversed the anhedonic and anxiogenic behaviors resulting from chronic unpredictable stress (CUS), suggesting that the synaptic actions of Ro25-6981 are dependent on mTOR signaling, since mTOR is required for fast rapid antidepressant activity (Nanxin et al., 2011). However, Wang et al. (2006) reported that the behavioral improvements from Ro25-6981 treatment in the FST do not persist as long as ketamine.

3.1.6 CGP-37849 as an Antidepressant

CGP-37849 (DL-(*E*)-2-amino-4-methyl-5-phosphono-3-pentonoic acid) is a competitive antagonist of the NMDA receptor (Fagg et al., 1990). It is a potent, orally active anticonvulsant in animal models (Schmutz et al., 1990) and has beneficial effects on the treatment of epilepsy (Fujikawa, Daniels, & Kim, 1994). CGP-37849 also exhibits neuroprotective activity (Gutnikov & Gaffan, 1996), antidepressant activity (Maj, Klimek, Gołembiowska, Rogóz, & Skuza, 1993; Papp & Moryl, 1994), and anxiolytic effects (Jessa, Nazar, Bidzinski, & Plaznik, 1996; Przegaliński, Tatarczyńska, & Chojnacka-Wójcik, 2000). Papp and Moryl (1994) demonstrated that chronic treatment for 4/5 weeks with CGP-37849 reversed the deficit in sucrose intake induced by stress in rats. Moreover, Maj et al. (1993) showed that chronic treatment with CGP-37849 administered twice daily for 14 days did not affect locomotor activity, enhanced D-amphetamine-induced hyperactivity, and increased the density of dopamine D1 receptors labeled with [3H]-SCH 23390 in the rat striatum. Furthermore, CGP-37849 retained its anxiolytic-like potency in the Vogel test, which measures the ability of drugs to release the drinking behavior of water-deprived rats exposed to a mild aversive stimulus ("punishment") (Jessa et al., 1996). The obtained results indicate that in some experimental paradigms, CGP-37849 given repeatedly produces effects similar to those of typical antidepressant drugs.

3.1.7 ACPC as an Antidepressant

ACPC (1-aminocyclopropanecarboxylic acid) is a high-affinity partial agonist at strychnine-insensitive glycine receptors (Marvizon, Lewin, & Skolnick, 1989). There is also evidence that ACPC might act directly as a low-affinity antagonist at the NMDA receptor (Nahum-Levy, Fossom, Skolnick, & Benveniste, 1999). ACPC has been found to produce anticonvulsant, neuroprotective, anxiolytic, and antidepressant-like effects in rats and mice (Anthony & Nevins, 1993; Papp & Moryl, 1994; Trullas et al., 1991; Von Lubitz et al., 1992) and was reported to block the acquisition of morphine on a conditioned place preference (Kotlinska & Biala, 2000). Chronic treatment for 5 weeks with ACPC gradually reversed CMS-induced reductions in sucrose consumption and the magnitude of this effect was comparable to that observed following similar administration of imipramine (10 mg/kg) (Papp & Moryl, 1994). Trullas et al. (1991) showed that ACPC administrated intraperitoneally and orally were equipotent in reducing immobility in the FST for at least 6 h. In the elevated plus maze, the ACPC increased both the % time in and % entries into the open arms, indicating an anxiolytic effect. These findings suggest that ACPC may constitute a novel class of antidepressant/anxiolytic agents.

3.1.8 AP-7 as an Antidepressant

AP-7 (2-amino-7-phosphonoheptanoic acid) is a selective NMDA receptor antagonist that competitively inhibits the glutamate binding site and thus activation of the NMDA receptor (Meldrum, Millan, Patel, & de Sarro, 1988). AP-7 functions specifically as an NMDA recognition site blocker, in contrast with 7-chlorokynurenate, which acts as a glycine site modulation blocker (Guillemin, 2012). Studies have shown that AP-7 has anticonvulsant (Meldrum et al., 1988) and antidepressant effects (Maj, Rogoz, Skuza, & Sowinska, 1992; Tan et al., 2015). In fact, AP-7 reduced immobility in the FST in rats, like many clinically effective antidepressants (Trullas & Skolnick, 1990).

3.1.9 Eliprodil as an Antidepressant

Eliprodil ((rac)-9) is a ligand that is selective for NMDA receptors containing the NR2B subunit (DeKeyser, Sulter, & Luiten, 1999). This compound given in subclinical doses potentiated the anticonvulsant effects of CGP-37849. Also, in normal mice, eliprodil decreased spontaneous locomotor activity, did not affect the locomotor hyperactivity induced by MK-801, and attenuated CGP-37849-induced locomotor hyperactivity (Dereń-Wesołek & Maj, 1993). In addition, eliprodil also produces antidepressant effects. In fact, Layer, Popik, Olds, and Skolnick (1995) demonstrated that chronic treatment with eliprodil, at dose of 5 and 20 mg/kg, produced a dose-dependent reduction in immobility in the FST and led to a down-regulation of β-adrenoceptors. β-Adrenoceptor downregulation has also been observed following chronic treatment with dizocilpine and ACPC (Klimek & Papp, 1994; Paul, Trullas, Skolnick, & Nowak, 1992), and the ability of eliprodil to produce this effect is consistent with the hypothesis that NMDA antagonists possess antidepressant actions (Layer et al., 1995).

3.2 Glycine Modulators

Sufficiently high synaptic concentrations of glycine appear to allow activation of NMDA receptors by presynaptic neurotransmitter release (Kutsuwada et al., 1992; Nahum-Levy, Tam, Shavit, & Benveniste, 2002; Yu et al., 2015). Thus, modulation of glycine levels has also been considered as a possible regulatory mechanism of NMDA receptor function. In addition, certain NMDA receptors demonstrate lower sensitivity to glycine and Mg^{2+}, which suggests variable sensitivity to endogenous receptor modulators (Kutsuwada et al., 1992; Monyer et al., 1992; Stein, Gray, & Zito, 2015). Glycine promotes NMDA receptor activity and reduces desensitization of the NMDA receptor (Benveniste & Mayer, 1993; Vyklický, Benveniste, & Mayer, 1990). Binding of an agonist to the NDMA receptor reduces its affinity for glycine (Yu et al., 2015), so the desensitizing currents activated by NMDA are determined by glycine interaction rates. This role desensitization by glycine plays in the brain is not fully understood, although it may contribute to deterioration of excitatory synaptic currents if their concentrations are sufficiently low (Zhang, Shuttleworth, Moskal, & Stanton, 2015).

Activation of the ionotropic channel requires binding of glycine and glutamate agonists and pre-depolarization, which can occur by depolarization of non-NMDA receptors on adjacent regions of the plasma membrane (Cunha et al., 2015; Zhang et al., 2014). The NMDA receptor is unique among glutamate receptors in that it contains additional sites for the regulatory modulators glycine, D-serine, and D-alanine. In the presence of glutamate, these modulators increase the duration of channel opening, thereby amplifying the transmission signal (Huang et al., 2013). In the absence of glycine, the NMDA receptor channel opening attenuates, even with an adequate amount of glutamate (Chen et al., 2011; Cunha et al., 2015; Zhang et al., 2014). Glycine transporter and D-amino acid oxidase inhibitors are currently under development. These compounds modulate brain levels of glycine and D-serine indirectly and can be used as potent psychopharmacological agent (Huang et al., 2013). Studies have shown that activation of the glycine B site of the NMDA receptor by D-serine inhibits zinc antidepressant activity (Poleszak et al., 2011; Zhang, Sullivan, Moskal, & Stanton, 2008). Chronic treatment with zinc reduced the capacity of glycine to bind and interact with NMDA receptors in the cortex of rats subjected to CMS (Cichy et al., 2009; Nowak, Ossowska, Jopek, & Papp, 1998).

3.3 Zinc and Magnesium

Studies have highlighted the involvement of zinc and magnesium (inorganic antagonists of the NMDA receptors) in the pathophysiology of MDD and in possible therapeutic treatment (Murck, 2013; Swardfager et al., 2013). In preclinical studies, zinc and magnesium produce antidepressant activity in some animal models of depression (Decollogne, Tomas, Lecerf, Adamowicz, & Seman, 1997; Kroczka, Branski, Palucha, Pilc, & Nowak, 2001; Kroczka, Zieba, Dudek, Pilc, & Nowak, 2000; Nowak, Siwek, Dudek, Zieba, & Pilc, 2003; Poleszak et al., 2004, 2005; Rosa, Lin, Calixto, Santos, & Rodrigues, 2003; Sowa-Kućma et al., 2008).

Zinc is found in large concentrations in the CNS, particularly in the hippocampal region (Barañano, Ferris, & Snyder, 2001). This metal is also found in synaptic vesicles and is released in the central nerve endings upon depolarization (Barañano et al., 2001; Nam, Lee, Hinton, & Choi, 2010). One of the most prominent effects of zinc is inhibition of the glutamatergic NMDA receptor by a voltage-dependent mechanism similar to magnesium in addition to an independent voltage mechanism at a separate site (Christine & Choi, 1990; Nam et al., 2010).

Zinc can also regulate the activity of PKC (Beyersmann & Haase, 2001). Zinc, in nanomolar concentrations, can activate translocation of PKC to the cell membrane and stimulate the autonomous activity of PKC in the absence of activating cofactors (Beyersmann & Haase, 2001). Zinc also plays an important role in the regulation of cell death in the CNS. Experimental studies show a neurotoxic effect of zinc at high levels (Horning, Blakemore, & Trombley, 2000; Park & Koh, 1999) and its involvement in diseases such as Alzheimer's disease (Bush, 2003). Some studies have shown a neuroprotective role for zinc in relation to neuronal death. It appears to have protective effects on neurons by inhibiting NMDA receptors and modulating the neurotoxicity of glutamate (Domínguez, Blasco-Ibáñez, Crespo, Marqués-Marí, & Martínez-Guijarro, 2003).

Experimental studies show that zinc deficiency induces a behavior similar to depression in animals (Młyniec, Budziszewska, Reczyński, Sowa-Kućma, & Nowak, 2013; Młyniec & Nowak, 2012; Tamano, Kan, Kawamura, Oku, & Takeda, 2009; Tassabehji, Corniola, Alshingiti, & Levenson, 2008; Whittle, Lubec, & Singewald, 2009). Zinc has shown antidepressant properties in clinical and preclinical models, such as in the FST (Kroczka et al., 2000, 2001; Rosa et al., 2003; Szewczyk, Peterson, & Jacobson, 2002) and with CMS (Cieślik et al., 2007). Zinc levels were reduced in the serum of patients with MDD, and these levels were regulated with antidepressant therapy (Maes, De Vos, Demedts, Wauters, & Neels, 1999; Maes et al., 1994; Mc Loughlin & Hodge, 1990; Nowak et al., 1998; Siwek et al., 2010). Thus, these results suggest an important role for zinc in the modulation of MDD.

Magnesium is an inhibitor of the NMDA receptor with high affinity for the NR1/NR2A, NR1/NR2B, NR1/NR2, and NR1/NR2D subunits. The affinity of magnesium for the NR2A and NR2B subunits is relatively lower than that of zinc (Paoletti & Neyton, 2007). In fact, some studies have shown that depression or depressive behavior may be related to adaptations that skew receptor subunit proportions (Feyissa, Chandran, Stockmeier, & Karolewicz, 2009; Tokita, Fujita, Yamaji, & Hashimoto, 2012).

Magnesium functions as a voltage-dependent constitutive channel blocker that is released after excitation and depolarization of the postsynaptic cell. Electrostatic repulsion acts as a voltage sensor. Asparagine residues in GluN1 and GluN2 are responsible for the conformational change of the NDMA receptor that promotes the opening of ion channels, allowing the influx of sodium, calcium, and potassium. The calcium permeability is responsible for regulating synaptic plasticity, cellular mechanism involved in memory formation, and learning processes. NMDA receptors formed by GluN2A or GluN2B subunits show a higher affinity for magnesium than those containing the other two isoforms of the same subunit (C or D) (Blanke & Van Dongen, 2009; Zito, Scheuss, Knott, Hill, & Svoboda, 2009). Inhibition of NMDA responses in heteromeric receptors GluN1/GluN2A and GluN1/GluN2B by zinc is voltage dependent. Magnesium and oxidant agents act as NMDA receptor inhibitors; on the other hand, polyamines and redox agents potentiate the NMDA receptor by increasing opening time, opening frequency, or affinity (Ruljancic, Mihanovic, Cepelak, Bakliza, & Curkovic, 2013; Zieba, Kata, Dudek, Schlegel-Zawadzka, & Nowak, 2000).

Several studies have shown a relationship between magnesium and MDD. In studies of patients with MDD, measurement of magnesium concentrations has been inconsistent. For example, blood concentration of magnesium was increased in some studies (Kirov, Birch, Steadman, & Ramsey, 1994; Widmer et al., 1992, 1995) and decreased in others (Banki, Vojnik, Papp, Balla, & Arató, 1985; Hashizume & Mori, 1990; Ruljancic et al., 2013; Zieba et al., 2000). Magnesium deficiency in the brain induced depressive-like behavior in animal models (Muroyama et al., 2009; Singewald, Sinner, Hetzenauer, Sartori, & Murck, 2004). Furthermore, NMDA receptor antagonist activity is enhanced by coadministration with magnesium (Poleszak, 2007; Poleszak et al., 2007).

A considerable percentage of suicide victims have been attributed to depression (Pawlak et al., 2013), and one hypothesis is an alteration in the homeostasis of zinc and magnesium in the brain (Nowak, Siwek, et al., 2003; Nowak, Szewczyk, et al., 2003; Takeda, 2011). Zinc and magnesium are potent NMDA receptor antagonists and deficiency of both can lead to hyperactivity of the NMDA receptor. These studies suggest the importance of zinc and magnesium in MDD via changes in the affinity of the NMDA receptor subunits.

3.4 Ascorbic Acid

Ascorbic acid (vitamin C) occurs physiologically as the ascorbate anion. It is an antioxidant vitamin that plays a role in interneuronal communication, acting as a neuromodulator. It is released from glutamatergic neurons attaining a relatively high extracellular level in the brain (Rebec & Pierce, 1994; Rice, 2000). Neurons maintain relatively high intracellular concentrations of ascorbate due to the activity of the sodium-dependent transporter SVCT2. It was reported that hippocampal cultures from SVCT2-deficient mice showed increased susceptibility to oxidative damage and NMDA-induced excitotoxicity, indicating that ascorbate may exert neuroprotective effects. The first preclinical study that showed that ascorbic acid may elicit antidepressant-like effects was published in 2009. Administration of ascorbic acid to mice caused an antidepressant-like effect in tail suspension test (TST) due to its interaction with the monoaminergic systems. In addition, a synergistic antidepressant-like effect was obtained when ascorbic acid was administered in combination with conventional antidepressants (Binfaré, Rosa, Lobato, Santos, & Rodrigues, 2009). In subsequent studies, the mechanism of the antidepressant-like effect of ascorbic acid was shown to involve the inhibition of NMDA receptors, the L-arginine-NO-cyclic guanosine 3,5-monophosphate pathway (Moretti et al., 2011), and potassium channels (Moretti, Budni, et al., 2012). More recently, its antidepressant effect was demonstrated to be dependent on activation of phosphatidylinositol-3 kinase (PI3K) and mTOR, inhibition of GSK-3β, and induction of heme oxygenase-1 (Moretti, Budni, Freitas, Rosa, & Rodrigues, 2014).

In addition, ascorbic acid exerted an antidepressant-like effect in mice submitted to CUS (Moretti, Colla, et al., 2012), acute restraint stress models of depression (Moretti et al., 2013), and depressive-like behavior induced by the administration of the proinflammatory cytokine TNF-α (Moretti et al., 2015). Clinical studies have also indicated that ascorbic acid may have a beneficial effect for the treatment of depression. In 1980, a case report showed that high doses of intravenous ascorbic acid (50 mg/kg/day) relieved adrenocorticotropic hormone-induced depression in a child (Cocchi, Silenzi, Calabri, & Salvi, 1980). Ascorbic acid administration was also effective in decreasing scores on Beck Depression Inventories in healthy young adults (Brody, 2002). The notion that ascorbic acid may exert antidepressant effects was confirmed in subsequent studies. It was reported to lower depressive symptoms in an elderly population on high dietary intake of vitamin C (Oishi, Doi, & Kawakami, 2009). In addition, it was demonstrated that pediatric patients treated for 6 months with FLX and ascorbic acid presented a significant decrease in depressive symptoms in comparison to the FLX plus placebo group (Amr, El-Mogy, Shams, Vieira, & Lakhan, 2013).

3.5 Guanosine

The purine nucleoside guanosine is an endogenous modulator of glutamatergic excitotoxicity and has been shown to promote neuroprotection in *in vivo* and *in vitro* models of neurotoxicity. Guanosine was reported to prevent seizures induced by the NMDA receptor agonist quinolinic acid (Lara et al., 2001; Schmidt, Lara, de Faria Maraschin, da Silveira Perla, & Souza, 2000) and cell death in the rat striatum (Malcon et al., 1997). Guanosine is also able to modulate glutamate transporter activity by increasing astrocytic glutamate uptake (Frizzo et al., 2001) and decreasing glutamate uptake into synaptic vesicles (Tasca et al., 2004). Guanosine was also reported to protect against glutamate-induced cell death in rat hippocampal slices by a mechanism dependent on PI3K/Akt pathway activation, inhibition of GSK-3β, and inducible nitric oxide synthase inhibition (Molz et al., 2011). In human neuroblastoma SH-SY5Y cells, guanosine protects against mitochondrial oxidative stress through the PI3K/Akt/GSK-3β signaling pathway and induction of the antioxidant enzyme heme oxygenase-1 (Dal-Cim et al., 2012). Of note, the administration of low doses of guanosine causes antidepressant-like effects in the FST and TST in mice, and its effect in the TST is dependent on the inhibition of either NMDA receptors or nitric oxide synthesis and on the activation of PI3K and mTOR signaling (Bettio et al., 2012). It also prevents the behavioral alterations in the FST and hippocampal oxidative damage induced by acute restraint stress in mice (Bettio et al., 2014).

4. CONCLUSION

There has been much interest in targeting the glutamatergic system when developing new and promising antidepressants. The basis for this interest is the evidence supporting ketamine, an NMDA receptor agonist, administered at subanesthetic doses producing rapidly effective treatment of refractory depression. However, due to the abuse potential of this drug and the fact that its chronic use may cause neurotoxic effects, the investigation of novel glutamatergic modulators that may be used for treatment-resistant depression is needed.

ACKNOWLEDGMENTS

Laboratory of Neurosciences (Brazil) is a center within the National Institute for Translational Medicine (INCT-TM). Laboratory of Neurobiology of Depression and Laboratory of Neurosciences are members of the Center of Excellence in Applied Neurosciences of Santa Catarina (NENASC). Work in the authors' laboratories has been funded by grants from CNPq (J.Q., A.L.S.R., and G.Z.R.), FAPESC (J.Q.), InstitutoCérebro e Mente, and UNESC (J.Q.). J.Q. and A.L.S.R. are CNPq Research Fellows. H.M.A is also grateful for CAPES for their research fellowship. Center for Translational Psychiatry (USA) is funded by Department of Psychiatry and Behavioral Sciences, The University of Texas Medical School at Houston. We thank Leandro D.V. Soares for making Figs. 1 and 2.

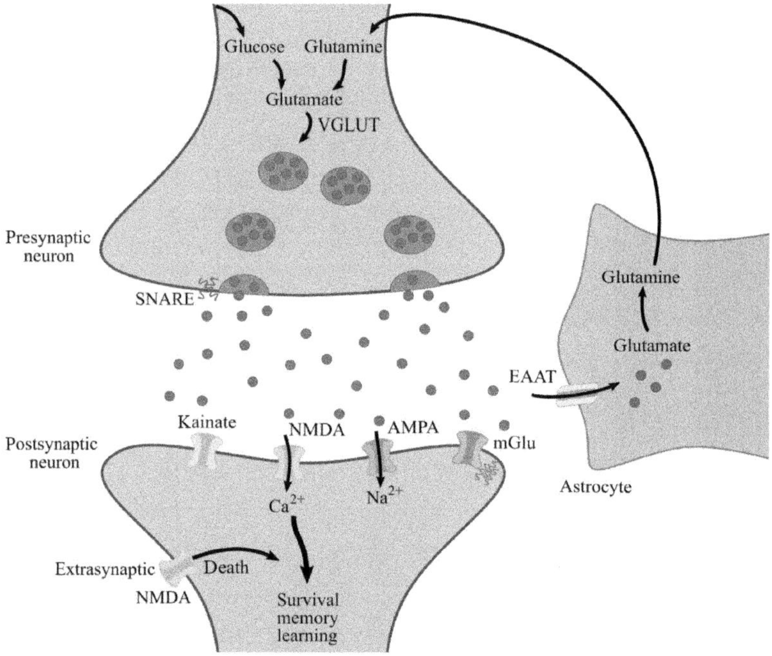

Figure 1 *Glutamatergic synapse*: Glutamate is synthetized in the presynaptic neuron from glucose metabolism or glutamine from astrocytes. Glutamate is transported to vesicles by VGLUT transporters and is released into the synapse by exocytosis. Glutamate acts on postsynaptic neuron at metabotropic (mGlu) and ionotropic receptors (NMDA, AMPA, and kainate). The activation of postsynaptic receptors leads to a survival signaling pathway activation that is involved with neuroplasticity, memory, and learning. On the other hand, excess levels of glutamate could activate extrasynaptic NMDA receptors that can activate signaling cascades involved with cell death. After activating the receptor, glutamate is transported by EAAT to astrocytes where it is converted to glutamine.

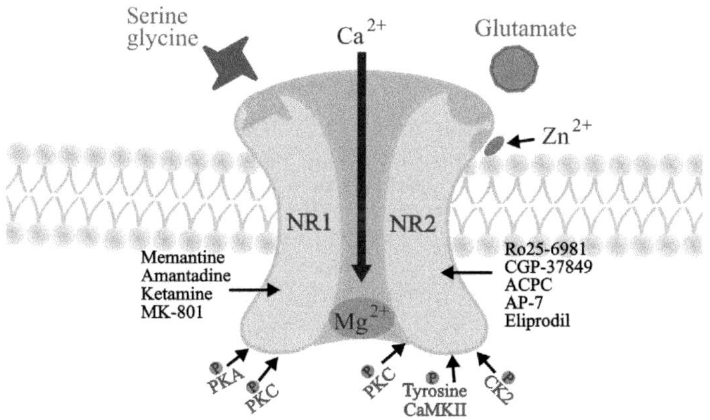

Figure 2 *NMDA receptor as target for antidepressant*: The complex NMDA receptor has one site binding for glutamate and another for co-agonists serine and glycine. Zinc (Zn^{2+}) also has a site binding in the receptor. Magnesium (Mg^{2+}) blocks the receptor channel. The phosphorylation of proteins such as PKA, PCK, tyrosine, CaMKII, and CK2 are important to NMDA receptor formation and function. The NMDA receptor may be an important target for antidepressant drugs. Modulators or antagonists (for example, ketamine, memantine, amantadine, MK-801, Ro25-6981, CGP-37849, ACPC, AP-7, and eliprodil) of this channel have demonstrated antidepressant properties in animal models of depression and in humans with MDD who are antidepressant treatment resistant.

REFERENCES

aan het Rot, M., Collins, K. A., Murrough, J. W., Perez, A. M., Reich, D. L., Charney, D. S., et al. (2010). Safety and efficacy of repeated-dose intravenous ketamine for treatment-resistant depression. *Biological Psychiatry*, *67*, 139–145.

Abe, T., Sugihara, H., Nawa, H., Shigemoto, R., Mizuno, N., & Nakanishi, S. (1992). Molecular characterization of a novel metabotropic glutamate receptor mGluR5 coupled to inositol phosphate/Ca2+ signal transduction. *The Journal of Biological Chemistry*, *267*, 13361–13368.

Abelaira, H. M., Réus, G. Z., Neotti, M. V., & Quevedo, J. (2014). The role of mTOR in depression and antidepressant responses. *Life Science*, *101*, 10–14.

Abelaira, H. M., Réus, G. Z., Petronilho, F., Barichello, T., & Quevedo, J. (2014). Neuroimmunomodulation in depression: A review of inflammatory cytokines involved in this process. *Neurochemical Research*, *39*, 1634–1639.

Abelaira, H. M., Réus, G. Z., & Quevedo, J. (2013). Animal models as tools to study the pathophysiology of depression. *Revista Brasileira de Psiquiatria*, *2*, S112–S120.

Akinfiresoye, L., & Tizabi, Y. (2013). Antidepressant effects of AMPA and ketamine combination: Role of hippocampal BDNF, synapsin, and mTOR. *Psychopharmacology*, *230*, 291–298.

Almeida, R. C., Felisbino, C. S., López, M. G., Rodrigues, A. L., & Gabilan, N. H. (2006). Evidence for the involvement of L-arginine-nitric oxide-cyclic guanosine monophosphate pathway in the antidepressant-like effect of memantine in mice. *Behavioural Brain Research*, *168*, 318–322.

Almeida, R. C., Souza, D. G., Soletti, R. S., López, M. G., Rodrigues, A. N. R., & Gabilan, N. H. (2006). Involvement of PKA, MAPK/ERK and CaMKII, but not

PKC in the acute antidepressant-like effect of memantine in mice. *Neuroscience Letters*, *395*, 93–97.

Amr, M., El-Mogy, A., Shams, T., Vieira, K., & Lakhan, S. E. (2013). Efficacy of vitamin C as an adjunct to fluoxetine therapy in pediatric major depressive disorder: A randomized, double-blind, placebo-controlled pilot study. *Nutrition Journal*, *12*, 31.

Anderson, I. M. (2000). Selective serotonin reuptake inhibitors versus tricyclic antidepressants: A meta-analysis of efficacy and tolerability. *Journal of Affective Disorders*, *58*(1), 19–36.

Anthony, E. W., & Nevins, M. E. (1993). Anxiolytic-like effects of N-methyl-D-aspartate-associated glycine receptor ligands in the rat potentiated startle test. *European Journal of Pharmacology*, *250*, 317–324.

Banki, C. M., Vojnik, M., Papp, Z., Balla, K. Z., & Arató, M. (1985). Cerebrospinal fluid magnesium and calcium related to amine metabolites, diagnosis, and suicide attempts. *Biological Psychiatry*, *20*, 163–171.

Barañano, D. E., Ferris, C. D., & Snyder, S. H. (2001). Atypical neural messengers. *Trends in Neurosciences*, *24*, 99–106.

Basso, A. M., Gioino, G., Molin, V. A., & Cancela, L. M. (1999). Chronic amphetamine facilitates immunosuppression in response to a novel aversive stimulus: Reversal by haloperidol pretreatment. *Pharmacology, Biochemistry, and Behavior*, *62*, 307–314.

Beal, M. F., Kowall, N. W., Ellison, D. W., Mazurek, M. F., Swartz, K. J., & Martin, J. B. (1986). Replication of the neurochemical characteristics of Huntington's disease by quinolinic acid. *Nature*, *321*, 168–171.

Benveniste, M., & Mayer, M. L. (1993). Multiple effects of spermine on N-methyl-D-aspartic acid receptor responses of rat cultured hippocampal neurones. *The Journal of Physiology*, *464*, 131–163.

Berman, R. M., Cappiello, A., Anand, A., Oren, D. A., Heninger, G. R., Charney, D. S., et al. (2000). Antidepressant effects of ketamine in depressed patients. *Biological Psychiatry*, *47*, 351–354.

Berton, O., & Nestler, E. J. (2006). New approaches to antidepressant drug discovery: Beyond monoamines. *Nature Reviews Neuroscience*, *7*, 137–151.

Bettio, L. E., Cunha, M. P., Budni, J., Pazini, F. L., Oliveira, A., Colla, A. R., et al. (2012). Guanosine produces an antidepressant-like effect through the modulation of NMDA receptors, nitric oxide-cGMP and PI3K/mTOR pathways. *Behavioural Brain Research*, *234*, 137–148.

Bettio, L. E., Freitas, A. E., Neis, V. B., Santos, D. B., Ribeiro, C. M., Rosa, P. B., et al. (2014). Guanosine prevents behavioral alterations in the forced swimming test and hippocampal oxidative damage induced by acute restraint stress. *Pharmacology, Biochemistry, and Behavior*, *127*, 7–14.

Beurel, E., Song, L., & Jope, R. S. (2011). Inhibition of glycogen synthase kinase-3 is necessary for the rapid antidepressant effect of ketamine in mice. *Molecular Psychiatry*, *16*, 1068–1070.

Beyersmann, D., & Haase, H. (2001). Functions of zinc in signaling, proliferation and differentiation of mammalian cells. *Biometals*, *14*, 331–341.

Binfaré, R. W., Rosa, A. O., Lobato, K. R., Santos, A. R. S., & Rodrigues, A. L. S. (2009). Ascorbic acid administration produces an antidepressant-like effect: Evidence for the involvement of monoaminergic neurotransmission. *Progress in Neuropsychopharmacology and Biological Psychiatry*, *33*, 530–540.

Björkholm, C., Jardemark, K., Schilström, B., & Svensson, T. H. (2015). Ketamine-like effects of a combination of olanzapine and fluoxetine on AMPA and NMDA receptor-mediated transmission in the medial prefrontal cortex of the rat. *European Neuropsychopharmacology*, *25*, 1842–1847.

Blanke, M. L., & Van Dongen, A. M. J. (2009). Activation mechanisms of the NMDA receptor. In *Frontiers in neuroscience*. CRC Press/Taylor & Francis. Chapter 13.

Bliss, T. V., & Collingridge, G. L. (1993). A synaptic model of memory: Long-term potentiation in the hippocampus. *Nature, 361*, 31–39.

Brody, S. (2002). High-dose ascorbic acid increases intercourse frequency and improves mood: A randomized controlled clinical trial. *Biological Psychiatry, 52*, 371–374.

Bush, A. I. (2003). Copper, zinc, and the metallobiology of Alzheimer disease. *Alzheimer Disease and Associated Disorders, 17*, 147–150.

Chaturvedi, H. K., Bapna, J. S., & Chandra, D. (2001). Effect of fluvoxamine and N-methyl-D-aspartate receptor antagonists on shock-induced depression in mice. *Indian Journal of Physiology and Pharmacology, 45*, 199–207.

Chen, S. L., Tao, P. L., Chu, C. H., Chen, S. H., Wu, H. E., Tseng, L. F., et al. (2012). Low-dose memantine attenuated morphine addictive behavior through its anti-inflammation and neurotrophic effects in rats. *Journal of Neuroimmune Pharmacology*, 7, 444–453.

Chen, R. Q., Wang, S. H., Yao, W., Wang, J. J., Ji, F., Yan, J. Z., et al. (2011). Role of glycine receptors in glycine-induced LTD in hippocampal CA1 pyramidal neurons. *Neuropsychopharmacology, 36*, 1948–1958.

Chittajallu, R., Vignes, M., Dev, K. K., Barnes, J. M., Collingridge, G. L., & Henley, J. M. (1996). Regulation of glutamate release by presynaptic kainate receptors in the hippocampus. *Nature, 379*, 78–81.

Choi, D. W., Koh, J. Y., & Peters, S. (1988). Pharmacology of glutamate neurotoxicity in cortical cell culture: Attenuation by NMDA antagonists. *Journal of Neuroscience, 8*, 185–196.

Choi, D. W., & Rothman, S. M. (1990). The role of glutamate neurotoxicity in hypoxic-ischemic neuronal death. *Annual Reviews of Neuroscience, 13*, 171–182.

Christine, C. W., & Choi, D. W. (1990). Effect of zinc on NMDA receptor-mediated channel currents in cortical neurons. *Journal of Neuroscience, 10*, 108–116.

Cichy, A., Sowa-Kućma, M., Legutko, B., Pomierny-Chamioło, L., Siwek, A., Piotrowska, A., et al. (2009). Zinc-induced adaptive changes in NMDA/glutamatergic and serotonergic receptors. *Pharmacological Reports, 61*, 1184–1191.

Cieślik, K., Klenk-Majewska, B., Danilczuk, Z., Wróbel, A., Łupina, T., & Ossowska, G. (2007). Influence of zinc supplementation on imipramine effect in a chronic unpredictable stress (CUS) model in rats. *Pharmacological Reports, 59*, 46–52.

Clineschmidt, B. V., Martin, G. E., Bunting, P. R., & Papp, N. L. (1982). Central sympathomimetic activity of (+)-5-methyl-10,11-dihydro-5 H-dibenzo[a,d]cyclohepten-5,10-imine (MK-801), a substance with potent anticonvulsant central sympathomimetic and apparent anxiolytic properties. *Drug Development Research, 2*, 135.

Cocchi, P., Silenzi, M., Calabri, G., & Salvi, G. (1980). Antidepressant effect of vitamin C. *Pediatrics, 65*, 862–863.

Collingridge, G., & Lester, R. A. (1989). Excitatory amino acid receptors in the vertebrate central nervous system. *Pharmacological Reviews, 41*, 143–210.

Cunha, M. P., Pazini, F. L., Ludka, F. K., Rosa, J. M., Oliveira, Á., Budni, J., et al. (2015). The modulation of NMDA receptors and L-arginine/nitric oxide pathway is implicated in the anti-immobility effect of creatine in the tail suspension test. *Amino Acids, 47*, 795–811.

Dal-Cim, T., Molz, S., Egea, J., Parada, E., Romero, A., Budni, J., et al. (2012). Guanosine protects human neuroblastoma SH-SY5Y cells against mitochondrial oxidative stress by inducing heme oxigenase-1 via PI3K/Akt/GSK-3β pathway. *Neurochemistry International, 61*, 397–404.

Danbolt, N. C. (2001). Glutamate uptake. *Progress in Neurobiology, 65*, 1–105.

Decollogne, S., Tomas, A., Lecerf, C., Adamowicz, E., & Seman, M. (1997). NMDA receptor complex blockade by oral administration of magnesium: Comparison with MK-801. *Pharmacology, Biochemistry, and Behavior, 58,* 261–268.

DeKeyser, J., Sulter, G., & Luiten, P. G. (1999). Clinical trials with neuroprotective drugs in acute ischaemic stroke: Are we doing the right thing? *Trends in Neurosciences, 22,* 535–540.

Dereń-Wesołek, A., & Maj, J. (1993). Central effects of SL 82.0715, an antagonist of polyamine site of the NMDA receptor complex. *Polish Journal of Pharmacology, 45,* 467–480.

Dietrich, D. E., Bode, L., Spannhuth, C. W., Lau, T., Huber, T. J., Brodhun, B., et al. (2000). Amantadine in depressive patients with Borna disease virus (BDV) infection: An open trial. *Bipolar Disorders, 2,* 65–70.

Dingledine, R. (1983). N-methyl aspartate activates voltage-dependent calcium conductance in rat hippocampal pyramidal cells. *The Journal of Physiology, 343,* 385–405.

Dingledine, R., Borges, K., Bowie, D., & Traynelis, S. F. (1999). The glutamate receptor ion channels. *Pharmacological Reviews, 51,* 7–61.

Dingledine, R., McBain, C. J., & McNamara, J. O. (1990). Excitatory amino acid receptors in epilepsy. *Trends in Pharmacological Sciences, 11,* 334–338.

Domínguez, M. I., Blasco-Ibáñez, J. M., Crespo, C., Marqués-Marí, A. I., & Martínez-Guijarro, F. J. (2003). Zinc chelation during non-lesioning overexcitation results in neuronal death in the mouse hippocampus. *Neuroscience, 116,* 791–806.

Fagg, G. E., Olpe, H. R., Pozza, M. F., Baud, J., Steinmann, M., Schmutz, M., et al. (1990). CGP 37849 and CGP 39551: Novel and potent competitive N-methyl-D-aspartate receptor antagonists with oral activity. *British Journal of Pharmacology, 99,* 791–797.

Ferguson, J. M., & Shingleton, R. N. (2007). An open-label, flexible-dose study of memantine in major depressive disorder. *Clinical Neuropharmacology, 30,* 136–144.

Feyissa, A. M., Chandran, A., Stockmeier, C. A., & Karolewicz, B. (2009). Reduced levels of NR2A and NR2B subunits of NMDA receptor and PSD-95 in the prefrontal cortex in major depression. *Progress in Neuropsychopharmacology & Biological Psychiatry, 33,* 70–75.

Frizzo, M. E., Lara, D. R., Dahm, K. C., Prokopiuk, A. S., Swanson, R. A., & Souza, D. O. (2001). Activation of glutamate uptake by guanosine in primary astrocyte cultures. *Neuroreport, 12,* 879–881.

Fujikawa, D. G., Daniels, A. H., & Kim, J. S. (1994). The competitive NMDA receptor antagonist CGP 40116 protects against status epilepticus-induced neuronal damage. *Epilepsy Research, 17,* 207–219.

Garcia, L. S., Comim, C. M., Valvassori, S. S., Réus, G. Z., Andreazza, A. C., Stertz, L., et al. (2008a). Chronic administration of ketamine elicits antidepressant-like effects in rats without affecting hippocampal brain-derived neurotrophic factor protein levels. *Basic Clinical Pharmacology and Toxicology, 103,* 502–506.

Garcia, L. S., Comim, C. M., Valvassori, S. S., Réus, G. Z., Barbosa, L. M., Andreazza, A. C., et al. (2008b). Acute administration of ketamine induces antidepressant-like effects in the forced swimming test and increases BDNF levels in the rat hippocampus. *Progress in Neuropsychopharmacology & Biological Psychiatry, 32,* 140–144.

Garcia, L. S., Comim, C. M., Valvassori, S. S., Réus, G. Z., Stertz, L., Kapczinski, F., et al. (2009). Ketamine treatment reverses behavioral and physiological alterations induced by chronic mild stress in rats. *Progress in Neuropsychopharmacology & Biological Psychiatry, 33,* 450–455.

Ghasemi, M., Kazemi, M. H., Yoosefi, A., Ghasemi, A., Paragomi, P., Amini, H., et al. (2014). Rapid antidepressant effects of repeated doses of ketamine compared with electroconvulsive therapy in hospitalized patients with major depressive disorder. *Psychiatry Research, 215,* 355–361.

Gideons, E. S., Kavalali, E. T., & Monteggia, L. M. (2014). Mechanisms underlying differential effectiveness of memantine and ketamine in rapid antidepressant responses. *Proceedings of the National Academy of Sciences of the United States of America*, *111*, 8649–8654.

Gill, R., Foster, A. C., & Woodruff, G. N. (1987). Systemic administration of MK-801 protects against ischemia-induced hippocampal neurodegeneration in the gerbil. *Journal of Neuroscience*, *7*, 3343.

Guan, X. T., Lin, W. J., & Tang, M. M. (2015). Comparison of stress-induced and LPS-induced depressive-like behaviors and the alterations of central proinflammatory cytokines mRNA in rats. *The American Journal of Psychiatry*, *4*, 113–122.

Guillemin, G. J. (2012). Quinolinic acid, the inescapable neurotoxin. *FEBS Journal*, *279*, 1356–1365.

Gutnikov, S. A., & Gaffan, D. (1996). Systemic NMDA receptor antagonist CGP-40116 does not impair memory acquisition but protects against NMDA neurotoxicity in rhesus monkeys. *Journal of Neuroscience*, *16*, 4041–4045.

Hashizume, N., & Mori, M. (1990). An analysis of hypermagnesemia and hypomagnesemia. *Journal of Medicinal Chemistry*, *29*, 368–372.

Henneberger, C., Papouin, T., Oliet, S. H., & Rusakov, D. A. (2010). Long-term potentiation depends on release of D-serine from astrocytes. *Nature*, *463*, 232–236.

Hoeffer, C. A., Tang, W., Wong, H., Santillan, A., Patterson, R. J., Martinez, L. A., et al. (2008). Removal of FKBP12 enhances mTOR-Raptor interactions, LTP, memory, and perseverative/repetitive behavior. *Neuron*, *60*, 832–845.

Horning, M. S., Blakemore, L. J., & Trombley, P. Q. (2000). Endogenous mechanisms of neuroprotection: Role of zinc, copper, and carnosine. *Brain Research*, *852*, 56–61.

Huang, C. C., Wei, I. H., Huang, C. L., Chen, K. T., Tsai, M. H., Tsai, P., et al. (2013). Inhibition of glycine transporter-I as a novel mechanism for the treatment of depression. *Biological Psychiatry*, *74*, 734–741.

Hustveit, O., Maurset, A., & Oye, I. (1995). Interaction of the chiral forms of ketamine with opioid, phencyclidine, sigma and muscarinic receptors. *Pharmacology and Toxicology*, *77*, 355–359.

Izquierdo, I. (1994). Pharmacological evidence for a role of long-term potentiation in memory. *The FASEB Journal*, *8*, 1139–1145.

Jahr, C. E., & Stevens, C. F. (1993). Calcium permeability of the N-methyl-D-aspartate receptor channel in hippocampal neurons in culture. *Proceedings of the National Academy of Sciences of the United States of America*, *90*, 11573–11577.

Jansen, K. L. (2000). A review of the nonmedical use of ketamine: Use, users and consequences. *Journal of Psychoactive Drugs*, *32*, 419–433.

Jessa, M., Nazar, M., Bidzinski, A., & Plaznik, A. (1996). The effects of repeated administration of diazepam, MK-801 and CGP 37849 on rat behavior in two models of anxiety. *European Neuropsychopharmacology*, *6*, 55–61.

Kessler, R. C., et al. (2005). Prevalence, severity, and comorbidity of 12-month DSM-IV disorders in the National Comorbidity Survey Replication. *Archives of General Psychiatry*, *62*, 617–627.

Kirov, G. K., Birch, N. J., Steadman, P., & Ramsey, R. G. (1994). Plasma magnesium levels in a population of psychiatric patients: Correlations with symptoms. *Neuropsychobiology*, *30*, 73–78.

Klimek, V., & Papp, M. (1994). The effect of MK-801 and imipramine on beta-adrenergic and 5-HT2 receptors in the chronic mild stress model of depression in rats. *Polish Journal of Pharmacology*, *46*, 67–69.

Kohrs, R., & Durieux, M. E. (1998). Ketamine: Teaching an old drug new tricks. *Anesthesia and Analgesia*, *87*, 1186–1193.

Koike, H., & Chaki, S. (2014). Requirement of AMPA receptor stimulation for the sustained antidepressant activity of ketamine and LY341495 during the forced swim test in rats. *Behavioural Brain Research*, *271*, 111–115.
Kollmar, R., Markovic, K., Thurauf, N., Schmitt, H., & Kornhuber, J. (2008). Ketamine followed by memantine for the treatment of major depression. *The Australian and New Zealand Journal of Psychiatry*, *42*, 170.
Kotlinska, J., & Biala, G. (2000). Memantine and ACPC affect conditioned place preference induced by cocaine in rats. *Polish Journal of Pharmacology*, *52*, 179–185.
Kouassi, E., Boukhris, W., Descotes, J., Zukervar, P., Li, Y. S., & Revillard, J. P. (1987). Selective T cell defects induced by dopamine. Administration in mice. *Immunopharmacology and Immunotoxicology*, *9*, 477–487.
Koukopoulos, A., Reginaldi, D., Serra, G., Koukopoulos, A., Sani, G., & Serra, G. (2010). Antimanic and mood-stabilizing effect of memantine as an augmenting agent in treatment-resistant bipolar disorder. *Bipolar Disorders*, *12*, 348–349.
Kroczka, B., Branski, P., Palucha, A., Pilc, A., & Nowak, G. (2001). Antidepressant-like properties of zinc in rodent forced swim test. *Brain Research Bulletin*, *55*, 297–300.
Kroczka, B., Zieba, A., Dudek, D., Pilc, A., & Nowak, G. (2000). Zinc exhibits an antidepressant-like effect in the forced swimming test in mice. *Polish Journal of Pharmacology*, *52*, 403–406.
Kubera, M., Basta-Kaim, A., Budziszewska, B., Rogóz, Z., Skuza, G., Leœkiewicz, M., et al. (2006). Effect of amantadine and imipramine on immunological parameters of rats subjected to a forced swimming test. *International Journal of Neuropsychopharmacology*, *9*, 297–305.
Kutsuwada, T., Kashiwabuchi, N., Mori, H., Sakimura, K., Kushiya, E., Araki, K., et al. (1992). Molecular diversity of the NMDA receptor channel. *Nature*, *358*, 36–41.
Lapidus, K. A., Soleimani, L., & Murrough, J. W. (2013). Novel glutamatergic drugs for the treatment of mood disorders. *Journal of Neuropsychiatric Disease and Treatment*, *9*, 1101–1112.
Lara, D. R., Schmidt, A. P., Frizzo, M. E., Burgos, J. S., Ramírez, G., & Souza, D. O. (2001). Effect of orally administered guanosine on seizures and death induced by glutamatergic agents. *Brain Research*, *912*, 176–180.
Lau, C. G., & Zukin, R. S. (2007). NMDA receptor trafficking in synaptic plasticity and neuropsychiatric disorders. *Nature Review Neuroscience*, *8*, 413–426.
Laube, B., Hirai, H., Sturgess, M., Betz, H., & Kuhse, J. (1997). Molecular determinants of agonist discrimination by NMDA receptor subunits: Analysis of the glutamate binding site on the NR2B subunit. *Neuron*, *18*, 493–503.
Layer, R. T., Popik, P., Olds, T., & Skolnick, P. (1995). Antidepressant-like actions of the polyamine site NMDA antagonist, eliprodil (SL-82.0715). *Pharmacology, Biochemistry, and Behavior*, *52*, 621–627.
Lee, S. Y., Chen, S. L., Chang, Y. H., Chen, S. H., Chu, C. H., Huang, S. Y., et al. (2014). Genotype variant associated with add-on memantine in bipolar II disorder. *The International Journal of Neuropsychopharmacology*, *17*, 189–197.
Lee, S. Y., Chen, S. L., Chang, Y. H., Chen, P. S., Huang, S. Y., Tzeng, N. S., et al. (2013). Add-on memantine to valproate treatment increased HDL-C in bipolar II disorder. *Journal of Psychiatric Research*, *47*, 1343–1348.
Lesch, K. P., & Schmitt, A. (2002). Antidepressants and gene expression profiling: How to SNARE novel drug targets. *The Pharmacogenomics Journal*, *2*, 346–348.
Lester, R. A., Clements, J. D., Westbrook, G. L., & Jahr, C. E. (1990). Channel kinetics determine the time course of NMDA receptor-mediated synaptic currents. *Nature*, *346*, 565–567.
Li, N., Lee, B., Liu, R. J., Banasr, M., Dwyer, J. M., Iwata, M., et al. (2010). mTOR-dependent synapse formation underlies the rapid antidepressant effects of NMDA antagonists. *Science*, *20*, 959–964.

Liebrenz, M., Stohler, R., & Borgeat, A. (2009). Repeated intravenous ketamine therapy in a patient with treatment-resistant major depression. *World Journal of Biological Psychiatry*, *10*, 640–643.

Lipton, S. A., & Rosenberg, P. A. (1994). Excitatory amino acids as a final common pathway for neurologic disorders. *The New England Journal of Medicine*, *330*, 613–622.

Liu, R. J., Fuchikami, M., Dwyer, J. M., Lepack, A. E., Duman, R. S., & Aghajanian, G. K. (2013). GSK-3 inhibition potentiates the synaptogenic and antidepressant-like effects of subthreshold doses of ketamine. *Neuropsychopharmacology*, *38*, 2268–2277.

Liu, M. Y., Wang, S., Yao, W. F., Zhang, Z. J., Zhong, X., Sha, L., et al. (2014). Memantine improves spatial learning and memory impairments by regulating NGF signaling in APP/PS1 transgenic mice. *Neuroscience*, *273*, 141–151.

Luft, A., & Mendes, F. F. (2005). Low S(+) ketamine doses: A review. *Revista Brasileira de Anestesiologia*, *55*, 460–469.

Ma, X. C., Dang, Y. H., Jia, M., Ma, R., Wang, F., Wu, J., et al. (2013). Long-lasting antidepressant action of ketamine, but not glycogen synthase kinase-3 inhibitor SB216763, in the chronic mild stress model of mice. *PLoS One*, *8*, e56053.

MacDermott, A. B., Mayer, M. L., Westbrook, G. L., Smith, S. J., & Barker, J. L. (1986). NMDA-receptor activation increases cytoplasmic calcium concentration in cultured spinal cord neurones. *Nature*, *321*, 519–522.

Maeng, S., Zarate, C. A., Jr., Du, J., Schloesser, R. J., McCammon, J., Chen, G., et al. (2008). Cellular mechanisms underlying the antidepressant effects of ketamine: Role of alpha-amino-3-hydroxy-5-methylisoxazole-4-propionic acid receptors. *Biological Psychiatry*, *63*, 349–352.

Maes, M., De Vos, N., Demedts, P., Wauters, A., & Neels, H. (1999). Lower serum zinc in major depression in relation to changes in serum acute phase proteins. *Journal of Affective Disorders*, *56*, 189–194.

Maes, M., D'Haese, P. C., Scharpé, S., D'Hondt, P., Cosyns, P., & De Broe, M. E. (1994). Hypozincemia in depression. *Journal of Affective Disorders*, *31*, 135–140.

Maj, J., Klimek, V., Gołembiowska, K., Rogóz, Z., & Skuza, G. (1993). Central effects of repeated treatment with CGP 37849, a competitive NMDA receptor antagonist with potential antidepressant activity. *Polish Journal of Pharmacology*, *45*, 455–466.

Maj, J., Rogoz, Z., Skuza, G., & Sowinska, H. (1992). The effects of MK-801 and antidepressant drugs in the forced swimming test in rats. *Neuropsychopharmacology*, *2*, 37–41.

Malcon, C., Achaval, M., Komlos, F., Partata, W., Sauressig, M., Ramírez, G., et al. (1997). GMP protects against quinolinic acid-induced loss of NADPH-diaphorase-positive cells in the rat striatum. *Neuroscience Letters*, *225*, 145–148.

Marvizon, J. C., Lewin, A. H., & Skolnick, P. (1989). 1-Aminocyclopropane carboxylic acid: A potent and selective ligand for the glycine modulatory site of the N-methyl-D-aspartate receptor complex. *Journal of Neurochemistry*, *52*, 992–994.

Masu, M., Tanabe, Y., Tsuchida, K., Shigemoto, R., & Nakanishi, S. (1991). Sequence and expression of a metabotropic glutamate receptor. *Nature*, *349*, 760–765.

Mc Loughlin, I. J., & Hodge, J. S. (1990). Zinc in depressive disorder. *Acta Psychiatrica Scandinavica*, *82*, 451–453.

Meldrum, B., & Garthwaite, J. (1990). Excitatory amino acid neurotoxicity and neurodegenerative disease. *Trends in Pharmacological Sciences*, *11*, 379–387.

Meldrum, B., Millan, M., Patel, S., & de Sarro, G. (1988). Anti-epileptic effects of focal micro-injection of excitatory amino acid antagonists. *Journal of Neural Transmission*, *72*, 191–200.

Meloni, D., Gambarana, C., De Montis, M. G., Dal Prá, P., Taddei, I., & Tagliamonte, A. (1993). Dizocilpine antagonizes the effect of chronic imipramine on learned helplessness in rats. *Pharmacology, Biochemistry, and Behavior*, *46*, 423–426.

Młyniec, K., Budziszewska, B., Reczyński, W., Sowa-Kućma, M., & Nowak, G. (2013). The role of the GPR39 receptor in zinc deficient-animal model of depression. *Behavioural Brain Research*, *238*, 30–35.
Młyniec, K., & Nowak, G. (2012). Zinc deficiency induces behavioral alterations in the tail suspension test in mice. Effect of antidepressants. *Pharmacological Reports*, *64*, 249–255.
Molz, S., Dal-Cim, T., Budni, J., Martín-de-Saavedra, M. D., Egea, J., Romero, A., et al. (2011). Neuroprotective effect of guanosine against glutamate-induced cell death in rat hippocampal slices is mediated by the phosphatidylinositol-3 kinase/Akt/glycogen synthase kinase 3β pathway activation and inducible nitric oxide synthase inhibition. *Journal of Neuroscience Research*, *89*, 1400–1408.
Monyer, H., Sprengel, R., Schoepfer, R., Herb, A., Higuchi, M., Lomeli, H., et al. (1992). Heteromeric NMDA receptors: Molecular and functional distinction of subtypes. *Science*, *256*, 1217–1221.
Moretti, M., Budni, J., Freitas, A. E., Neis, V. B., Ribeiro, C. M., Balen, G. O., et al. (2015). TNF-α-induced depressive-like phenotype and p38(MAPK) activation are abolished by ascorbic acid treatment. *European Neuropsychopharmacology*, *25*, 902–912.
Moretti, M., Budni, J., Freitas, A. E., Rosa, P. B., & Rodrigues, A. L. S. (2014). Antidepressant-like effect of ascorbic acid is associated with the modulation of mammalian target of rapamycin pathway. *Journal of Psychiatric Research*, *48*, 16–24.
Moretti, M., Budni, J., Ribeiro, C. M., & Rodrigues, A. L. S. (2012). Involvement of different types of potassium channels in the antidepressant-like effect of ascorbic acid in the mouse tail suspension test. *European Journal of Pharmacology*, *687*, 21–27.
Moretti, M., Budni, J., Santos, D., Antunes, A., Daufenbach, J., Manosso, L., et al. (2013). Protective effects of ascorbic acid on behavior and oxidative status of restraint-stressed mice. *Journal of Molecular Neuroscience*, *49*, 68–79.
Moretti, M., Colla, A., de Oliveira Balen, G., dos Santos, D. B., Budni, J., de Freitas, A. E., et al. (2012). Ascorbic acid treatment, similarly to fluoxetine, reverses depressive-like behavior and brain oxidative damage induced by chronic unpredictable stress. *Journal of Psychiatric Research*, *46*, 331–340.
Moretti, M., Freitas, A. E., Budni, J., Fernandes, S. C., Balen, G. O., & Rodrigues, A. L. S. (2011). Involvement of nitric oxide-cGMP pathway in the antidepressant-like effect of ascorbic acid in the tail suspension test. *Behavioural Brain Research*, *225*, 328–333.
Moryl, E., Danysz, W., & Quack, G. (1993). Potential antidepressive properties of amantadine, memantine and bifemelane. *Pharmacology and Toxicology*, *72*, 394–397.
Murck, H. J. (2013). Ketamine, magnesium and major depression—From pharmacology to pathophysiology and back. *Psychiatry Research*, *47*, 955–965.
Muroyama, A., Inaka, M., Matsushima, H., Sugino, H., Marunaka, Y., & Mitsumoto, Y. (2009). Enhanced susceptibility to MPTP neurotoxicity in magnesium-deficient C57BL/6N mice. *Neuroscience Research*, *63*, 72–75.
Musazzi, L., Treccani, G., Mallei, A., & Popoli, M. (2013). The action of antidepressants on the glutamate system: Regulation of glutamate release and glutamate receptors. *Biological Psychiatry*, *73*, 1180–1188.
Nahum-Levy, R., Fossom, L. H., Skolnick, P., & Benveniste, M. (1999). Putative partial agonist 1-aminocyclopropanecarboxylic acid acts concurrently as a glycine-site agonist and a glutamate-site antagonist at N-methyl-D-aspartate receptors. *Molecular Pharmacology*, *56*, 1207–1218.
Nahum-Levy, R., Tam, E., Shavit, S., & Benveniste, M. (2002). Glutamate but not glycine agonist affinity for NMDA receptors is influenced by small cations. *Journal of Neuroscience*, *22*, 2550–2560.

Nakajima, Y., Iwakabe, H., Akazawa, C., Nawa, H., Shigemoto, R., Mizuno, N., et al. (1993). Molecular characterization of a novel retinal metabotropic glutamate receptor mGluR6 with a high agonist selectivity for L-2-amino-4-phosphonobutyrate. *The Journal of Biological Chemistry, 268*, 11868–11873.

Nam, H. W., Lee, M. R., Hinton, D. J., & Choi, D. S. (2010). Reduced effect of NMDA glutamate receptor antagonist on ethanol-induced ataxia and striatal glutamate levels in mice lacking ENT1. *Neuroscience Letters, 479*, 277–281.

Nanxin, L., Rong-Jian, L., Jason, M. D., Mounira, B., Boyoung, L., Hyeon, S., et al. (2011). Glutamate NMDA receptor antagonists rapidly reverse behavioral and synaptic deficits caused by chronic stress exposure. *Biological Psychiatry, 69*, 754–761.

Nemeroff, C. B., & Owens, M. J. (2002). Treatment of mood disorders. *Nature Neuroscience, 5*, 1068–1070.

Nishitani, N., Nagayasu, K., Asaoka, N., Yamashiro, M., Shirakawa, H., Nakagawa, T., et al. (2014). Raphe AMPA receptors and nicotinic acetylcholine receptors mediate ketamine-induced serotonin release in the rat prefrontal cortex. *International Journal of Neuropsychopharmacology, 17*, 1321–1326.

Nowak, G., Ossowska, G., Jopek, R., & Papp, M. (1998). Strychnine-insensitive glycine/NMDA sites are altered in two stress models of depression. *Polish Journal of Pharmacology, 50*, 365–369.

Nowak, G., Siwek, M., Dudek, D., Zieba, A., & Pilc, A. (2003). Effect of zinc supplementation on antidepressant therapy in unipolar depression: A preliminary placebo-controlled study. *Polish Journal of Pharmacology, 55*, 1143–1147.

Nowak, G., Szewczyk, B., Sadlik, K., Piekoszewski, W., Trela, F., Florek, E., et al. (2003). Reduced potency of zinc to interact with NMDA receptors in hippocampal tissue of suicide victims. *Polish Journal of Pharmacology, 55*, 455–459.

Oishi, J., Doi, H., & Kawakami, N. (2009). Nutrition and depressive symptoms in community-dwelling elderly persons in Japan. *Acta Medica Okayama, 63*, 9–17.

Okamoto, N., Hori, S., Akazawa, C., Hayashi, Y., Shigemoto, R., Mizuno, N., et al. (1994). Molecular characterization of a new metabotropic glutamate receptor mGluR7 coupled to inhibitory cyclic AMP signal transduction. *The Journal of Biological Chemistry, 269*, 1231–1236.

Olney, J. W. (1990). Excitotoxicity: An overview. *Canada Diseases Weekly Report, 16*, 47–57.

Olney, J. W., & de Gubareff, T. (1978). Glutamate neurotoxicity and Huntington's chorea. *Nature, 271*, 557–559.

Olney, J. W., Labruyere, J., & de Gubareff, T. (1980). Brain damage in mice from voluntary ingestion of glutamate and aspartate. *Neurobehavioral Toxicology, 2*, 125–129.

Owolabi, R. A., Akanmu, M. A., & Adeyemi, O. I. (2014). Effects of ketamine and N-methyl-D-aspartate on fluoxetine-induced antidepressant-related behavior using the forced swimming test. *Neuroscience Letters, 566*, 172–176.

Panatier, A., Theodosis, D. T., Mothet, J. P., Touquet, B., Pollegioni, L., Poulain, D. A., et al. (2006). Glia-derived D-serine controls NMDA receptor activity and synaptic memory. *Cell, 125*, 775–784.

Paoletti, P., & Neyton, J. (2007). NMDA receptor subunits: Function and pharmacology. *Current Opinion in Pharmacology, 7*, 39–47.

Papp, M., & Moryl, E. (1994). Antidepressant activity of non-competitive and competitive NMDA receptor antagonists in a chronic mild stress model of depression. *European Journal of Pharmacology, 263*, 1–7.

Park, J. A., & Koh, J. Y. (1999). Induction of an immediate early gene egr-1 by zinc through extracellular signal-regulated kinase activation in cortical culture: Its role in zinc-induced neuronal death. *Journal of Neurochemistry, 73*, 450–456.

Park, S. W., Lee, J. G., Seo, M. K., Lee, C. H., Cho, H. Y., Lee, B. J., et al. (2014). Differential effects of antidepressant drugs on mTOR signalling in rat hippocampal neurons. *International Journal of Neuropsychopharmacology*, *17*, 1831–1846.

Parpura, V., Basarsky, T. A., Liu, F., Jeftinija, K., Jeftinija, S., & Haydon, P. G. (1994). Glutamate-mediated astrocyte-neuron signalling. *Nature*, *369*, 744–747.

Parsons, M. P., & Raymond, L. A. (2014). Extrasynaptic NMDA receptor involvement in central nervous system disorders. *Neuron*, *82*, 279–293.

Paul, I. A., & Skolnick, P. (2003). Glutamate and depression: Clinical and preclinical studies. *Annals of the New York Academy of Sciences*, *1003*, 250–272.

Paul, I. A., Trullas, R., Skolnick, P., & Nowak, G. (1992). Downregulation of cortical beta-adrenoceptors by chronic treatment with functional NMDA antagonists. *Psychopharmacology*, *106*, 285–287.

Pawlak, J., Dmitrzak-Węglarz, M., Skibińska, M., Szczepankiewicz, A., Leszczyńska-Rodziewicz, A., Rajewska-Rager, A., et al. (2013). Suicide attempts and clinical risk factors in patients with bipolar and unipolar affective disorders. *General Hospital Psychiatry*, *35*, 427–432.

Peeters, M., Page, G., Maloteaux, J. M., & Hermans, E. (2002). Hypersensitivity of dopamine transmission in the rat striatum after treatment with the NMDA receptor antagonist amantadine. *Brain Research*, *949*, 32–41.

Pfenninger, E. G., Durieux, M. E., & Himmelseher, S. (2002). Cognitive impairment after small-dose ketamine isomers in comparison to equianalgesic racemic ketamine in human volunteers. *Anesthesiology*, *96*, 357–366.

Poleszak, E. (2007). Modulation of antidepressant-like activity of magnesium by serotonergic system. *Journal of Neural Transmission*, *114*, 1129–1134.

Poleszak, E., Szewczyk, B., Kedzierska, E., Wlaź, P., Pilc, A., & Nowak, G. (2004). Antidepressant- and anxiolytic-like activity of magnesium in mice. *Pharmacology, Biochemistry, and Behavior*, *78*, 7–12.

Poleszak, E., Wlaź, P., Kedzierska, E., Nieoczym, D., Wróbel, A., Fidecka, S., et al. (2007). NMDA/glutamate mechanism of antidepressant-like action of magnesium in forced swim test in mice. *Pharmacology, Biochemistry, and Behavior*, *88*, 158–164.

Poleszak, E., Wlaź, P., Kedzierska, E., Radziwon-Zaleska, M., Pilc, A., Fidecka, S., et al. (2005). Effects of acute and chronic treatment with magnesium in the forced swim test in rats. *Pharmacological Reports*, *57*, 654–658.

Poleszak, E., Wlaź, P., Szewczyk, B., Wlaź, A., Kasperek, R., Wróbel, A., et al. (2011). A complex interaction between glycine/NMDA receptors and serotonergic/noradrenergic antidepressants in the forced swim test in mice. *Journal of Neural Transmission*, *118*, 1535–1546.

Price, D. L. (1999). New order from neurological disorders. *Nature*, *399*, A3–A5.

Pruus, K., Rudissaar, R., Allikmets, L., & Harro, J. (2010). The effect of the NMDA receptor antagonist dizocilpine on behavioral manifestations of serotonin and adrenergic antidepressants in rats. *Methods & Findings in Experimental & Clinical Pharmacology*, *32*, 123–128.

Przegaliński, E., Tatarczyńska, E., & Chojnacka-Wójcik, E. (2000). The influence of the benzodiazepine receptor antagonist flumazenil on the anxiolytic-like effects of CGP 37849 and ACPC in rats. *Neuropharmacology*, *39*, 1858–1864.

Raeder, J. C., & Stenseth, L. B. (2000). Ketamine: A new look at an old drug. *Current Opinion Anaesthesiology*, *13*, 463–468.

Rajkowska, G. (2000). Histopathology of the prefrontal cortex in major depression: What does it tell us about dysfunctional monoaminergic circuits? *Progress in Brain Research*, *126*, 397–412.

Ranju, V., Sathiya, S., Kalaivani, P., Priya, R. J., & Saravana Babu, C. (2015). Memantine exerts functional recovery by improving BDNF and GDNF expression in 3-nitropropionic acid intoxicated mice. *Neuroscience Letters*, *23*(586), 1–7.

Rebec, G. V., & Pierce, R. C. (1994). A vitamin as neuromodulator: Ascorbate release into the extracellular fluid of the brain regulates dopaminergic and glutamatergic transmission. Progress. *Neurobiology*, *43*, 537–565.

Réus, G. Z., Abelaira, H. M., dos Santos, M. A., Carlessi, A. S., Tomaz, D. B., Neotti, M. V., et al. (2013). Ketamine and imipramine in the nucleus accumbens regulate histone deacetylation induced by maternal deprivation and are critical for associated behaviors. *Behavioural Brain Research*, *256*, 451–456.

Réus, G. Z., Abelaira, H. M., Stringari, R. B., Fries, G. R., Kapczinski, F., & Quevedo, J. (2012). Memantine treatment reverses anhedonia, normalizes corticosterone levels and increases BDNF levels in the prefrontal cortex induced by chronic mild stress in rats. *Metabolic Brain Disease*, *27*, 175–182.

Réus, G. Z., Carlessi, A. S., Titus, S. E., Abelaira, H. M., Ignácio, Z. M., da Luz, J. R., et al. (2015). A single dose of S-ketamine induces long-term antidepressant effects and decreases oxidative stress in adulthood rats following maternal deprivation. *Developmental Neurobiology*, *75*(11), 1268–1281.

Réus, G. Z., Fries, G. R., Stertz, L., Badawy, M., Passos, I. C., Barichello, T., et al. (2015). The role of inflammation and microglial activation in the pathophysiology of psychiatric disorders. *Neuroscience*, *300*, 141–154.

Réus, G. Z., Jansen, K., Titus, S., Carvalho, A. F., Gabbay, V., & Quevedo, J. (2015). Kynurenine pathway dysfunction in the pathophysiology and treatment of depression: Evidences from animal and human studies. *Journal of Psychiatry Research*, *68*, 316–328.

Réus, G. Z., Nacif, M. P., Abelaira, H. M., Tomaz, D. B., dos Santos, M. A., Carlessi, A. S., et al. (2015a). Ketamine ameliorates depressive-like behaviors and immune alterations in adult rats following maternal deprivation. *Neuroscience Letters*, *584*, 83–87.

Réus, G. Z., Nacif, M. P., Abelaira, H. M., Tomaz, D. B., dos Santos, M. A., Carlessi, A. S., et al. (2015b). Ketamine treatment partly reverses alterations in brain derived-neurotrophic factor, oxidative stress and energy metabolism parameters induced by an animal model of depression. *Current Neurovascular Research*, *12*, 73–84.

Réus, G. Z., Stringari, R. B., Kirsch, T. R., Fries, G. R., Kapczinski, F., Roesler, R., et al. (2010). Neurochemical and behavioural effects of acute and chronic memantine administration in rats: Further support for NMDA as a new pharmacological target for the treatment of depression? *Brain Research Bulletin*, *81*, 585–589.

Réus, G. Z., Stringari, R. B., Rezin, G. T., Fraga, D. B., Daufenbach, J. F., Scaini, G., et al. (2012). Administration of memantine and imipramine alters mitochondrial respiratory chain and creatine kinase activities in rat brain. *Journal of Neural Transmission*, *119*, 481–491.

Réus, G. Z., Stringari, R. B., Ribeiro, K. F., Ferraro, A. K., Vitto, M. F., Cesconetto, P., et al. (2011). Ketamine plus imipramine treatment induces antidepressant-like behavior and increases CREB and BDNF protein levels and PKA and PKC phosphorylation in rat brain. *Behavioural Brain Research*, *221*, 166–171.

Réus, G. Z., Vieira, F. G., Abelaira, H. M., Michels, M., Tomaz, D. B., dos Santos, M. A., et al. (2014). MAPK signaling correlates with the antidepressant effects of ketamine. *Journal of Psychiatry Research*, *55*, 15–21.

Rice, M. E. (2000). Ascorbate regulation and its neuroprotective role in the brain. *Trends in Neurosciences*, *23*, 209–216.

Rogóz, Z., Skuza, G., Daniel, W. A., Wójcikowski, J., Dudek, D., & Wróbel, A. (2007). Amantadine as an additive treatment in patients suffering from drug-resistant unipolar depression. *Pharmacology Reports*, *59*, 778–784.

Rogóz, Z., Skuza, G., Maj, J., & Danysz, W. (2002). Synergistic effect of uncompetitive NMDA receptor antagonists and antidepressant drugs in the forced swimming test in rats. *Neuropharmacology*, *42*, 1024–1030.

Rosa, A. O., Lin, J., Calixto, J. B., Santos, A. R., & Rodrigues, A. L. (2003). Involvement of NMDA receptors and L-arginine-nitric oxide pathway in the antidepressant-like effects of zinc in mice. *Behavioural Brain Research*, *144*, 87–93.

Rudy, C. C., Hunsberger, H. C., Weitzner, D. S., & Reed, M. N. (2015). The role of the tripartite glutamatergic synapse in the pathophysiology of Alzheimer's disease. *Aging Disease*, *6*, 131–148.

Ruljancic, N., Mihanovic, M., Cepelak, I., Bakliza, A., & Curkovic, K. D. (2013). Platelet serotonin and magnesium concentrations in suicidal and non-suicidal depressed patients. *Magnesium Research*, *26*, 9–17.

Sabino, V., Narayan, A. R., Zeric, T., Steardo, L., & Cottone, P. (2013). mTOR activation is required for the anti-alcohol effect of ketamine, but not memantine, in alcohol-preferring rats. *Behavioural Brain Research*, *247*, 9–16.

Sattler, R., & Rothstein, J. D. (2007). Targeting an old mechanism in a new disease—Protection of glutamatergic dysfunction in depression. *Biological Psychiatry*, *6*, 137–138.

Schmidt, A. P., Lara, D. R., de Faria Maraschin, J., da Silveira Perla, A., & Souza, D. O. (2000). Guanosine and GMP prevent seizures induced by quinolinic acid in mice. *Brain Research*, *864*, 40–43.

Schmutz, M., Portet, C., Jeker, A., Klebs, K., Vassout, A., Allgeier, H., et al. (1990). The competitive NMDA receptor antagonists CGP 37849 and CGP 39551 are potent, orally-active anticonvulsants in rodents. *Naunyn-Schmiedebergs Archives of Pharmacology*, *342*, 61–66.

Serafini, G., Pompili, M., Innamorati, M., Dwivedi, Y., Brahmachari, G., & Girardi, P. (2013). Pharmacological properties of glutamatergic drugs targeting NMDA receptors and their application in major depression. *Current Pharmaceutical Design*, *19*, 1898–1922.

Singewald, N., Sinner, C., Hetzenauer, A., Sartori, S. B., & Murck, H. (2004). Magnesium-deficient diet alters depression- and anxiety-related behavior in mice—Influence of desipramine and Hypericum perforatum extract. *Neuropharmacology*, *47*, 1189–1197.

Siwek, M., Dudek, D., Schlegel-Zawadzka, M., Morawska, A., Piekoszewski, W., Opoka, W., et al. (2010). Serum zinc level in depressed patients during zinc supplementation of imipramine treatment. *Journal of Affective Disorders*, *126*, 447–452.

Skuza, G., & Rogóz, Z. (2003). Sigma1 receptor antagonists attenuate antidepressant-like effect induced by co-administration of 1,3 di-o-tolylguanidine (DTG) and memantine in the forced swimming test in rats. *Polish Journal of Pharmacology*, *55*, 1149–1152.

Sowa-Kućma, M., Legutko, B., Szewczyk, B., Novak, K., Znojek, P., Poleszak, E., et al. (2008). Antidepressant-like activity of zinc: Further behavioral and molecular evidence. *Journal of Neural Transmission*, *115*, 1621–1628.

Stein, I. S., Gray, J. A., & Zito, K. (2015). Non-ionotropic NMDA receptor signaling drives activity-induced dendritic spine shrinkage. *Journal of Neuroscience*, *35*, 12303–12308.

Swardfager, W., Herrmann, N., Mazereeuw, G., Goldberger, K., Harimoto, T., & Lanctôt, K. L. (2013). Zinc in depression: A meta-analysis. *Biological Psychiatry*, *74*, 872–878.

Szewczyk, N. J., Peterson, B. K., & Jacobson, L. A. (2002). Activation of Ras and the mitogen-activated protein kinase pathway promotes protein degradation in muscle cells of Caenorhabditis elegans. *Molecular and Cellular Biology*, *22*, 4181–4188.

Takeda, A. (2011). Zinc signaling in the hippocampus and its relation to pathogenesis of depression. *Molecular Neurobiology*, *44*, 166–174.

Tamano, H., Kan, F., Kawamura, M., Oku, N., & Takeda, A. (2009). Behavior in the forced swim test and neurochemical changes in the hippocampus in young rats after 2-week zinc deprivation. *Neurochemistry International*, *55*, 536–541.

Tan, L., Ge, H., Tang, J., Fu, C., Duanmu, W., Chen, Y., et al. (2015). Amantadine preserves dopamine level and attenuates depression-like behavior induced by traumatic brain injury in rats. *Behavioural Brain Research*, *279*, 274–282.

Tang, C. M., Dichter, M., & Morad, M. (1989). Quisqualate activates a rapidly inactivating high conductance ionic channel in hippocampal neurons. *Science (New York, NY)*, *243*, 1474–1477.

Tasca, C. I., Santos, T. G., Tavares, R. G., Battastini, A. M., Rocha, J. B., & Souza, D. O. (2004). Guanine derivatives modulate L-glutamate uptake into rat brain synaptic vesicles. *Neurochemistry International*, *44*, 423–431.

Tassabehji, N. M., Corniola, R. S., Alshingiti, A., & Levenson, C. W. (2008). Zinc deficiency induces depression-like symptoms in adult rats. *Physiology & Behavior*, *95*, 365–369.

Tokita, K., Fujita, Y., Yamaji, T., & Hashimoto, K. (2012). Depressive-like behavior in adrenocorticotropic hormone-treated rats blocked by memantine. *Pharmacology, Biochemistry, and Behavior*, *102*, 329–334.

Trivedi, M. H. (2006). Major depressive disorder: Remission of associated symptoms. *Journal of Clinical Psychiatry*, *6*, 27–32.

Trullas, R., Folio, T., Young, A., Miller, R., Boje, K., & Skolnick, P. (1991). 1-Aminocyclopropanecarboxylates exhibit antidepressant and anxiolytic actions in animal models. *European Journal of Pharmacology*, *203*, 379–385.

Trullas, R., & Skolnick, P. (1990). Functional antagonists at the NMDA receptor complex exhibit antidepressant actions. *European Journal of Pharmacology*, *185*, 1–10.

Von Lubitz, D. K. J. E., Lin, R. C. S., McKenzie, R. J., Devlin, T. M., McCabe, T. R., & Skolnick, P. (1992). A novel treatment of global ischaemia with a glycine partial agonist. *European Journal of Pharmacology*, *219*, 153–158.

Vyklický, L., Jr., Benveniste, M., & Mayer, M. L. (1990). Modulation of N-methyl-D-aspartic acid receptor desensitization by glycine in mouse cultured hippocampal neurones. *The Journal of Physiology*, *428*, 313–331.

Walker, A. K., Budac, D. P., Bisulco, S., Lee, A. W., Smith, R. A., Beenders, B., et al. (2013). NMDA receptor blockade by ketamine abrogates lipopolysaccharide-induced depressive-like behavior in C57BL/6J mice. *Neuropsychopharmacology*, *38*, 1609–1616.

Wang, J. Q., Guo, M. L., Jin, D. Z., Xue, B., Fibuch, E. E., & Mao, L. M. (2014). Roles of subunit phosphorylation in regulating glutamate receptor function. *European Journal of Pharmacology*, *728*, 183–187.

Wang, T., Huang, X. J., Van, K. C., Went, G. T., Nguyen, J. T., & Lyeth, B. G. (2014). Amantadine improves cognitive outcome and increases neuronal survival after fluid percussion traumatic brain injury in rats. *Journal of Neurotrauma*, *31*, 370–377.

Wang, M., Yang, Y., Dong, Z., Cao, J., & Xu, L. (2006). NR2B-containing N-methyl-D-aspartate subtype glutamate receptors regulate the acute stress effect on hippocampal long-term potentiation/long-term depression in vivo. *Neuroreports*, *17*, 1343–1346.

Whittle, N., Lubec, G., & Singewald, N. (2009). Zinc deficiency induces enhanced depression-like behaviour and altered limbic activation reversed by antidepressant treatment in mice. *Amino Acids*, *36*, 147–158.

Widmer, J., Bovier, P., Karege, F., Raffin, Y., Hilleret, H., Gaillard, J. M., et al. (1992). Evolution of blood magnesium, sodium and potassium in depressed patients followed for three months. *Neuropsychobiology*, *6*, 173–179.

Widmer, J., Henrotte, J. G., Raffin, Y., Bovier, P., Hilleret, H., & Gaillard, J. M. (1995). Relationship between erythrocyte magnesium, plasma electrolytes and cortisol, and intensity of symptoms in major depressed patients. *Journal of Affective Disorders*, *34*, 201–209.

Wong, E. H. F., Knight, A. R., & Woodruff, G. N. (1988). [3H]MK-801 labels a site on the N-methyl-D-aspartate receptor channel complex in rat brain membranes. *Journal of Neurochemistry*, *50*, 274–281.

World Health Organization. (2012). *Sixty-fifth World Health Assembly*.

Wu, G. J., Chen, T. L., Ueng, Y. F., & Chen, R. M. (2008). Ketamine inhibits tumor necrosis factor-alpha and interleukin-6 gene expressions in lipopolysaccharide-stimulated macrophages through suppression of toll-like receptor 4-mediated c-Jun N-terminal kinase phosphorylation and activator protein-1 activation. *Toxicology and Applied Pharmacology*, *228*, 105–113.

Yang, C., Zhou, Z. Q., Gao, Z. Q., Shi, J. Y., & Yang, J. J. (2013). Acute increases in plasma mammalian target of rapamycin, glycogen synthase kinase-3β, and eukaryotic elongation factor 2 phosphorylation after ketamine treatment in three depressed patients. *Biological Psychiatry*, *73*, e35–e36.

Yu, J., Ding, C. P., Wang, J., Wang, T., Zhang, T., Zeng, X. Y., et al. (2015). Red nucleus glutamate facilitates neuropathic allodynia induced by spared nerve injury through non-NMDA and metabotropic glutamate receptors. *Journal of Neuroscience Research*, *93*, 1839–1848.

Yuan, T., & Bellone, C. (2013). Glutamatergic receptors at developing synapses: The role of GluN3A-containing NMDA receptors and GluA2-lacking AMPA receptors. *European Journal of Pharmacology*, *719*, 107–111.

Zarate, C. A., Jr., Singh, J. B., Carlson, P. J., Brutsche, N. E., Ameli, R., Luckenbaugh, D. A., et al. (2006). A randomized trial of an N-methyl-D-aspartate antagonist in treatment-resistant major depression. *Archives of General Psychiatry*, *63*(8), 856–864.

Zarate, C. A., Jr., Singh, J. B., Quiroz, J. A., De Jesus, G., Denicoff, K. K., Luckenbaugh, D. A., et al. (2006). A double-blind, placebo-controlled study of memantine in the treatment of major depression. *The American Journal of Psychiatry*, *163*, 153–155.

Zarrindast, M. R., Nasehi, M., Pournaghshband, M., & Yekta, B. G. (2012). Dopaminergic system in CA1 modulates MK-801 induced anxiolytic-like responses. *Pharmacology, Biochemistry, and Behavior*, *103*, 102–110.

Zhang, X. Y., Ji, F., Wang, N., Chen, L. L., Tian, T., & Lu, W. (2014). Glycine induces bidirectional modifications in N-methyl-D-aspartate receptor-mediated synaptic responses in hippocampal CA1 neurons. *The Journal of Biological Chemistry*, *289*, 31200–31211.

Zhang, X. L., Shuttleworth, C. W., Moskal, J. R., & Stanton, P. K. (2015). Suppression of spreading depolarization and stabilization of dendritic spines by GLYX-13, an NMDA receptor glycine-site functional partial agonist. *Experimental Neurology*, *273*, 312–321.

Zhang, X. L., Sullivan, J. A., Moskal, J. R., & Stanton, P. K. (2008). A NMDA receptor glycine site partial agonist, GLYX-13, simultaneously enhances LTP and reduces LTD at Schaffer collateral-CA1 synapses in hippocampus. *Neuropharmacology*, *55*, 1238–1250.

Zieba, A., Kata, R., Dudek, D., Schlegel-Zawadzka, M., & Nowak, G. (2000). Serum trace elements in animal models and human depression: Part III. Magnesium. Relationship with copper. *Human Psychopharmacology*, *15*, 631–635.

Zito, K., Scheuss, V., Knott, G., Hill, T., & Svoboda, K. (2009). Rapid functional maturation of nascent dendritic spines. *Neuron*, *61*, 247–258.

CHAPTER SIX

AMPA Receptors as Therapeutic Targets for Neurological Disorders

Kevin Lee*[,1], Lucy Goodman*[,1], Chantelle Fourie*, Susan Schenk[†], Beulah Leitch[‡], Johanna M. Montgomery*[,2]

*Department of Physiology, Centre for Brain Research, University of Auckland, Auckland, New Zealand
[†]School of Psychology, Victoria University, Wellington, New Zealand
[‡]Department of Anatomy, University of Otago, Dunedin, New Zealand
[2]Corresponding author: e-mail address: jm.montgomery@auckland.ac.nz

Contents

Abstract

Almost every neurological disease directly or indirectly affects synapse function in the brain. However, these diseases alter synapses through different mechanisms, ultimately resulting in altered synaptic transmission and/or plasticity. Glutamate is the major neurotransmitter that mediates excitatory synaptic transmission in the brain through activation of alpha-amino-3-hydroxy-5-methyl-4-isoxazole-propionate (AMPA) receptors.

[1] These authors contributed equally to this work.

Advances in Protein Chemistry and Structural Biology, Volume 103
ISSN 1876-1623
http://dx.doi.org/10.1016/bs.apcsb.2015.10.004

These receptors have therefore been identified as a target for the development of therapeutic treatments for neurological disorders including epilepsy, neurodegenerative diseases, autism, and drug addiction. The fact that AMPA receptors play a dominant role throughout the brain raises the significant challenge of selectively targeting only those regions affected by disease, and clinical trials have raised doubt regarding the feasibility of specifically targeting AMPA receptors for new therapeutic options. Benzamide compounds that act as positive allosteric AMPA receptor modulators, known as AMPAkines, can act on specific brain regions and were initially proposed to revolutionize the treatment of cognitive deficits associated with neurological disorders. Their therapeutic potential has since declined due to inconsistent results in clinical trials. However, recent advances in basic biomedical research are significantly increasing our knowledge of AMPA receptor structure, binding sites, and interactions with auxiliary proteins. In particular, the large complex of postsynaptic proteins that interact with AMPA receptor subunits have been shown to control AMPA receptor insertion, location, pharmacology, synaptic transmission, and plasticity. These proteins are now being considered as alternative therapeutic target sites for modulating AMPA receptors in neurological disorders.

Almost every neurological disorder either directly or indirectly alters synaptic function in the brain; however, the underlying subcellular changes that impair synaptic function are unique for each disorder. Example synaptopathologies occurring in neurological disorders include changes in surface postsynaptic receptor localization (Gladding et al., 2012; Milnerwood et al., 2010; Parsons & Raymond, 2014; Wasterlain, Naylor, Liu, Niquet, & Baldwin, 2013), postsynaptic protein synthesis (Huber, Gallagher, Warren, & Bear, 2002; Osterweil et al., 2013), excitatory–inhibitory balance (Baroncelli et al., 2011; Chen, Xiong, & Yan, 2013; Fernandez et al., 2007; Kaila, Ruusuvuori, Seja, Voipio, & Puskarjov, 2014; Na, Morris, Nelson, & Monteggia, 2014), presynaptic transmitter release (Arons et al., 2012; Usdin, Shelbourne, Myers, & Madison, 1999; Zhao et al., 2006), synaptic protein–protein interactions (Arons et al., 2012; Epis et al., 2010; Marcello et al., 2013; Sun, Savanenin, Reddy, & Liu, 2001), and synapse density (Belichenko et al., 2004; Grabrucker et al., 2011; Hamos, DeGennaro, & Drachman, 1989; Neuhofer et al., 2015; Nishijima et al., 2014; Terry et al., 1991) to name just a few. These synaptic changes manifest in altered synaptic transmission and/or synaptic plasticity. Therefore, proteins intricately involved in synapse function have been major targets for developing therapies for neurological disorders such as Alzheimer's disease (AD), autism spectrum disorders (ASD), Parkinson's disease (PD), epilepsy, amyotrophic lateral sclerosis (ALS), ischemia, and drug addiction. Here, we review the structure, function, and regulation of one major therapeutic target—the AMPA-type glutamate receptor, and how pharmacological targeting of this receptor has largely failed to translate to therapies for neurological disorders. Over recent years, however, our fundamental understanding of AMPA receptor dynamics and regulation at synapses has significantly increased, especially with regard to how AMPA receptor scaffold proteins control AMPA receptor dynamics, synaptic expression, and localization (Elias & Nicoll, 2007; Fitzjohn, Doherty, & Collingridge, 2006; Jeyifous et al., 2009; Li et al., 2011; Montgomery, Zamorano, & Garner, 2004; Sun & Turrigiano, 2011; Waites et al., 2009; Fig. 1). In combination with the low clinical success of AMPA receptor focussed drug trials, this raises the issue of whether AMPA receptor scaffold proteins are potential therapeutic targets due to their ability to alter multiple aspects of AMPA receptor physiology. Therefore in this review, we also summarize a large body of work that has revealed how these scaffold proteins control AMPA receptor targeting, localization, and dynamics, and whether they are a viable avenue for therapeutic treatments of neurological disorders.

1. AMPA RECEPTOR STRUCTURE AND FUNCTION

In the brain, glutamate is the major excitatory neurotransmitter involved in regulating neuronal activity through fast excitatory synaptic transmission. When released from presynaptic terminals, glutamate signaling activates two distinctly different families of receptors consisting of ionotropic glutamate receptors (iGluRs) and metabotropic glutamate receptors (mGluRs) (Julio-Pieper, Flor, Dinan, & Cryan, 2011; Traynelis et al., 2010). The iGluRs are further classified into α-amino-3-hydroxy-5-methyl-4-isoxazolepropinoic acid receptor (AMPAR), *N*-methyl-D-aspartate receptor (NMDAR), and kainate receptor subfamilies, which all share a common voltage-gated ion channel function (Traynelis et al., 2010).

AMPA receptors mediate fast excitatory neurotransmission (Colquhoun, Jonas, & Sakmann, 1992), and are the targets for multiple signaling pathways that regulate the strength of glutamatergic excitatory synapses (i.e., synaptic plasticity). In the mammalian brain, most AMPA receptors are tetramers comprised of varying combinations of four different glutamate receptor subunits, termed GluA1–4 (Hollmann, O'Shea-Greenfield, Rogers, & Heinemann, 1989; Keinanen et al., 1990; Nakanishi, Shneider, & Axel, 1990). The subunit composition of AMPA receptors varies depending on the brain region and specific developmental stage: for example, at the mature

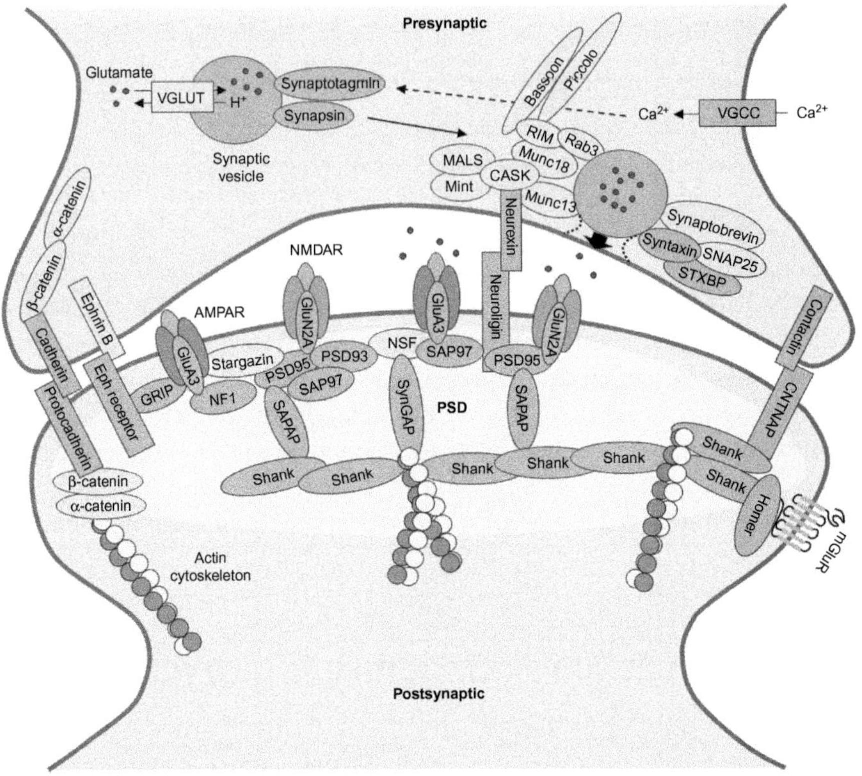

Figure 1 Schematic representation of pre- and postsynaptic proteins located at glutamatergic excitatory synapses. Glutamate is stored in synaptic vesicles via vesicular glutamate transporter (VGLUT). Synaptotagmin on synaptic vesicles senses the Ca^{2+} transient from the opening of voltage-gated Ca^{2+} channels (VGCCs) induced by action potentials and this leads to subsequent targeting, docking, priming, and fusion of synaptic vesicles mediated by groups of presynaptic proteins. The pre- and postsynaptic sides of the synapse are structurally and functionally stabilized by a number of cell adhesion molecules. Glutamate, once released, bind to glutamate receptors such as AMPA receptors, NMDA, and metabotropic glutamate receptors (mGluRs). Scaffolding proteins at the postsynaptic density (PSD) are involved in trafficking and stabilizing the glutamate receptors at the synapse, and mediating intracellular signaling stimulated by the activation of the receptors.

glutamatergic synapses of the hippocampus, a brain region known to be critically involved in learning and memory, AMPA receptors consist primarily of heteromeric GluA1/GluA2 subunits or GluA2/GluA3 subunits, with a very small fraction of homomeric GluA1 receptors (Wenthold, Petralia, Blahos, & Niedzielski, 1996). In the striatum, a brain region which plays a major role in movement disorders such as PD, the parvalbumin containing interneurons express the GluA1, GluA2/3, and GluA2/3/4c AMPA receptor subunits (Chen, Veenman, & Reiner, 1996; Tallaksen-Greene & Albin, 1994), while the somatostatin-containing and cholinergic interneurons express GluA5/6/7 and GluN1 (Chen et al., 1996) but not GluA1 or GluA2/3/4c (Tallaksen-Greene & Albin, 1994).

The functional properties of AMPA receptors largely depend on subunit composition. Although the extracellular and transmembrane regions of all AMPA receptor subunits are highly homologous, their intracellular carboxyl (C) termini are structurally and functionally divergent: The C-termini of AMPA receptor subunits contain regulatory domains that undergo posttranslational modifications including protein phosphorylation, ubiquitination, and palmitoylation. They are also the sites of interaction with scaffold proteins that bind signaling proteins and cytoskeletal proteins (Anggono & Huganir, 2012; Shepherd & Huganir, 2007). Among AMPA receptor subunits, the presence of a GluA2 subunit exerts a profound effect on the biophysical property of AMPA receptor heteromeric complexes. The GluA2 subunit is subject to RNA editing where the genomic glutamine (Q) codon for residue 607 is replaced by the arginine (R) codon (Lomeli et al., 1994; Sommer, Kohler, Sprengel, & Seeburg, 1991). The state of RNA editing in GluA2 controls multiple AMPA receptor properties including Ca^{2+} permeability, channel conductance, kinetics, receptor affinity for glutamate, and subunit assembly into a functional receptor (Burnashev, Monyer, Seeburg, & Sakmann, 1992; Geiger et al., 1995; Greger, Khatri, Kong, & Ziff, 2003; Hume, Dingledine, & Heinemann, 1991; Mansour, Nagarajan, Nehring, Clements, & Rosenmund, 2001; Swanson, Kamboj, & Cull-Candy, 1997). GluA2-containing AMPA receptors exhibit a lower channel conductance, open probability, and Ca^{2+} permeability, whereas GluA2-lacking AMPA receptors have a high channel conductance, open channel probability, and Ca^{2+} permeability (Burnashev et al., 1992; Swanson et al., 1997). Thus, the presence or absence of the GluA2 subunit within AMPA receptor subunit assembly can significantly influence AMPA receptor properties and thereby synaptic transmission.

2. SYNAPTIC PLASTICITY AND AMPA RECEPTOR DYNAMICS

2.1 Synaptic Plasticity

There are several forms of synaptic plasticity expressed in the brain but a large body of the research has focussed on NMDA receptor-dependent long-term potentiation (LTP) at hippocampal CA1 synapses (Bliss & Collingridge, 1993; Malenka, 2003; Malinow & Malenka, 2002; Park et al., 2013; Poncer, 2003). LTP is measured as a persistent increase in the postsynaptic AMPA receptor-mediated evoked response and is induced by tetanic or theta-burst stimulation of presynaptic inputs (Bliss & Lømo, 1973; Larson & Munkácsy, 2015; Park et al., 2013), or by pairing presynaptic action potentials with postsynaptic depolarization (Larson & Lynch, 1986; Montgomery, Pavlidis, & Madison, 2001; Park et al., 2013). Conversely, synaptic strength can be decreased by low-frequency stimulation, as measured by a decrease in the AMPA receptor-mediated synaptic currents. This is termed long-term depression (LTD) (Dudek & Bear, 1992), and together LTP and LTD enable bidirectional changes in synaptic strength.

The expression and maintenance of multiple forms of LTP relies on AMPA receptors that are rapidly inserted to the synapse upon LTP induction or rapidly removed from the synapse upon LTD induction (Fig. 2). Using recombinant AMPA receptors (GluA1-GFP) expressed in hippocampal CA1 neurons, it has been shown that the GluA1 subunit seems to be the main component for activity-dependent delivery of AMPA receptors to synapses during LTP (Hayashi et al., 2000; Horton & Ehlers, 2004; Lu et al., 2001; Man et al., 2003; Park, Penick, Edwards, Kauer, & Ehlers, 2004; Passafaro, Piech, & Sheng, 2001; Pickard et al., 2001; Shi, Hayashi, Esteban, & Malinow, 2001; Shi et al., 1999). Adult knockout mice lacking GluA1 could not generate LTP in the CA1 region of hippocampus, and insertion of GFP-tagged GluA1 containing receptors to these mice restored LTP in CA1 pyramidal neurons (Mack et al., 2001; Zamanillo et al., 1999).

Protein kinases and AMPA receptor phosphorylation are also critical for increasing synaptic strength (Fig. 2). During LTP, Ca^{2+} influx through NMDA receptors activates calcium calmodulin protein kinase II (CaMKII) and triggers its autophosphorylation (Barria, Muller, Derkach, Griffith, & Soderling, 1997). Activated CaMKII then phosphorylates the GluA1 subunit of AMPA receptors and enhances AMPA receptor conductance (Barria et al., 1997; Derkach, Barria, & Soderling, 1999; Lee, Barbarosie, Kameyama, Bear, & Huganir, 2000; McGlade-McCulloh, Yamamoto, Tan, Brickey, & Soderling, 1993). AMPA receptor phosphorylation state also influences surface receptor diffusion. cAMP-dependent PKA phosphorylates Ser845 in the GluA1 subunit to promote receptor insertion (Ehlers, 2000; He et al., 2009; Man, Sekine-aizawa, & Huganir, 2007) and directs the receptor to extrasynaptic sites (Oh, Derkach, Guire, & Soderling, 2006). Mobile pools of AMPA receptors are recruited

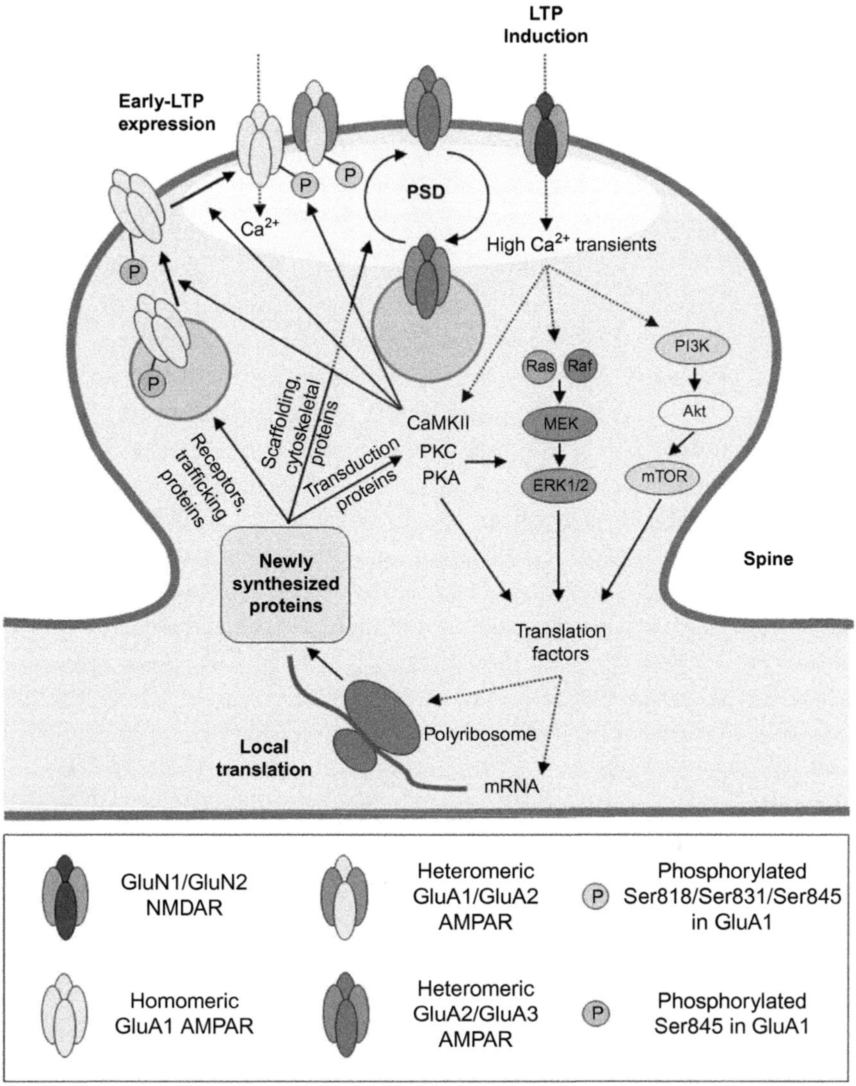

Figure 2 Expression pathways of LTP. Stimulation of NMDA receptors leads to Ca^{2+} influx that in turn activates multiple protein kinases in dendritic spines. These kinases phosphorylate the AMPA receptors at the synapse to change their properties or the AMPA receptors in the intracellular reserve pool to induce insertion of more AMPA receptors to the cell surface. Furthermore, protein kinases activate translation factors that are required for the stabilization of local mRNAs and for the initiation of local translation of mRNAs. Local protein synthesis provides a feedforward mechanism to persistently increase receptor numbers, receptor trafficking, levels of scaffolding and cytoskeleton proteins, and lateral diffusion and stabilization of AMPA receptors.

and trapped at activated synapses when NMDA receptors are activated due to Ca^{2+} influx and subsequent activation of CaMKII (Opazo et al., 2010). CaMKII activation leads to phosphorylation of Stargazin (see Section 3.1) and its binding to postsynaptic scaffold proteins, and thus trapping AMPA receptors at the synapse (Opazo et al., 2010). The NMDAR Ca^{2+} influx-dependent synaptic incorporation and/or stabilization of AMPA receptors may also require PKC phosphorylation of Ser818 in the GluA1 subunit and CaMKII-dependent phosphorylation of a PDZ domain-containing protein such as SAP97 (see Section 3.2; Boehm et al., 2006; Hayashi et al., 2000). Thus, LTP is controlled by changes in the properties, localization, and number of AMPA receptors, which in turn require protein to protein interactions and protein phosphorylation by protein kinases.

In contrast, AMPA receptor dephosphorylation via protein phosphatases and their subsequent removal from the synapse is critical for the expression of LTD (Carroll et al., 1999; Man et al., 2000; Mulkey & Malenka, 1992). Following low-frequency stimulation of NMDA receptors, low-level calcium influx activates the Ca^{2+}-sensitive protein phosphatase 2B (PP2B), also known as calcineurin, which dephosphorylates inhibitor 1 (I1) and activation of PP1, which in turn dephosphorylates CaMKII (Lisman, 1989; Mulkey, Endo, Shenolikar, & Malenka, 1994; Mulkey, Herron, & Malenka, 1993). Our understanding of the molecular mechanisms of AMPA receptor endocytosis during NMDA receptor-dependent LTD has been possible by the identification of proteins that bind directly to and regulate the synaptic function of AMPA receptors. *N*-ethylmaleimide-sensitive factor (NSF), an ATPase, binds directly to the GluA2 subunit to stabilize AMPA receptors on the membrane (Nishimune et al., 1998; Osten et al., 1998). AMPA receptors are internalized by clathrin-mediated endocytosis during LTD. The overlapping region on GluA2 that both clathrin adaptor protein AP2 and NSF interact with, is indicative of a possible mechanism for the regulation of AMPA receptor surface expression during LTD (Man et al., 2000). The LTD-induced replacement of NSF by AP2 to initiate the AMPA receptor endocytosis is mediated by a number of proteins including hippocalcin, which acts as a neuronal calcium sensor and translocates to the plasma membrane upon a small rise in Ca^{2+}, forming a complex with AP2 and GluA2 to initiate the clathrin-dependent endocytosis of AMPA receptor (Palmer et al., 2005). Other AP2 targeting molecules include small GTPase Ras-related protein (RalA), Rab5, and also RalA-binding protein 1 (RaBP1) (Brown, Tran, Backos, & Esteban, 2005).

2.2 AMPA Receptor Dynamics

AMPA receptors are localized within the postsynaptic density (PSD) and also on the neuronal membrane outside of synapses (Fig. 3). Electron microscopy and more recent fluorescent super resolution imaging techniques have revealed that AMPA receptors are not distributed uniformly across the synaptic membrane, and are observed at central, peripheral, and extrasynaptic locations (Dani, Huang, Bergan, Dulac, & Zhuang, 2010). Perisynaptic glutamate receptors are those located ~100 nm from the edge of the PSD, and extrasynaptic glutamate receptors are located further from the synapse, along the spine neck, dendritic shafts, and soma

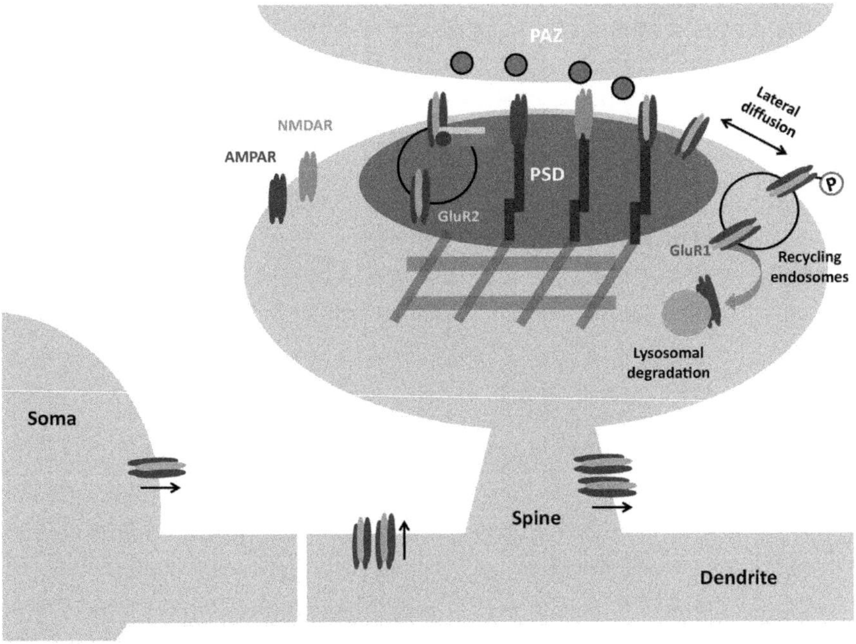

Figure 3 AMPA receptor incorporation at synapses. AMPA receptor exocytosis has been observed directly at synapses, perisynaptic to the PSD, on the dendritic spine, spine shaft, and soma. GluA2-containing AMPA receptors continually exchange with existing synaptic AMPA receptors under basal conditions, although their exocytosis has also been observed on dendrites. Surface expression of GluA2-containing receptors is maintained by the intracellular scaffolding proteins ABP/GRIP and PICK1, and NSF. In contrast, GluA1-containing AMPA receptors are inserted into the membrane via recycling endosomes outside of the PSD and diffuse laterally into the synapse where they are scaffolded by PSD95 proteins. AMPA receptor incorporation is upregulated in response to synaptic activity and involves phosphorylation of GluA1. NMDA receptors do not appear to undergo extensive endosomal recycling at the synapse.

(Newpher & Ehlers, 2008). Several recent studies have confirmed the presence of AMPA receptor "nanodomains" on the synaptic surface (Fukata et al., 2013; Hoze et al., 2012; MacGillavry, Song, Raghavachari, & Blanpied, 2013; Nair et al., 2013). The average synapse contains 1–2 AMPA receptor nanodomains each approximately 70 nm in size that are separated by a few hundred nanometres of synaptic membrane, with the number and density of nanodomains increasing with larger PSD size (Fukata et al., 2013; Nair et al., 2013). Live cell super resolution imaging has revealed the highly dynamic nature of these nanodomains, which can form and disappear in minutes to hours (Nair et al., 2013). AMPA receptor movements are highly heterogeneous, moving freely within the synapse, and diffusing laterally into these nanodomains where they become more stably integrated into the membrane (Nair et al., 2013).

The density of AMPA receptors within the synapse can directly determine synaptic strength (Raghavachari & Lisman, 2004). Estimates of absolute AMPA receptors densities suggest that a single synapse may contain anywhere from 0 to 150 AMPA receptors (Chen et al., 2008; Masugi-Tokita et al., 2007; Matsuzaki et al., 2001; Nusser et al., 1998; Tanaka et al., 2005). The number of AMPA receptors clustered at synapses is correlated with spine type and synaptic maturity (Matsuzaki et al., 2001). Whereas NMDAR-only "silent synapses" are more common during early development, AMPA receptors are recruited to synapses with increasing maturation (Gomperts, Carroll, Malenka, & Nicoll, 2000; Liao, Zhang, O'Brien, Ehlers, & Huganir, 1999; Montgomery et al., 2001). The location of AMPA receptors at the synapse is also critical to synaptic function, as AMPA receptors have a lower affinity for glutamate, they must be positioned near to the site of presynaptic release in order to be activated by glutamate (Lisman, Raghavachari, & Tsien, 2007). The observation that increased exocytosis of AMPA receptors does not necessarily lead to increased excitatory postsynaptic currents (Schnell et al., 2002) suggests that newly exocytosed receptors do not always have direct access to glutamate. While insertion of AMPA receptors has been observed directly at the synapse (Adesnik, Nicoll, & England, 2005; Gerges, Backos, Rupasinghe, Spaller, & Esteban, 2006), current thinking suggests a multistep incorporation of receptors into synapses (Choquet, 2010; Opazo & Choquet, 2011; Chater & Goda, 2014). AMPA receptors have been shown to be exocytosed at extrasynaptic sites away from synapses (Kennedy, Davison, Robinson, & Ehlers, 2010; Patterson, Szatmari, & Yasuda, 2010; Tanaka & Hirano, 2012; Tao-Cheng et al., 2011). This includes locations at or on dendritic spines very near the synapse (Adesnik et al., 2005; Gerges et al., 2006; Kennedy et al., 2010; Park et al., 2006; Wang et al., 2009), on the dendritic shaft (Jaskolski, Mayo-Martin, Jane, & Henley, 2009; Lin, Makino, Sharma, & Hayashi, 2009; Makino & Malinow, 2009; Patterson et al., 2010; Yudowski, 2007), or the cell soma (Lin et al., 2009; Yudowski, 2007).

Newly exocytosed AMPA receptors can diffuse laterally from the extrasynaptic membrane to the synapse (Fig. 3). In fact, the population of surface receptors appears to be in dynamic equilibrium between extrasynaptic and synaptic compartments (Borgdorff & Choquet, 2002; Tardin, Cognet, Bats, Lounis, & Choquet, 2003). Under basal conditions, about 70% of surface AMPA receptors are mobile, with half of these freely exchanging between the synaptic and extrasynaptic membrane (Heine et al., 2008). Molecular uncoupling of the endocytic zone from the synapse reduces the number and mobility of AMPA receptors at synapses and the ability to undergo synaptic potentiation (Lu, Helton, Blanpied, & Rácz, 2007; Petrini et al., 2009). Furthermore, inhibiting the GTPase dynamin that is involved in endocytosis reduces the lateral diffusion of AMPA receptors into synapses via membrane drift (Jaskolski et al., 2009). Mobile surface GluA2 subunits frequently stop just outside synapses (Borgdorff & Choquet, 2002), and the mobility of receptors in areas of clathrin assembly are greatly reduced compared to the extrasynaptic membrane (Petrini et al., 2009).

Receptor movement across the membrane surface is an efficient method of receptor incorporation into synapses. While fast receptor exocytosis from internal stores to the synaptic membrane takes place in a matter of minutes (Adesnik et al., 2005; Ehlers, 2000; Passafaro et al., 2001), lateral movement of receptors into synapses occurs within seconds (Adesnik et al., 2005; Bats, Groc, & Choquet, 2007; Tardin et al., 2003). AMPA receptors can be grouped by their diffusive properties in the membrane: those with fast mobility are located mainly at extrasynaptic locations where they display free, unrestrained movements reaching velocities of up to several hundred nanometres per second (Tardin et al., 2003; Ashby, Maier, Nishimune, & Henley, 2006; Tomita et al., 2005; Heine et al., 2008; Petrini et al., 2009; Bats et al., 2007; Nair et al., 2013; Groc et al., 2004; Borgdorff & Choquet, 2002). However, over time these newly exocytosed, freely moving AMPA receptors become immobile at synapses (Petrini et al., 2009; Tardin et al., 2003). Furthermore, slowly mobile or immobile receptors are more likely to be found at synaptic locations where their movements are more restricted to domains of less than 150 nm radius (Tardin et al.,

2003; Nair et al., 2013). However, measurements of AMPA receptor diffusion rates using large latex beads (Borgdorff & Choquet, 2002), antibodies (Groc et al., 2004; Tardin et al., 2003), or quantum dots (QDs) (Bats et al., 2007; Heine et al., 2008; Petrini et al., 2009) may be affected by the very small space between opposing synaptic terminals. Indeed, a recent study showed that AMPA receptors labeled with smaller (9 nm) QDs visited synapses more often than those labeled with conventional QDs (Cai et al., 2014). Therefore, technological advances in AMPA receptor labeling and imaging resolution will help to confirm our understanding of AMPA receptor dynamics at synapses.

The insertion of intracellular AMPA receptors into the extrasynaptic membrane under basal conditions may provide a reserve pool of receptors that can move into the synapse during periods of increased synaptic activity. Measurements of AMPA receptor exocytosis during synaptic activity has suggested that the majority (70–90%) of receptors reach the synapse via lateral diffusion whereas the remainder are exocytosed from internal stores (Patterson et al., 2010). LTP induced by synaptic NMDA receptor stimulation increases receptor diffusion into spines (Jaskolski et al., 2009). The overall effect may be to enhance the population of immobile receptors at synapses, as observed after 30 min of LTP treatment (Szepesi et al., 2014). Furthermore, enhancing the reserve pool of extrasynaptic AMPA receptors increases the response to electrical LTP stimulation (Oh et al., 2006), whereas uncoupling the endocytic zone from the synapse prevents LTP (Petrini et al., 2009). In contrast, LTD induced by global NMDA receptor stimulation may stabilize AMPA receptors and prevent their movement into spines (Jaskolski et al., 2009).

AMPA and NMDA receptors undergo clathrin-mediated endocytosis at endocytic zones located a few hundred nanometres from the edge of the PSD (Blanpied, Scott, & Ehlers, 2002; Man et al., 2000; Petralia, Wang, & Wenthold, 2003; Rácz et al., 2004), suggesting a coupling between endocytosis and exocytosis machinery lateral to the synapse (Fig. 3). Recycling of AMPA receptors between the membrane surface and the intracellular compartments also provides a pool of receptors ready to respond to changes in synaptic excitability. The dynamics of this receptor recycling can be manipulated by synaptic activity. Under basal conditions, continual recycling of receptors between intracellular stores and the plasma membrane occurs over a relatively slow timescale from minutes to hours (Adesnik et al., 2005; Lüscher et al., 1999; Oh et al., 2006; Park et al.,

2006). This occurs at both the synaptic (Adesnik et al., 2005; Park et al., 2006) and extrasynaptic membrane (Lüscher et al., 1999; Oh et al., 2006; Yudowski, 2007), and can occur independently from clathrin activity (Glebov, Tigaret, Mellor, & Henley, 2015; Zheng, Jeyifous, Munro, Montgomery, & Green, 2015).

3. AMPA RECEPTOR INTERACTIONS WITH POSTSYNAPTIC DENSITY PROTEINS

What molecular interactions occur at the synapse to cluster AMPA receptors into different surface locations? A large body of research has shown that AMPA receptors are stabilized at synapses by interactions with intracellular scaffolding proteins (Montgomery et al., 2004; Zheng, Seabold, Horak, & Petralia, 2011). At synapses, AMPA receptors are densely clustered at the PSD, a postsynaptic disk-like membrane specialization approximately 200–800 nm wide and 30–50 nm thick, located directly opposed to the active zone (Carlin, Grab, Cohen, & Siekevitz, 1980). The PSD is a specialized megaorganelle composed of many cytoplasmic scaffold proteins, signaling enzymes and cytoskeletal elements which anchor glutamate receptors and cell adhesion molecules (CAMs), recruit intracellular signaling molecules and provide a structural link to the cytoskeletal network of the spine (Sheng & Hoogenraad, 2007; Fig. 1). Among the many PSD proteins, two important families are *m*embrane *a*ssociated *gu*anylate *k*inases (MAGUKs) and *s*rc *h*omology 3 (SH3) domain and *ank*yrin repeat proteins (Shanks), which are known as key organizers of excitatory glutamatergic synapses. The *Drosophila* tumor suppressor gene (DLG) subfamily of MAGUKs is composed of the synaptic proteins PSD93, PSD95, SAP97, and SAP102, which all contain three PDZ domains, an SH3 domain, and a GUK domain (Cho, Hunt, & Kennedy, 1992; Kim, Cho, Rothschild, & Sheng, 1996; Müller et al., 1995). Through their binding domains, MAGUKs interact with glutamate receptors and other ion channels to regulate their trafficking, targeting, and insertion at the synapse (Chen et al., 2000; El-Husseini Ael et al., 2002; Gardoni et al., 2003; Jeyifous et al., 2009; Kim, Niethammer, Rothschild, Jan, & Sheng, 1995; Kim & Sheng, 1996; Kornau, Schenker, Kennedy, & Seeburg, 1995; Leonard, Davare, Horne, Garner, & Hell, 1998; Migaud et al., 1998; Nehring et al., 2000; Wu, Nash, Zamorano, & Garner, 2002).

3.1 PSD95 and Stargazin Play a Primary Role in AMPA Receptor Insertion into the Surface Membrane

PSD95 binds to the C-terminus of the GluN2A/B subunit of the NMDA receptor (Kornau et al., 1995; Niethammer, Kim, & Sheng, 1996) and was initially thought to play a major role in regulating NMDA receptors at synapses. Surprisingly, mice lacking PSD95 still show normal clustering of NMDA receptors, suggesting that PSD95 is not necessary for the scaffolding of NMDA receptors at the PSD (Migaud et al., 1998). Overexpressing PSD95 also does not affect the number of postsynaptic NMDA receptors but increases the number of AMPA receptors at the PSD and enhances synaptic neurotransmission through AMPA receptors (Ehrlich & Malinow, 2004; El-Husseini Ael et al., 2002; Elias et al., 2006; Schnell et al., 2002). Thus, PSD95's ability to promote glutamatergic function without directly binding to the AMPA receptor suggests that it must interact with intermediate adaptor proteins in order to achieve its scaffolding functions.

The transmembrane AMPA receptor regulatory proteins (TARPs) provide the intermolecular link between surface AMPA receptors and intracellular PDZ containing proteins such as PSD95 (Chen et al., 2000). The most well-known TARP protein, Stargazin (also known as γ-2) is the first member of the Type 1 family of TARP proteins (Tomita et al., 2003), and was identified as the mutant protein in the epileptic *Stargazer* mouse (Letts et al., 1997, 1998). Stargazin is widely expressed throughout the brain, particularly in the cerebellum, cerebral cortex, and hippocampus (Tomita et al., 2003), and is enriched in the PSD fraction of excitatory synapses where it is strongly associated with the AMPA receptor (Fukata et al., 2005; Tomita, Fukata, Nicoll, & Bredt, 2004; Tomita et al., 2003). Stargazin can modulate the properties of surface AMPA receptors via its transmembrane domains, slowing channel deactivation and desensitization, and increasing the rate of channel opening (Tomita et al., 2005). In addition, Stargazin acts as an adaptor protein, binding via its C-terminal intracellular tail to PDZ domain-containing proteins, thus linking surface AMPA receptors to the intracellular scaffold (Chen et al., 2000; Schnell et al., 2002). For example, Stargazin overexpression promotes AMPA receptor expression at synapses, but can only achieve this by directly binding to the first two PDZ domains of PSD95 (Bats et al., 2007; Chen et al., 2000; Schnell et al., 2002). Thus, Stargazin regulates both the surface expression and function of the AMPA receptor.

PSD95 and Stargazin proteins appear to direct the subsynaptic surface distribution of AMPA receptors. PSD95 proteins have been observed with super resolution microscopy to form distinct subdomains of similar size as the AMPA receptor nanodomains on the synaptic surface (Fukata et al., 2013; MacGillavry et al., 2013). Overexpression of PSD95 concentrates AMPA receptors in nanodomains, whereas knockdown of PSD95 or preventing the interaction between PSD95 and Stargazin reduces the number of AMPA receptor nanodomains and increases the diffusion rate of the receptor (Hoze et al., 2012; Nair et al., 2013). The formation of these PSD95 nanodomains is dynamically modified by synaptic activity driven by the palmitoylation state of PSD95's N-terminal domain (Fukata et al., 2013). In addition, the dynamic clustering of synaptic scaffolding proteins and surface receptors appears to be complimentary, as disrupting the interaction between PSD95 and GluA1 in turn reduces the clustering of PSD95 (MacGillavry et al., 2013). Considering that these nanodomains are separated by a few hundred nanometres of synaptic membrane, these imaging experiments suggest that the PSD structure itself is dynamic and can remodel to cluster surface glutamate receptors into functionally distinct compartments driven by PSD95 (Kerr & Blanpied, 2012; Nair et al., 2013).

PSD95 and Stargazin are unlikely to be the only scaffolding proteins responsible for the formation of these AMPA receptor nanodomains. Knockdown of PSD95 or Stargazin, or disrupting their interaction does not completely abolish AMPA receptor nanodomains (Nair et al., 2013). Furthermore, PSD95 and AMPA receptor nanodomains do not always align with each other. For example, whereas live cell super resolution imaging observed a high degree of colocalization between AMPA receptor nanodomains with subdomains of overexpressed PSD95 protein (MacGillavry et al., 2013), dSTORM imaging of endogenous GluA1 subunits detected only moderate colocalization between AMPA receptors and PSD95 subdomains (Nair et al., 2013). Other major scaffolding proteins are likely to support the PSD95 scaffold, as Shank3, Homer1c, and GKAP also form these subdomains (MacGillavry et al., 2013), and may initiate clustering of other scaffolding proteins.

3.2 SAP97 Plays a Major Role in AMPA Receptor Trafficking and Surface Location

SAP97 directly binds to the extreme C-terminus as well as the upstream SSG tripeptide sequence of the GluA1 subunit of the AMPA receptor via its second PDZ domain (Cai, Coleman, Niemi, & Keinänen, 2002; Leonard et al., 1998; Von Ossowski et al., 2006). Both SAP97 and PSD95 undergo alternative N-terminal domain splicing that gives rise to both a palmitoylated α isoform and a β isoform that contains an L27 domain (Chetkovich et al., 2002; Cho et al., 1992; Müller et al., 1995; Schlüter, Xu, & Malenka, 2006). βSAP97 is the dominant form of SAP97 in the brain (Chetkovich et al., 2002; Müller et al., 1995). Within α and βSAP97 there are five other alternative splice variants, called I1–I5 that arise from different insertions located at the N-terminus or between the SH3 and GUK domains (Fig. 4; Lue, Marfatia, Branton, & Chishti, 1994; McLaughlin et al., 2002; Mori et al., 1998).

The N-terminus of SAP97 has a myosin VI binding site and direct binding between myosin VI, SAP97 and GluA1 may be involved in trafficking AMPA receptors to and from the postsynaptic plasma membrane (Wu et al., 2002). SAP97 also plays a key role in sorting and independently trafficking glutamate receptors from the soma to the synaptic membrane. βSAP97 associates with immature, incompletely glycosylated AMPA receptors in the endoplasmic reticulum or *cis*-golgi (Sans et al., 2001). In the ER, βSAP97 sorts AMPA and NMDA receptors from each other and directs them along different trafficking routes toward the synapse (Jeyifous et al., 2009). βSAP97 transports GluA1-containing AMPA receptors to the synapse via the conventional golgi network (Jeyifous et al., 2009; Sans et al., 2001). This is achieved by assuming a compact formation that preferentially binds AMPA receptors (Lin, Jeyifous, & Green, 2013). Considering that GluA1 can also bind directly to protein 4.1 via its C-terminal tail (Shen, Liang, Walensky, & Huganir, 2000), a protein complex consisting of compact

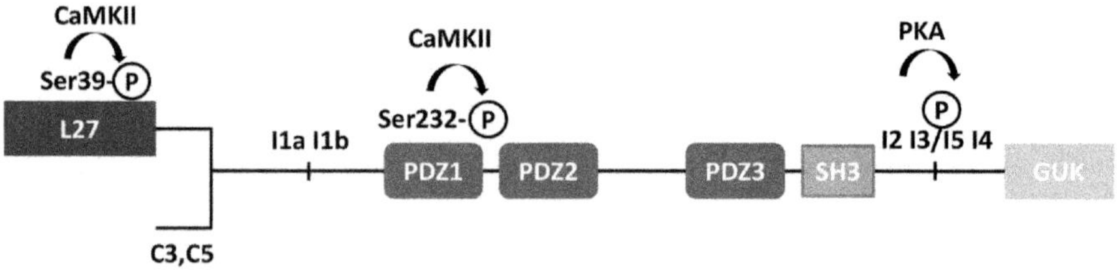

Figure 4 Shank proteins, a family of multidomain postsynaptic scaffolding proteins. Shank proteins are characterized by a set of protein binding domains, including Shank/ProSAP N-terminal domain (SPN), ankyrin repeats (ANK), Src homology 3 (SH3) domain, PDZ domain, a proline-rich/homer, and contactin binding domain within proline-rich region and a sterile alpha motif (SAM) domain. Depending on the isoform of Shank proteins, the combination of protein–protein interacting domains is different.

βSAP97, protein 4.1, and GluA1-containing AMPA receptors may associate with the cytoskeleton *en route* to the synapse. Interestingly, one lab reported that a GluA1 C-terminal mutant that cannot bind to SAP97 can still reach the plasma membrane whereas βSAP97 itself was retained in the cytosol (Zhou et al., 2008), suggesting that GluA1 can reach the plasma membrane via other transport mechanisms whereas βSAP97 does not execute its transport function in the absence of GluA1.

Conflicting data has been reported on the ability of SAP97 to alter AMPA receptor currents: SAP97 has been shown to increase (Schlüter et al., 2006), decrease (Li et al., 2011; Waites et al., 2009), or not affect (Schnell et al., 2002) excitatory glutamatergic neurotransmission. These conflicts may reflect the difficulty in assigning a single function to a protein with multiple sites of alternative splicing, as the presence of many SAP97 isoforms in the PSD suggests that this protein holds a range of functional responsibilities. While most research has focused on the dominant N-terminal βSAP97 isoform, the use of variable or unstated regions of alternative splicing throughout the rest of the protein makes comparisons between studies difficult. Indeed, direct comparisons between α- and βSAP97 expressing neurons suggest that α- and βSAP97 induce opposing effects on surface AMPA receptor expression (Li et al., 2011; Waites et al., 2009). It is proposed that αSAP97 scaffolds receptors directly at synapses, whereas βSAP97 holds surface receptors away from functional synapses at extrasynaptic sites where they do not contribute to neurotransmission (Li et al., 2011; Waites et al., 2009).

3.3 Shank Proteins

The family of *SH*3 and multiple *ank*yrin repeat domains proteins (Shank1, Shank2, and Shank3), are postsynaptic scaffolding proteins characterized by a set of protein to protein interaction domains: ankyrin repeats, an Src homology 3 (SH3) domain, a PDZ domain, a proline-rich/homer and contactin binding domain, and a sterile alpha motif (SAM) domain (Fig. 5; Boeckers et al., 1999). Shank proteins interact directly and indirectly with glutamate receptors, ion channels, CAMs, signaling molecules, and cytoskeletal actin. Through these interactions, Shank proteins provide a fundamental platform for structural and functional formation, maturation, and maintenance of excitatory synapses and for synaptic plasticity; thus Shanks are regarded as "master regulators" of the glutamatergic excitatory

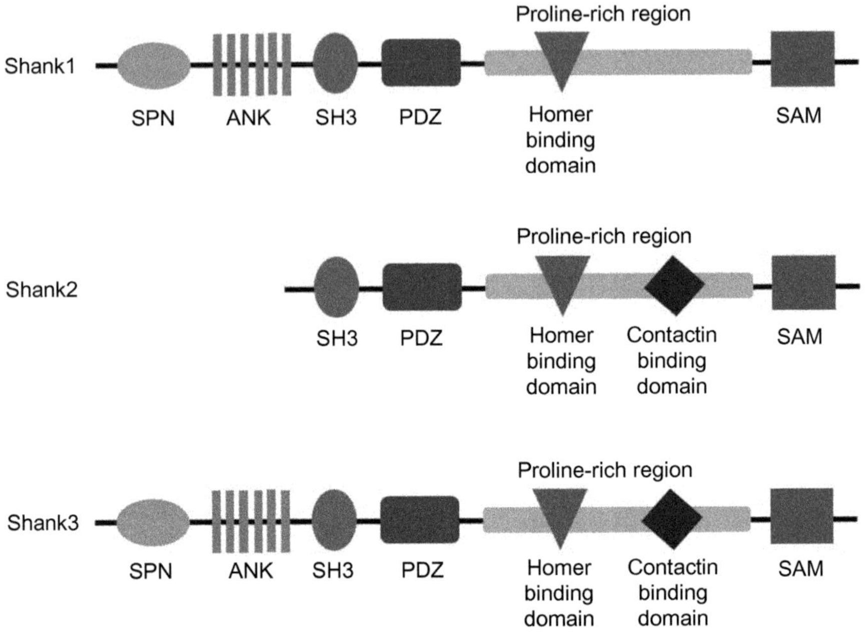

Figure 5 Structural domains of SAP97, including sites of alternative splicing and post-translational modifications. SAP97 contains three PDZ domains, an SH3 domain, and a GK domain. The protein is alternatively spliced at its N-terminus with either α inserts (two cysteine resides that can be palmitoylated) or βSAP97 inserts (L27 domain). Additional regions of alternative splicing are found at the N-terminus prior to the first PDZ domain (I1a, I1b), and in the HOOK region between SH3 and GK domains (I2, I3, I4, I5). SAP97 can be phosphorylated at three known sites, two of which are alternatively spliced inserts, suggesting functional divergence between the SAP97 variants.

postsynapse (Berkel et al., 2012; Grabrucker et al., 2011; Roussignol et al., 2005; Sala, Roussignol, Meldolesi, & Fagni, 2005; Sala et al., 2001).

Immunostaining of Shank proteins in hippocampal neuronal cultures has revealed the exclusive expression of Shank proteins at the PSD of excitatory synapses (Boeckers et al., 1999; Lim et al., 1999; Naisbitt et al., 1999; Yao et al., 1999; Zitzer, Honck, Bachner, Richter, & Kreienkamp, 1999). A detailed analysis by immunogold electron microscopy demonstrated that the majority of Shank is concentrated in the deeper part of the PSD (Valtschanoff & Weinberg, 2001). In addition, Shank proteins were observed to be intensely labeled in a region immediately below the PSD, forming a filamentous network approximately 120 nm below the postsynaptic membrane, which is contiguous to the PSD but distinct from the cytoplasm in the rest of the spine (Tao-Cheng, Dosemeci, Gallant, Smith, & Reese, 2010).

The protein interactions mediated by individual Shank domains enable Shank proteins to produce diverse functional roles at excitatory synapses. The GluA1 subunit of AMPA receptors directly binds to the PDZ domain of Shank3 via its C-terminal PDZ-binding motif (Uchino et al., 2006). Shank contains a SH3 domain that binds to glutamate receptor-interacting protein (GRIP), which is involved in stabilization of AMPA receptors during receptor cycling (Sheng & Kim, 2000; Xia, Zhang, Staudinger, & Huganir, 1999; see Section 3.4). Shank proteins also interact directly with neuroligins through the PDZ domain or indirectly via PSD95 (Meyer, Varoqueaux, Neeb, Oschlies, & Brose, 2004). Neuroligins are postsynaptic CAMs that interact with presynaptic neurexins (Chih, Engelman, & Scheiffele, 2005; Ichtchenko et al., 1995). The neurexin–neuroligin complex is an integral part of transsynaptic signaling between pre- and postsynapses: inducing presynaptic and postsynaptic differentiation and maturation (Dean et al., 2003; Graf, Kang, Hauner, & Craig, 2006; Scheiffele, Fan, Choih, Fetter, & Serafini, 2000). Arons et al. (2012) showed that overexpression of Shank3 in cultured hippocampal neurons enhanced both pre and postsynaptic protein expression, increased presynaptic transmitter release and postsynaptic AMPA receptor-mediated currents, revealing that Shanks strengthen glutamatergic synapses via the neurexin–neuroligin complex.

3.4 ABP/GRIP and PICK1

The GluA2 subunit of AMPA receptors also interacts with multiple intracellular AMPA receptor-binding proteins (ABPs). The extreme C-terminal tail of GluA2 can bind directly to the PDZ domains of protein interacting with C Kinase 1 (PICK1) (Xia et al., 1999), and the closely related proteins ABP and GRIP (Dong et al., 1997; Srivastava et al., 1998). ABP/GRIP are large proteins with six or seven PDZ domains, whereas PICK1 proteins contain only a single PDZ domain. However, both ABP/GRIP and PICK1 can multimerize via their PDZ domains to form part of a lipid and protein scaffold within the synapse (Perez et al., 2001; Srivastava et al., 1998).

ABP/GRIP proteins provide a molecular scaffold that maintains GluA2 expression on the synaptic surface or within intracellular stores (Daw et al., 2000; DeSouza, Fu, States, & Ziff, 2002; Dong et al., 1997; Osten et al., 2000). ABP/GRIP proteins are enriched in the PSD where they interact with GluA2/3 receptors (Wyszynski et al., 1999), and disrupting the interaction between ABP/GRIP and GluA2 reduces the surface expression of the receptor (Osten et al., 2000; Shi et al., 2001). However, some studies observed an increase in AMPA receptor function (Daw et al., 2000), or no change at all in AMPA receptor targeting or recycling (Braithwaite, Xia, & Malenka, 2002) after disrupting ABP/GRIP and GluA2 binding, suggesting that ABP/GRIP could play a more subtle role in regulating the intracellular pool of AMPA receptors. Indeed, considering that ABP/GRIP immunolabeling has been observed at both excitatory and inhibitory synapses, nonsynaptic membranes, and intracellular compartments (Wyszynski et al., 1999), this protein is likely to play multiple roles in AMPA receptor trafficking and recycling. Interestingly, PSD95 does not appear to interact with ABP/GRIP (Srivastava et al., 1998), and does not deliver GluA2/3 subunits to synapses (Ehrlich & Malinow, 2004), suggesting that ABP/GRIP offers GluA2 a protein scaffold unique from that of PSD95 MAGUK proteins that scaffold GluA1 receptors.

The synaptic scaffolding of GluA2/3 subunits by ABP/GRIP can be regulated by PICK1 as well as other signaling proteins that are capable of binding GluA2. Phosphorylation of GluA2's C-terminal serine 880 residue by PKCα within the ABP/GRIP and PICK1 binding site reduces the receptor's affinity for ABP/GRIP but not PICK1 (Matsuda, Mikawa, & Hirai, 1999; Xia, Chung, Wihler, Huganir, & Linden, 2000). PICK1 is a low calcium sensor that can bind directly to PKAα (Matsuda et al., 1999; Perez et al., 2001; Staudinger, Lu, & Olson, 1997), and activation of this kinase, or overexpression of PICK1 relocates phosphorylated GluA2 to an intracellular compartment (Perez et al., 2001). Together, this suggests that during LTD, PICK1 allows PKCα to phosphorylate GluA2, unbinding the receptor from the ABP/GRIP scaffold (Kim, Anwyl, Suh, Djamgoz, & Rowan, 2001; Kim, Chung, Lee, & Huganir, 2001; Matsuda et al., 1999), and directing it toward endocytotic pathways (Lu & Ziff, 2005; Perez et al., 2001). However, NMDA-induced internalization of AMPA receptors is not dependent on PICK1 or phosphorylation at Ser880, and inhibition of PKCα does not block NMDA receptor-dependent LTD (Oliet, Malenka, & Nicoll, 1997; Peineau et al., 2009). Rather ABP/GRIP bind to the GluA2 subunit to maintain a store of AMPA receptors at nonsynaptic sites (intracellular or extrasynaptic sites), and PICK1 may enable the dissociation of AMPA receptors from ABP/GRIP at these nonsynaptic sites, providing a pool near the synapses, ready for membrane insertion (Daw et al., 2000; Emond et al., 2010; Kim, Anwyl, et al., 2001; Kim, Chung, et al.,

2001). These studies suggest that the scaffold of surface GluA2-containing AMPA receptors and their recycling within internal stores is maintained by a complex interaction of GluA2 with the scaffolding proteins ABP/GRIP, PICK1, and NSF.

4. AMPA RECEPTORS IN NEURODEGENERATIVE DISEASES

Synaptic deficits in neurodegenerative disease have been a major focus of research over the past 20 years. Animal models of neurodegenerative diseases including Huntington's disease (HD), AD, and PD have shown deficits in synaptic transmission and plasticity, in combination with behavioral deficits in learning, memory, and motor function (e.g., Hodgson et al., 1999; Kolodziejczyk, Parsons, Southwell, Hayden, & Raymond, 2014; Murphy et al., 2000; Picconi et al., 2005; Usdin et al., 1999), suggesting that altered synaptic mechanisms play a significant role in the manifestation of the clinical symptoms associated with these diseases. Deciphering the mechanisms that contribute to altered synaptic transmission and plasticity has therefore been a major research focus.

The major feature of AD is the presence of amyloid beta (Aβ) plaques and neurofibrillary tangles within neurons, but another major feature is loss of synapses. One of the first signs of AD is cognitive decline, which has been shown to correlate with synapse loss in animal AD models and AD human brain (DeKosky & Scheff, 1990; DeKosky, Scheff, & Styren, 1996; Hamos et al., 1989; Penzes & Vanleeuwen, 2011; Terry et al., 1991), and decreased AMPA receptor-mediated synaptic transmission in animal AD models (DeKosky & Scheff, 1990; Penzes & Vanleeuwen, 2011). Addition of Aβ to neurons induces loss of synapses, impairs LTP, enhances LTD, and increases AMPA receptor internalization to depress transmission (Hsieh et al., 2006; Kim, Anwyl, et al., 2001; Kim, Chung, et al., 2001; Knobloch & Mansuy, 2008; Lau & Zukin, 2007; Selkoe, 2002).

The major effects of AD on AMPA receptor transmission led to clinical trials of drugs that modulate AMPA receptors to treat AD-related cognitive decline. AMPAkines are a class of pharmacological agents that act as positive modulators of AMPA receptors to enhance excitatory glutamatergic neurotransmission (Baumbarger, Muhlhauser, Yang, & Nisenbaum, 2001; Miu et al., 2001; Ornstein et al., 2000). They decrease the onset of AMPA receptor desensitization and deactivation to enhance glutamatergic synaptic transmission (Arai & Kessler, 2007; Baumbarger et al., 2001; Gates, Ogden, &

Bleakman, 2001; Vandergriff, Huff, Bond, & Lodge, 2001). Because AMPAkines exhibit subunit preference, it was thought that they could act on specific brain regions and therefore enable targeted treatment for specific regions exhibiting depressed synaptic transmission (Montgomery, Kessler, & Arai, 2009). This could be particularly important for different neurodegenerative disorders such as AD and PD, which depress glutamatergic synapse function in the hippocampus (AD) versus striatum (PD). Indeed, some AMPAkines were shown to significantly alter hippocampal glutamatergic synaptic currents more than thalamic currents (Xia, Kessler, & Arai, 2005). AMPAkines were also shown to facilitate the induction of LTP (Baudry et al., 2012; Lynch & Gall, 2006; Rex et al., 2006) and increase the release of BDNF (Jourdi et al., 2009; Lauterborn, Lynch, Vanderklish, Arai, & Gall, 2000; Legutko, Li, & Skolnick, 2001; Mackowiak, O'Neill, Hicks, Bleakman, & Skolnick, 2002; Murray et al., 2003; Rex et al., 2006; Simmons, Mehta, Lauterborn, Gall, & Lynch, 2011). These compounds raised much excitement and clinical interest for their therapeutic potential in treating cognitive decline associated with neurodegenerative disorders. Several studies showed AMPAkines could improve learning and memory in rodent and nonhuman primate tasks, and improve cognition in human patients (Hampson, Rogers, Lynch, & Deadwyler, 1998a, 1998b; Ingvar et al., 1997; Lynch, Rex, Chen, & Gall, 2008; Quirk & Nisenbaum, 2002; Staubli, Rogers, & Lynch, 1994). Based on their ability to increase BDNF, AMPAkines were also targeted as therapeutic options to protect against the loss of dopaminergic innervation to the caudate and putamen of the striatum that occurs in PD. Administration of LY503430, which induced dose-dependent increases in AMPA receptor-mediated currents, was found to protect against cell loss in the striatum and substantia nigra, and reduce behavioral effects in three different PD animal models (Murray et al., 2003; O'Neill et al., 2005). Unfortunately, interest in these compounds has since waned, as no study has progressed into large scale phase III trials for the treatment of neurological disorders (Chang, Verbich, & McKinney, 2012; Lynch et al., 2008; Ward, Bax, & Harries, 2010). For example, the positive AMPA receptor modulator LY451395, which reversed age-associated memory deficits in rodent models, did not improve cognitive function (Chappell et al., 2007) nor AD-associated agitation and aggression (Trzepacz et al., 2013). However, research still continues to develop new analogs of AMPAkines (Ward et al., 2010).

Animal models of PD have shown elevated levels of AMPA receptor expression and dephosphorylation (Ba et al., 2011; Silverdale et al.,

2010), raising the possibility that AMPA receptor antagonists may have therapeutic potential in PD. However, the efficacy of AMPA receptor antagonists to prevent PD symptoms in animal models has also produced variable results. AMPA receptor antagonists have been shown to elevate the antiparkinsonian effects of levadopa and resistance in levadopa-induced dyskinesia (Kobylecki, Cenci, Crossman, & Ravenscroft, 2010). Subsequently, the antidyskinetic potential of the noncompetitive AMPA receptor antagonist perampanel was examined in clinical trials; however, patients given perampanel alongside their levodopa did not show a significant reduction in levodopa-induced dyskinesias (Eggert et al., 2010; Lees et al., 2012).

In HD, glutamatergic synaptic changes have largely focussed on NMDA-type glutamate receptors as they contribute to excitotoxic mechanisms in striatal MSN death—a major neuropathological feature of HD (Fan & Raymond, 2007; Graham et al., 2009; Martin & Gusella, 1986). In particular, increased localization of NMDA receptors at extrasynaptic sites in HD is thought to activate cell death pathways (Hardingham, Fukunaga, & Bading, 2002; Kaufman et al., 2012; Xu et al., 2009), and blockade of these receptors can reverse motor deficits in HD animals (Milnerwood et al., 2010). However, AMPA receptor deficits are also observed in HD mice and in postmortem HD human brain (Fourie et al., 2014; Mandal et al., 2011). Expression of mutant huntingtin protein impairs AMPA receptor-mediated synaptic transmission by disrupting the transport of GluA2-containing AMPA receptors along microtubules (Mandal et al., 2011). AMPA receptor modulation through AMPAkines has also been investigated in HD where AMPAkine treatment has been shown to upregulate BDNF, rescue synaptic plasticity, reduce HD-associated decreases in striatal area and memory deficits, and significantly improve motor performance in HD mice (Simmons et al., 2009, 2011). The AMPAkine used in this work, CX929, has a short half-life, and it remains to be determined whether it will be of clinical use.

4.1 Synaptic Scaffold Proteins in Neurodegenerative Diseases

It is now very clear that PSD proteins such as MAGUKs, Shanks, and TARPs are major regulators of AMPA receptor synaptic targeting and stability in the synaptic membrane (Section 3), raising the possibility that these proteins hold significant potential as therapeutic targets to alter AMPA receptor-mediated synaptic transmission rather than pharmacologically targeting the receptors themselves. Considering the diverse roles that the

PSD scaffolding proteins play in glutamate receptor trafficking and scaffolding, perturbations to these proteins may underlie the pathology of many neurological diseases (Gardoni, 2008). Indeed, abnormalities in the subcellular distribution and expression levels of SAP97, PSD93, PSD95, and SAP102 have been observed in rodent models of PD, AD, and HD (Almeida et al., 2005; Gardoni et al., 2006; Luthi-Carter et al., 2003; Nash, Johnston, Collingridge, Garner, & Brotchie, 2005; Shao, Mirra, Sait, Sacktor, & Sigurdsson, 2011; Torres-Peraza et al., 2008), and in postmortem human AD, PD, and HD brain (Fourie et al., 2014; Proctor, Coulson, & Dodd, 2010; Yuki et al., 2014). Mutant huntingtin also disrupts its ability to bind PSD95 (Sun et al., 2001). Similarly, AD brains show impaired expression levels and interactions between SAP97 and GluA1 suggesting improper glutamate receptor trafficking during the course of the disease (Marcello, Gardoni, Di Luca, & Pérez-Otaño, 2010; Marcello, Epis, et al., 2010; Wakabayashi et al., 1999). Furthermore, SAP97's trafficking functions may impart improper glutamatergic function with the amyloid precursor protein (APP) cascade during the pathogenesis of AD (Marcello, Epis, et al., 2010; Marcello, Gardoni, et al., 2010). SAP97 can bind and traffick the A disintegrin and metalloproteinase 10 (ADAM10) enzyme to the synaptic membrane where it shifts APP metabolism toward the nonamyloidogenic protein products (Marcello et al., 2007). However, this interaction is disrupted in AD, potentially favoring the formation of amyloid protein plaques (Epis et al., 2010; Marcello et al., 2013; Musardo, Marcello, Gardoni, & Di Luca, 2014). These data show that abnormal PSD protein–protein interactions significantly contribute to glutamatergic synapse dysfunction in neurodegenerative diseases, and modulation of these interactions by small peptides is a strategy for new therapeutic options (Gardoni et al., 2006; Minnerup, Sutherland, Buchan, & Kleinschnitz, 2012; Zhou et al., 2010). Cell permeable peptides are small peptides that cross the blood–brain barrier and transport into neurons where they can disrupt PSD protein interactions with their binding partners (Aarts et al., 2002; Gardoni et al., 2006; Zhou et al., 2010). Data from animal studies show significant promise, with peptide treatment that disrupts PSD95 interactions with nNOS or with the NMDA receptor resulting in decreased excitotoxicity and ischemic damage in stroke models, without significant side effects (Luo et al., 2014; Zhou, Zhang, Zhang, & Zhu, 2012; Zhou et al., 2015). This strategy is also showing potential for developing treatments for neurodegenerative diseases, with peptides that uncouple GluN2A binding to PSD-MAGUKs resulting in a significant reduction in dyskinesias in PD rodent models (Gardoni et al., 2012; Mellone & Gardoni, 2013).

Recent neuropathological findings in both animal models and human AD brain suggest that Shank proteins also play a major role in the synaptopathology of AD (Gong, Lippa, Zhu, Lin, & Rosso, 2009; Grabrucker et al., 2011; Guilmatre, Huguet, Delorme, & Bourgeron, 2014; Pham et al., 2010; Roselli, Hutzler, Wegerich, Livrea, & Almeida, 2009). Aβ induces a reduction in Shank proteins at synapses in rodent and human brain (Grabrucker et al., 2011; Roselli et al., 2009), and experiments in hippocampal neurons have shown that this synaptic deficit can be rescued by zinc supplementation (Grabrucker et al., 2011). A zinc link in AD has been suggested for many years (e.g., Bush et al., 1994; Cuajungco & Lees, 1997; Danscher et al., 1997; Lovell, 2009; Lovell, Robertson, Teesdale, Campbell, & Markesbery, 1998; Vogler & Busciglio, 2014), as Aβ has a high binding affinity for zinc, promoting its oligomerization (Bush et al., 1994; Miller, Ma, & Nussinov, 2010), and high zinc concentrations have been measured in amyloid deposits and plaques in AD brain (Lovell et al., 1998; Suh et al., 2000). Because the organization of postsynaptic Shank platforms at the PSD is regulated by zinc (Baron et al., 2006), it is proposed that Aβ "traps" synaptic zinc and prevents it from binding Shank proteins, disrupting the Shank postsynaptic platform, and inducing major changes in glutamatergic synapse number and function that could underlie the synaptopathology of AD (Grabrucker et al., 2011).

Altogether, it is clear that changes in PSD protein localization, expression levels, and/or interactions with glutamate receptors occur in multiple neurodegenerative disorders. Basic biomedical research has shown us that these proteins have significant control over glutamate receptors at the synapse. Given that pharmaceutical targeting of AMPA receptors directly has yet to yield effective therapeutic options for neurodegenerative diseases, PSD proteins such as MAGUKs and Shanks may be feasible new pharmaceutical targets in the development of novel therapies (de Bartolomeis, Latte, Tomasetti, & Iasevoli, 2014; Gardoni, Marcello, & Di Luca, 2009; Grabrucker et al., 2011). However, much research needs to be done to determine how these proteins could be specifically targeted, determining how to rescue their expression to normal levels (e.g., zinc-dependent regulation of Shanks), and the efficacy and therapeutic potential of cell permeable peptides that alter the interactions of PSD proteins with their glutamate receptor-binding partners (Aarts et al., 2002; Gardoni et al., 2006; Zhou et al., 2010).

5. AMPA RECEPTORS AND ASD

ASD are a group of behaviorally defined neurodevelopmental disorders characterized by an inability to form reciprocal social interactions, impairments in verbal and nonverbal communication, and markedly repetitive behaviors (American Psychiatric Association, 2013; Bailey, Phillips, & Rutter, 1996; Bauman & Kemper, 2005a, 2005b). ASD comprises a complex and heterogeneous group of conditions that include autism, Rett's and Asperger's syndromes, and pervasive developmental disorder-otherwise nonspecified (PDD-NOS). Recent advances in genetic diagnostic technology have unveiled a remarkable convergence of several candidate genes on common molecular and cellular pathways including glutamatergic synapses (Gilman et al., 2011; Toro et al., 2010).

Human studies have identified genetic changes in glutamate receptors occur with ASD. A 19 mb deletion of chromosome 4q in a human ASD patient included the gene that encodes the AMPA receptor GluA2 subunit (Ramanathan et al., 2004). Chromosome 6q21 was also identified as a candidate region for ASD with a functional SNP in the glutamate receptor 6 (*GluR6* or *GRIK2*) gene associated with ASDs (Jamain et al., 2002). Postmortem brain from ASDs patients have also shown specific decreases in AMPA receptor density in the cerebellum (Purcell, Jeon, Zimmerman, Blue, & Pevsner, 2001).

In the late 1990s, infantile autism was proposed to be a "hypoglutamatergic disorder" (Carlsson, 1998), raising the possibility that glutamatergic agonists such as AMPAkines could be beneficial in the therapeutic treatment for ASD. A small double-blind placebo-controlled study of 40 ASD children showed that treatment with piracetam, a positive modulator of AMPA receptors, in combination with risperidone resulted in improved behavioral scores compared to those treated with risperidone alone (Akhondzadeh et al., 2008), suggesting that combining antipsychotic medications and a glutamate receptor modulator may increase synergistic effects in the treatment of autism. In an ASD mouse model, acute treatment with AMPAkines was found to reverse sociability and cognitive deficits, again supporting the AMPA receptor system as a pharmacological target for autism (Silverman et al., 2013). In human ASD patients, the glutamatergic system has also been targeted with riluzole and memantine. Riluzole treatment improved repetitive behaviors, and memantine showed significant improvements in behavior and memory (reviewed by Doyle &

McDougle, 2012). There have been no published double-blind, placebo-controlled clinical trials of either of these drugs, however. Currently, the major pharmaceutical treatments for ASD include serotonin reuptake inhibitors (however, recent work has shown no effect on improving repetitive behaviors; Williams et al., 2013) or antipsychotics (to control irritability, stereotypy, and hyperactivity; Doyle & McDougle, 2012). It is clear that there is a significant pharmaceutical need for the treatment of the core ASD symptoms and AMPA receptor modulators show some promise in this regard.

AMPA receptor modulators have been tested in Fragile X Syndrome (FXS) patients, a genetic syndrome that causes developmental problems including cognitive impairment and intellectual disabilities (Bagni & Oostra, 2013). FXS results from silencing of the *FMR1* gene and messenger RNA (mRNA), leading to the lack of production of fragile X mental retardation protein. This increases mGluR5 signaling and increases AMPA receptor internalization, resulting in deficits in LTP and LTD (Hu et al., 2008; Nakamoto et al., 2007; Suvrathan et al., 2010; Wang et al., 2010). Approximately 1/3 of FXS patients also have autism, with the Fragile X mutation being the most common single gene cause of ASD. Due to the deficits in AMPA signaling, AMPAkine compounds have been proposed as potential FXS treatments by potentiating AMPA receptor signaling. One randomized, double-blind, placebo-controlled clinical trial that has examined AMPAkine treatment (4 weeks with CX516) in adult FXS patients resulted in no significant changes in cognitive or behavioral function (Berry-Kravis et al., 2006). However, the lack of effect could be due to low potency of CX516 or the short treatment period.

5.1 Synaptic Changes in ASD Animal Models: Identification of Future Therapeutic Targets?

ASD-associated mutations occur in a number of proteins at the synapse, including neurexin, neuroligins, Shanks, synapsins, gephyrin, cadherins, GABA, and glutamate receptors (e.g., Arons et al., 2012; Chubykin et al., 2005; De Rubeis et al., 2014; Etherton et al., 2009; Fassio et al., 2011; Lionel et al., 2013; Ma et al., 2005; Pagnamenta et al., 2011; Peca et al., 2011; Shao et al., 2003; Yoo et al., 2013). The ASD-associated mutations that have been identified in these major synaptic proteins have led to a suggestion that "synaptopathy," alterations in synaptic structure and function, is a major determinant for ASD pathogenesis (Won et al., 2013).

Mutations in *SHANK* genes have been identified in ASD individuals (Leblond et al., 2014) and considering the fundamental role played by the

Shank proteins at the synapse (see Section 3.3), it is not surprising that ASD-associated mutations in *SHANK* could result in significant impairments in glutamatergic synaptic structure and function. Indeed multiple animal models of ASD have been created to characterize the structural, functional, and behavioral effects of Shank ASD-associated mutations that reduce Shank1, Shank2, or Shank3 expression (Bozdagi et al., 2010; Hung et al., 2008; Peca et al., 2011; Schmeisser et al., 2012; Wang et al., 2011; Won et al., 2012). Reductions in the expression levels of PSD proteins, glutamate receptors, synaptic transmission, synaptic plasticity, postsynaptic spine structure, and density were reported in these studies. Behavioral testing of these ASD animal models also revealed deficits in socialization, as well as increase in anxiety and repetitive behaviors (Bozdagi et al., 2010; Hung et al., 2008; Peca et al., 2011; Schmeisser et al., 2012; Wang et al., 2011; Won et al., 2012).

Multiple ASD-associated point mutations in Shanks have also been identified in ASD patients, particularly in Shank2 and Shank3 (Berkel et al., 2010; Durand et al., 2007; Leblond et al., 2012, 2014). To date, no data from animal models carrying these point mutations have been published; however, expression of Shank3 point mutations in cultured neurons show that they reduce AMPA and NMDA receptor-mediated synaptic transmission (Arons et al., 2012; Durand et al., 2012). Shank3 ASD-associated mutations also prevent Shank3 transsynaptic signaling through neurexin and neuroligin, revealing these mutations not only interfere with postsynaptic AMPA and NMDA receptors but also with presynaptic structure and function (Arons et al., 2012). Moreover, multiple ASD-associated mutations also occur in transsynaptic neuroligins and neurexins (Béna et al., 2013; Jamain et al., 2003; Kim et al., 2008; Laumonnier et al., 2004). ASD-associated point mutations in neuroligin enhance neuroligin-3 and AMPA receptor interactions, resulting in increased AMPA receptor internalization and consequently decreased postsynaptic AMPA receptor levels (Chanda et al., 2015). However, when the same point mutation was introduced into neuroligin-4, AMPA-receptor-mediated synaptic responses were increased, indicating that the same mutation can cause differential effects on AMPA receptor regulation. Another PSD protein recently associated with ASD is GRIP. Five missense variants have been identified in highly conserved amino acid residues between PDZ domains 4–6 in GRIP1 in human autistic patients (Mejias et al., 2011). Expression of these mutants in neurons altered GRIP1 binding to GluA2/3 and induced altered AMPA receptor recycling and surface expression. GRIP1/2 knockout mice also exhibit severe defects in social interactions, suggesting that GRIP proteins can modulate social behavior, a core feature of ASD.

Together these data show that ASD-associated mutations in glutamatergic synaptic proteins impact on core cellular pathways that control AMPA receptor synaptic expression and coordinate the function and maturation of excitatory synapses in the CNS. Further studies of mice expressing ASD-associated mutations in glutamatergic signaling may help elucidate how proteins localized to excitatory glutamatergic synapses regulate social and cognitive behaviors in ASD. These results highlight glutamate signaling pathways as promising candidates in the development of novel AMPA receptor-based therapeutic targets for ASD.

6. AMPA RECEPTORS IN THE TREATMENT OF DRUG ADDICTION

Persistent changes in brain circuitry are associated with pathological drug-seeking, so that drug-related stimuli initiate drug-seeking behavior while brain circuits that underlie inhibitory control and executive functioning become dysfunctional. An increase in synaptic dopamine, particularly in the nucleus accumbens shell, is a crucial initial response for coding the rewarding effect of a drug and enabling the development of associative learning that is critical for drug-seeking expressed in response to stimuli or contexts associated with the rewarding effect of a drug. Following repeated, intermittent, exposure to most psychostimulant drugs, this dopamine response becomes sensitized (Robinson & Becker, 1986) and this sensitized response might underlie the preferential attention to drugs of abuse over other, natural, reinforcers.

The nucleus accumbens also receives glutamatergic input from the prefrontal cortex and amygdala (Voorn et al., 2004). Thus, there are converging dopamine and glutamate inputs in the nucleus accumbens that modulate the response to psychostimulant drugs. Increased drug-produced DA in hippocampus and amygdala also strengthens the relationship between a response and the delivery of the drug reinforcer; i.e., learning stimulus-reward associations. Additionally, midbrain dopaminergic neurons project to cortical sites and pyramidal neurons of the mPFC send glutamatergic projections to the ventral tegmental area, the nucleus accumbens, and the amygdala (Gabbott et al., 2005; Hoover & Vertes, 2007), further modulating those neurochemical responses. Increased dopamine in prefrontal, anterior cingulate and orbitofrontal regions impacts cognitive control and executive functioning via modulation of glutamatergic outputs to limbic regions.

The role of AMPA receptors and plasticity in pathological drug-seeking, a characteristic of the shift from drug use to drug misuse, has been demonstrated (Pierce & Wolf, 2013; Wolf & Ferrario, 2010). AMPA receptors have been localized to the GABAergic medium spiny neurons (MSN) that comprise a large proportion of nucleus accumbens cells. An important role of these receptors was initially identified by local injections of AMPA (Cornish & Kalivas, 2000; Ping et al., 2008) or AMPA receptor antagonists (Famous et al., 2008; Ping et al., 2008) which produced or inhibited drug-seeking, respectively, following extinction of extensive exposure to cocaine self-administration. The mechanism underlying drug-produced plasticity has been studied using a range of behavioral and neurochemical assays. These studies have shown persistent increases in AMPA receptors that lack the GluA2 subunit, thereby increasing conductance and also inducing a cascade of calcium-dependent signaling. With the development of compulsive drug-seeking, AMPA receptor expression on the MSN in the nucleus accumbens increased (Conrad et al., 2008; Mameli et al., 2009) and it has been suggested that this reflects insertion of AMPA receptors into synapses, thereby increasing the surface expression of these receptors (Pierce & Wolf, 2013; Purgianto et al., 2013; Scheyer et al., 2014). This drug-induced upregulation of calcium-permeable AMPA receptors markedly increases the response to glutamate arising from prelimbic afferents and released in the nucleus accumbens. Because drugs of abuse increase synaptic dopamine levels well beyond what is produced following other, natural reinforcers (like food, water, or sex), stimulus–response relationships are acquired preferentially and more readily.

Based on this hypothesis, it should be possible to treat SUDs with drugs that counteract the neuroplastic changes in AMPA receptors induced by repeated exposure to drugs of abuse (Kalivas & Volkow, 2011; Rao et al., 2015; Schmidt & Pierce, 2010). Several drugs have been proposed as potential pharmacotherapies, based on the ability to limit effects of glutamate at selective receptors, but effectiveness of drugs that are selective for AMPA receptors have not been tested. Thus although both acamprosate and topiramate are both undergoing clinical trials for the treatment of SUDs, it is important that both of these drugs have myriad of effects in addition to modulation of AMPA responses. Acamprosate has been in use for more than 30 years but the pharmacology and behavioral pharmacology related to its putative "antiaddiction" properties are still poorly understood (Kalk &

Lingford-Hughes, 2014). Topiramate is an efficacious AMPA antagonist (Gryder & Rogawski, 2003). Clinical results of the effects of topiramate suggest that it is effective in the treatment of alcohol dependence, particularly with respect to decreasing craving (Guglielmo et al., 2015) and increasing abstinence (Blodgett et al., 2014), although substantive limitations to the various study designs have been noted (Shinn & Greenfield, 2010). Topiramate is also effective in maintaining cocaine abstinence and decreasing cocaine craving (Johnson et al., 2013; Kim & Lawrence, 2014). Less impressive results were produced when subjects with a methamphetamine SUD were treated with topiramate (Elkashef et al., 2012) although a follow-up study suggested effectiveness in a proportion of subjects who reduced methamphetamine use or achieved abstinence, although the basis for the different responses was not apparent (Ma et al., 2013).

It is noteworthy that most of the research on the effectiveness of drugs that might produce effects via glutamate receptor mechanisms has been conducted in human subjects, using compounds that have already received FDA approval for use to alleviate the symptoms of other disorders. These drugs are ubiquitous; they have a complex pharmacology that limits the ability to ascribe any beneficial effects to specific mechanisms. Because there has been limited preclinical evidence to provide either indices of effectiveness in relevant animal models of drug-taking and/or drug-seeking or to determine relevant pharmacological mechanisms, the important information that is required in order to understand whether these drugs are effective, under what conditions they might be effective, and what the mechanism for the effectiveness might be is lacking. Given the strong indication that dysregulation in AMPA receptor mechanisms underlies compulsive drug-seeking, approaches to identify more specific compounds are certainly warranted.

7. AMPA RECEPTORS AND EPILEPSY

Epilepsy affects approximately 50 million people worldwide making it one of the most common neurological diseases globally (WHO, 2015). It is characterized by recurrent seizures as a result of excessive and/or hypersynchronous activity within neural networks. The underlying pathophysiology is thought to be an imbalance between excitatory and inhibitory neurotransmission, hence many antiepileptic drugs (AEDs) act by modulating the excitatory–inhibitory balance either through reducing excitatory activity or by enhancing GABAergic inhibitory mechanisms. However, current AEDs are only successful in controlling seizures in approximately 70% of patients and have substantial side effects. In some cases, they can even exacerbate seizures (Chaves & Sander, 2005; Thomas et al., 2006); hence, there is a pressing need for more basic research to determine the precise underlying molecular and cellular mechanisms, which currently are not fully known and likely to be multifactorial. Recent research indicates that for some forms of epilepsy, enhanced inhibition may even be contributory to seizure generation (Cope et al., 2009; Seo & Leitch, 2014, 2015), contrary to the widely held view concerning the underlying pathophysiology.

The role of AMPA receptors in epilepsy and epileptogenesis and their potential as targets for drug treatment have been investigated for more than two decades (for reviews, see Rogawski, 2011; 2013; Hanada, 2014a). As AMPA receptors are responsible for fast glutamate-mediated excitation at synapses, AMPA receptor antagonists have been investigated for antiseizure activity in a broad range of animal seizure models for both focal and generalized epilepsies (De Sarro et al., 2005). They have also been tested in clinical investigations in humans (for reviews, see Citraro et al., 2014; Hanada, 2014a).

AMPARs antagonists fall into two broad categories: competitive AMPARs antagonists (e.g., quinoxalinediones, such as CNQX and NBQX), which compete with glutamate for the ligand binding site; and noncompetitive antagonists (e.g., talampanel and perampanel), which bind to an allosteric site and do not compete with glutamate. Early quinoxalinediones, despite showing promise as potent antagonists of glutamate in cultures (Drejer & Honoré, 1988), had poor *in vivo* activity due to problems in crossing the blood–brain barrier. Subsequently analogs were developed, such as NBQX, that had systemic and anticonvulsant activity in diverse seizure models (Chapman et al., 1991; Namba et al., 1994; Taylor & Vartanian, 1992; Yamaguchi et al., 1993). However, ultimately these competitive AMPA receptor antagonists proved to be of limited value for treating human epilepsy as they induced severe neurological side effects in animal models of epilepsy. Among the noncompetitive AMPA receptor antagonists developed to treat epileptic seizures, only perampanel has reached the market. Although talampanel (GYKI 53773, LY300164) was evaluated in epilepsy and showed positive signals in a Phase II clinical trial in patients with refractory partial-onset seizures, its further clinical development was suspended due to its pharmacokinetic profile and short half-life (3 h), necessitating multiple daily doses to be effective (Langan et al.,

2003). In fact all selective AMPA receptor antagonists to date, with the exception of perampanel, have proven unsuitable for use in treatment of epileptic patients for reasons linked to tolerability and pharmacokinetic profile, despite being very effective in preclinical trials on animal models.

Perampanel (trade name: Fycompa) is the first selective AMPA receptor antagonist to be approved for epilepsy treatment (Hanada, 2014b; Hanada et al., 2011). In Phase III trials it proved effective, when administered orally (in a single daily dose), as an adjunctive AED in adult and adolescent patients from age 12 years who had ongoing focal epileptic seizures despite receiving one to three AEDs (French et al., 2012; Kramer et al., 2014; Krauss et al., 2012; Steinhoff, 2015). It received marketing authorization approval by the European Commission for the treatment of partial epilepsy on 27 July 2012, and by the Food and Drug Administration (FDA) in the United States on 22 October 2012. Perampanel has since been widely licensed and introduced in over 30 countries. The marketing of the perampanel has once again focussed attention on AMPA receptors as potential drug targets for epilepsy therapy, and the use of antagonist of AMPA receptors as potent AED with broad efficacy against a range of epilepsy syndromes.

However, an exception to the broad efficacy of AMPA receptor antagonists as AEDs is their failure to inhibit seizures in animal models of absence (petit mal) epilepsy, such as the genetic absence epilepsy rat from Strasbourg [GAERS], indicating that they will not be of use in the treatment of absence seizures in humans (Hanada et al., 2011; Jakus et al., 2004; Rogawski, 2011). Absence epilepsy is a generalized form of epilepsy known to arise from disturbances within the thalamocortical system (Crunelli & Leresche, 2002; McCormick & Contreras, 2001). Alteration of AMPA receptor expression has been shown to contribute to the absence epilepsy phenotype by altering cortical (Kennard et al., 2011) and thalamocortical network oscillations (Beyer et al., 2008). In fact, in several animal models of absence epilepsy (e.g., stargazer mouse model of absence epilepsy and GRIA 4 knockout), absence seizures are associated with neuron and synapse-specific loss of AMPA receptors (Barad et al., 2012; Menuz & Nicoll, 2008; Paz et al., 2011). Hence, an AED, which is a selective antagonist of AMPA receptors, is unlikely to be of therapeutic value for this type of epilepsy.

Future basic research into AMPA receptor auxiliary proteins such as the TARPs (see Section 3.1) may allow for a more targeted modulation of AMPA receptors expression and function. As TARPs are differentially localized in different brain regions and even in different neurons, targeting individual isoforms may allow specific neuronal circuits to be selectively modulated without globally affecting synaptic transmission. Several human genetic analyses suggest that TARP pathology can underlie familial epilepsy syndromes. Linkage and mutational analysis studies have shown a link between childhood absence epilepsy and the *CACNG3* (TARP γ-3) mutation (Everett et al., 2007). Furthermore, homozygosity mapping has identified a 4 cM region within 22q13.1, which contains the *CACNG2* (human stargazin, TARP γ-2) locus in a family with epilepsy (Knight et al., 2008). However, whether specific AMPA receptor complexes such as TARPs can be uniquely targeted is yet to be established.

8. TRANSLATING DATA FROM ANIMAL MODELS TO THE HUMAN BRAIN

Animal models of neurodevelopmental and neurodegenerative disease have shed light on the changes in synapse structure and function that occur in response to diseases such as ASD, AD, HD and PD. Although no single animal model entirely reflects the human state of the disease, they have significantly increased our understanding of the molecular mechanisms that potentially govern synapse function, structure and ultimately behavior.

The major shortfall of animal models is that many therapies that have shown promise and significant therapeutic effects in animal models have failed to translate to therapies in humans (Section 4; Dragunow, 2008). The reasons for this are extremely complex, but likely reflect the significant difficulty in capturing the genetic diversity associated with human disease. For example, cellular studies in the human brain show significant inherent variability which is not evident in experiments using genetically similar animal models. Human cellular studies are also largely restricted to end-stage studies, making it difficult to determine whether the cellular changes that occur at presymptomatic and symptomatic stages of neurological disease are similar to those in animal models. In order to develop new therapies that prevent disease progression rather than target symptoms, this information is paramount. Animal models may also not accurately mimic disease processes occurring in human cells: for example, differential changes in glutamatergic synaptic proteins occur in the human hippocampus in human HD versus PD brain, and these changes were not observed in the rodent HD brain (Fourie et al., 2014). Further confounding this issue is the observation that animal models of the same disorder can also show conflicting data: with regard to glutamatergic synapses, animal models of AD, PD, and HD have shown conflicting results with regard to the changes in expression in glutamatergic receptor subunits and PSD proteins (Ariano et al., 2005; Cepeda et al., 2001; Dunah et al., 2000; Fan et al., 2007; Fourie et al., 2014; Gardoni et al., 2006; Jarabek et al., 2004; Nash et al., 2005; Torres-Peraza et al., 2008), likely due to genetic differences between animal models, the different stages of the disease (i.e., symptomatic vs. presymptomatic vs. end-stage), and the different experimental approaches taken to assess these changes. As a result, significant research remains to be done to determine why therapeutic treatments in animal studies are failing to result in therapies to repair the human brain.

9. CONCLUSIONS AND FUTURE DIRECTIONS

There is no doubt that the pathological signature of almost all neurological disorders involves altered glutamatergic synapse structure and function. As a result, AMPA receptors have been a major target in the development of pharmaceutical therapies for neurological disorders such as AD, HD, PD, ASD, epilepsy, and drug addiction, but little has translated into clinically effective treatment options and it appears that there is not the bright future for AMPAkines as once thought. While research on AMPA receptor structure, function, localization and binding proteins has progressed significantly over recent years, it remains to be determined whether AMPA receptor-interacting proteins are therapeutic options in human patients. The advent of super-resolution microscopy to examine the precise regulation of AMPA receptors at synapses will undoubtedly reveal a new magnitude of information on how neurological diseases alter AMPA receptors and their PSD binding partners. It is also clear that more research needs to be performed in human tissue to identify human-specific cellular changes, as these may underlie the lack of translation of therapies from animal models to human patients. In this regard, the expansion iPS technology and the use of human brain banks will be imperative.

REFERENCES

Aarts, M., Liu, Y., Liu, L., Besshoh, S., Arundine, M., Gurd, J. W., et al. (2002). Treatment of ischemic brain damage by perturbing NMDA receptor-PSD-95 protein interactions. *Science*, *298*(5594), 846–850.

Adesnik, H., Nicoll, R. A., & England, P. M. (2005). Photoinactivation of native AMPA receptors reveals their real-time trafficking. *Neuron*, *48*(6), 977–985.

Akhondzadeh, S., Tajdar, H., Mohammadi, M. R., Mohammadi, M., Nouroozinejad, G. H., et al. (2008). A double-blind placebo controlled trial of piracetam added to risperidone in patients with autistic disorder. *Child Psychiatry and Human Development*, *39*(3), 237–245.

Almeida, C. G., Tampellini, D., Takahashi, R. H., Greengard, P., Lin, M. T., et al. (2005). Beta-amyloid accumulation in APP mutant neurons reduces PSD-95 and GluR1 in synapses. *Neurobiology of Disease*, *20*(2), 187–198.

American Psychiatric Association (Ed.), (2013). *Diagnostic and statistical manual of mental disorders: DSM-V* (p. 1994). Washington, DC: American Psychiatric Association.

Anggono, V., & Huganir, R. L. (2012). Regulation of AMPA receptor trafficking and synaptic plasticity. *Current Opinion in Neurology*, *22*(3), 461–469.

Arai, A. C., & Kessler, M. (2007). Pharmacology of ampakine modulators: From AMPA receptors to synapses and behavior. *Current Drug Targets*, *8*(5), 583–602.

Ariano, M. A., Wagle, N., & Grissell, A. E. (2005). Neuronal vulnerability in mouse models of Huntington's disease: Membrane channel protein changes. *Journal of Neuroscience Research*, *80*(5), 634–645.

Arons, M. H., Thynne, C. J., Grabrucker, A. M., Li, D., Schoen, M., et al. (2012). Autism-associated mutations in ProSAP2/Shank3 impair synaptic transmission and neurexin-neuroligin-mediated transsynaptic signaling. *Journal of Neuroscience*, *32*(43), 14966–14978.

Ashby, M. C., Maier, S. R., Nishimune, A., & Henley, J. M. (2006). Lateral diffusion drives constitutive exchange of AMPA receptors at dendritic spines and is regulated by spine morphology. *Journal of Neuroscience*, *26*(26), 7046–7055.

Ba, M., Kong, M., Yu, G., Sun, X., Liu, Z., et al. (2011). GluR1 phosphorylation and persistent expression of levodopa-induced motor response alterations in the Hemi-Parkinsonian rat. *Neurochemical Research*, *36*(6), 1135–1144.

Bagni, C., & Oostra, B. A. (2013). Fragile X syndrome: From protein function to therapy. *American Journal of Medical Genetics Part A*, *161A*, 2809–2821.

Bailey, A., Phillips, W., & Rutter, M. (1996). Autism: towards an integration of clinical, genetic, neuropsychological, and neurobiological perspectives. *Journal of Child Psychology and Psychiatry*, *37*(1), 89–126.

Barad, Z., Shevtsova, O., Arbuthnott, G. W., & Leitch, B. (2012). Selective loss of AMPA receptors at corticothalamic synapses in the epileptic stargazer mouse. *Neuroscience*, *217*, 19–31.

Baron, M. K., Boeckers, T. M., Vaida, B., Faham, S., Gingery, M., et al. (2006). An architectural framework that may lie at the core of the postsynaptic density. *Science*, *311*(5760), 531–535.

Baroncelli, L., Braschi, C., Spolidoro, M., Begenisic, T., Maffei, L., et al. (2011). Brain plasticity and disease: A matter of inhibition. *Neural Plasticity*, *2011*, 286073.

Barria, A., Muller, D., Derkach, V., Griffith, L. C., & Soderling, T. R. (1997). Regulatory phosphorylation of AMPA-type glutamate receptors by CaM-KII during long-term potentiation. *Science*, *276*(5321), 2042–2045.

Bats, C., Groc, L., & Choquet, D. (2007). The interaction between Stargazin and PSD-95 regulates AMPA receptor surface trafficking. *Neuron*, *53*(5), 719–734.

Baudry, M., Kramar, E., Xu, X., Zadran, H., Moreno, S., Lynch, G., et al. (2012). Ampakines promote spine actin polymerization, long-term potentiation, and learning in a mouse model of Angelman syndrome. *Neurobiology of Disease*, *47*, 210–215.

Bauman, M. L., & Kemper, T. L. (2005a). Neuroanatomic observations of the brain in autism: A review and future directions. *International Journal of Developmental Neuroscience*, *23*(2–3), 183–187.

Bauman, M. L., & Kemper, T. L. (2005b). In *The neurobiology of autism*. Baltimore: Johns Hopkins University Press.

Baumbarger, P., Muhlhauser, M., Yang, C. R., & Nisenbaum, E. S. (2001). LY392098, a novel AMPA receptor potentiator: Electrophysiological studies in prefrontal cortical neurons. *Neuropharmacology*, *40*(8), 992–1002.

Belichenko, P. V., Masliah, E., Kleschevnikov, A. M., Villar, A. J., Epstein, C. J., et al. (2004). Synaptic structural abnormalities in the Ts65Dn mouse model of down syndrome. *Journal of Comparative Neurology*, *480*(3), 281–298.

Béna, F., Bruno, D. L., Eriksson, M., van Ravenswaaij-Arts, C., Stark, Z., et al. (2013). Molecular and clinical characterization of 25 individuals with exonic deletions of NRXN1 and comprehensive review of the literature. *American Journal of Medical Genetics. Part B, Neuropsychiatric Genetics*, *162B*(4), 388–403.

Berkel, S., Marshall, C. R., Weiss, B., Howe, J., Roeth, R., et al. (2010). Mutations in the SHANK2 synaptic scaffolding gene in autism spectrum disorder and mental retardation. *Nature Genetics*, *42*(6), 489–491.

Berkel, S., Tang, W., Treviño, M., Vogt, M., Obenhaus, H. A., Gass, P., et al. (2012). Inherited and de novo SHANK2 variants associated with autism spectrum disorder impair neuronal morphogenesis and physiology. *Human Molecular Genetics*, *21*(2), 344–357.

Berry-Kravis, E., Krause, S. E., Block, S. S., Guter, S., Wuu, J., et al. (2006). Effect of CX516, an AMPA-modulating compound, on cognition and behavior in fragile X syndrome: A controlled trial. *Journal of Child and Adolescent Psychopharmacology*, *16*(5), 525–540.

Beyer, B., Deleuze, C., Letts, V. A., Mahaffey, C. L., Boumil, R. M., Lew, T. A., et al. (2008). Absence seizures in C3H/HeJ and knockout mice caused by mutation of the AMPA receptor subunit Gria4. *Human Molecular Genetics*, *17*, 1738–1749.

Blanpied, T. A., Scott, D. B., & Ehlers, M. D. (2002). Dynamics and regulation of clathrin coats at specialized endocytic zones of dendrites and spines. *Neuron*, *36*(3), 435–449.

Bliss, T. V., & Collingridge, G. L. (1993). A synaptic model of memory: Long-term potentiation in the hippocampus. *Nature*, *361*(6407), 31–39.

Bliss, T. V. P., & Lømo, T. (1973). Long-lasting potentiation of synaptic transmission in the dentate area of the anaesthetized rabbit following stimulation of the perforant path. *The Journal of Physiology*, *232*(2), 331–356.

Blodgett, J. C., Del Re, A. C., Maisel, N. C., & Finney, J. W. (2014). A meta-analysis of topiramate's effects for individuals with alcohol use disorders. *Alcoholism: Clinical and Experimental Research*, *38*(6), 1481–1488.

Boeckers, T. M., Kreutz, M. R., Winter, C., Zuschratter, W., Smalla, K. H., et al. (1999). Proline-rich synapse-associated protein-1/cortactin binding protein 1 (ProSAP1/CortBP1) is a PDZ-domain protein highly enriched in the postsynaptic density. *Journal of Neuroscience*, *19*(15), 6506–6518.

Boehm, J., Kang, M. G., Johnson, R. C., Esteban, J., & Huganir, R. L. (2006). Synaptic incorporation of AMPA receptors during LTP is controlled by a PKC phosphorylation site on GluR1. *Neuron*, *51*(2), 213–225.

Borgdorff, A. J., & Choquet, D. (2002). Regulation of AMPA receptor lateral movements. *Nature*, *417*(6889), 649–653.

Bozdagi, O., Sakurai, T., Papapetrou, D., Wang, X., Dickstein, D., Takahashi, N., et al. (2010). Haploinsufficiency of the autism-associated Shank3 gene leads to deficits in synaptic function, social interaction, and social communication. *Molecular Autism*, *1*(1), 15.

Braithwaite, S. P., Xia, H., & Malenka, R. C. (2002). Differential roles for NSF and GRIP/-ABP in AMPA receptor cycling. *Proceedings of the National Academy of Sciences*, *99*(10), 7096–7101.

Brown, T. C., Tran, I. C., Backos, D. S., & Esteban, J. A. (2005). NMDA receptor-dependent activation of the small GTPase Rab5 drives the removal of synaptic AMPA receptors during hippocampal LTD. *Neuron*, *45*(1), 81–94.

Burnashev, N., Monyer, H., Seeburg, P. H., & Sakmann, B. (1992). Divalent ion permeability of AMPA receptor channels is dominated by the edited form of a single subunit. *Neuron*, *8*(1), 189–198.

Bush, A. I., Pettingell, W. H., Multhaup, G., Paradis, M. D., Vonsattel, J. P., et al. (1994). Rapid induction of Alzheimer A beta amyloid formation by zinc. *Science*, *265*(5177), 1464–1467.

Cai, C., Coleman, S. K., Niemi, K., & Keinänen, K. (2002). Selective binding of synapse-associated protein 97 to GluR-A alpha-amino-5-hydroxy-3-methyl-4-isoxazole propionate receptor subunit is determined by a novel sequence motif. *Journal of Biological Chemistry*, *277*(35), 31484–31490.

Cai, E., Ge, P., Lee, S. H., Jeyifous, O., Wang, Y., Liu, Y., et al. (2014). Stable small quantum dots for synaptic receptor tracking on live neurons. *Angewandte Communications*, *53*, 12484–12488.

Carlin, R. K., Grab, D. J., Cohen, R. S., & Siekevitz, P. (1980). Isolation and characterization of postsynaptic densities from various brain regions: Enrichment of different types of postsynaptic densities. *Journal of Cell Biology*, *86*(3), 831–845.

Carlsson, M. L. (1998). Hypothesis: Is infantile autism a hypoglutamatergic disorder? Relevance of glutamate—Serotonin interactions for pharmacotherapy. *Journal of Neural Transmission*, *105*(4-5), 525–535.

Carroll, R. C., Beattie, E. C., Xia, H., Luscher, C., Altschuler, Y., Nicoll, R. A., et al. (1999). Dynamin-dependent endocytosis of ionotropic glutamate receptors. *Proceedings of the National Academy of Sciences of the United States of America*, *96*(24), 14112–14117.

Cepeda, C., Ariano, M. A., Calvert, C. R., Flores-Hernández, J., Chandler, S. H., Leavitt, B. R., et al. (2001). NMDA receptor function in mouse models of Huntington disease. *Journal of Neuroscience Research*, *66*(4), 525–539.

Chanda, S., Aoto, J., Lee, S. J., Wernig, M., & Südhof, T. C. (2015). Pathogenic mechanism of an autism-associated neuroligin mutation involves altered AMPA-receptor trafficking. *Molecular Psychiatry*, Mar 17. http://dx.doi.org/10.1038/mp.2015.20.

Chang, P. K., Verbich, D., & McKinney, R. A. (2012). AMPA receptors as drug targets in neurological disease—advantages, caveats, and future outlook. *European Journal of Neuroscience*, *35*(12), 1908–1916.

Chapman, A. G., Smith, S. E., & Meldrum, B. S. (1991). The anticonvulsant effect of the non-NMDA antagonists NBQX and GYKI 52466 in mice. *Epilepsy Research*, *9*, 92–96.

Chappell, A. S., Gonzales, C., Williams, J., Witte, M. M., Mohs, R. C., et al. (2007). AMPA potentiator treatment of cognitive deficits in Alzheimer disease. *Neurology*, *68*(13), 1008–1012.

Chater, T. E., & Goda, Y. (2014). The role of AMPA receptors in postsynaptic mechanisms of synaptic plasticity. *Frontiers in Cellular Neuroscience*, *8*(November), 1–14.

Chaves, J., & Sander, J. W. (2005). Seizure aggravation in idiopathic generalized epilepsies. *Epilepsia*, *46*(Suppl. 9), 133–139.

Chen, L., Chetkovich, D. M., Petralia, R. S., Sweeney, N. T., Kawasaki, Y., Wenthold, R. J., et al. (2000). Stargazin regulates synaptic targeting of AMPA receptors by two distinct mechanisms. *Nature*, *408*(6815), 936–943.

Chen, Q., Veenman, C. L., & Reiner, A. (1996). Cellular expression of ionotropic glutamate receptor subunits on specific striatal neuron types and its implication for striatal vulnerability in glutamate receptor-mediated excitotoxicity. *Neuroscience*, *73*(3), 715–731.

Chen, X., Winters, C. A., Azzam, R., Li, X., Galbraith, J. A., Leapman, R. D., et al. (2008). Organization of the core structure of the postsynaptic density. *Proceedings of the National Academy of Sciences*, *105*(11), 4453–4458.

Chen, G. J., Xiong, Z., & Yan, Z. (2013). Aβ impairs nicotinic regulation of inhibitory synaptic transmission and interneuron excitability in prefrontal cortex. *Molecular Neurodegeneration*, *8*, 3.

Chetkovich, D. M., Bunn, R. C., Kuo, S. H., Kawasaki, Y., Kohwi, M., & Bredt, D. S. (2002). Postsynaptic targeting of alternative postsynaptic density-95 isoforms by distinct mechanisms. *Journal of Neuroscience*, *22*(15), 6415–6425.

Chih, B., Engelman, H., & Scheiffele, P. (2005). Control of excitatory and inhibitory synapse formation by neuroligins. *Science*, *307*(5713), 1324–1328.

Cho, K. O., Hunt, C. A., & Kennedy, M. B. (1992). The rat brain postsynaptic density fraction contains a homolog of the Drosophila discs-large tumor suppressor protein. *Neuron*, *9*(5), 929–942.

Choquet, D. (2010). Fast AMPAR trafficking for a high-frequency synaptic transmission. *European Journal of Neuroscience*, *32*(March), 250–260.

Chubykin, A. A., Liu, X., Comoletti, D., Tsigelny, I., Taylor, P., et al. (2005). Dissection of synapse induction by neuroligins: Effect of a neuroligin mutation associated with autism. *Journal of Biological Chemistry*, *280*, 22365–22374.

Citraro, R., Aiello, R., Franco, V., De Sarro, G., & Russo, E. (2014). Targeting α-amino-3-hydroxyl-5-methyl-4-isoxazole-propionate receptors in epilepsy. *Expert Opinion on Therapeutic Targets*, *18*, 319–334.

Colquhoun, D., Jonas, P., & Sakmann, B. (1992). Action of brief pulses of glutamate on AMPA/kainate receptors in patches from different neurones of rat hippocampal slices. *Journal of Physiology*, *458*, 261–287.

Conrad, K. L., Tseng, K. Y., Uejima, J. L., Reimers, J. M., Heng, L. J., Shaham, Y., et al. (2008). Formation of accumbens GluR2-lacking AMPA receptors mediates incubation of cocaine craving. *Nature*, *454*(7200), 118–121.

Cope, D. W., Di Giovanni, G., Fyson, S. J., Orbán, G., Errington, A. C., Lorincz, M. L., et al. (2009). Enhanced tonic $GABA_A$ inhibition in typical absence epilepsy. *Nature Medicine*, *15*, 1392–1399.

Cornish, J. L., & Kalivas, P. W. (2000). Glutamate transmission in the nucleus accumbens mediates relapse in cocaine addiction. *Journal of Neuroscience*, *20*(15), Rc89.

Crunelli, V., & Leresche, N. (2002). Childhood absence epilepsy, genes, channels, neurons and networks. *Nature Review Neuroscience*, *3*, 371–382.

Cuajungco, M. P., & Lees, G. J. (1997). Zinc and Alzheimer's disease: Is there a direct link? *Brain Research. Brain Research Reviews*, *23*(3), 219–236.

Dani, A., Huang, B., Bergan, J., Dulac, C., & Zhuang, X. (2010). Superresolution imaging of chemical synapses in the brain. *Neuron*, *68*(5), 843–856.

Danscher, G., Jensen, K. B., Frederickson, C. J., Kemp, K., Andreasen, A., et al. (1997). Increased amount of zinc in the hippocampus and amygdala of Alzheimer's diseased brains: A proton-induced X-ray emission spectroscopic analysis of cryostat sections from autopsy material. *Journal of Neuroscience Methods*, *76*(1), 53–59.

Daw, M. I., Chittajallu, R., Bortolotto, Z. A., Dev, K. K., Duprat, F., Henley, J. M., et al. (2000). PDZ proteins interacting with C-terminal GluR2/3 are involved in a PKC-dependent regulation of AMPA receptors at hippocampal synapses. *Neuron*, *28*(3), 873–886.

de Bartolomeis, A., Latte, G., Tomasetti, C., & Iasevoli, F. (2014). Glutamatergic postsynaptic density protein dysfunctions in synaptic plasticity and dendritic spines morphology: Relevance to schizophrenia and other behavioral disorders pathophysiology, and implications for novel therapeutic approaches. *Molecular Neurobiology*, *49*(1), 484–511.

De Rubeis, S., He, X., Goldberg, A. P., Poultney, C. S., Samocha, K., et al. (2014). Synaptic, transcriptional and chromatin genes disrupted in autism. *Nature*, *515*(7526), 209–215.

De Sarro, G., Gitto, R., Russo, E., Ibbadu, G. F., Barreca, M. L., De Luca, L., et al. (2005). AMPA receptor antagonists as potential anticonvulsant drugs. *Current Topics in Medicinal Chemistry*, *5*, 31–42.

Dean, C., Scholl, F. G., Choih, J., DeMaria, S., Berger, J., et al. (2003). Neurexin mediates the assembly of presynaptic terminals. *Nature Neuroscience*, *6*(7), 708–716.

DeKosky, S. T., & Scheff, S. W. (1990). Synapse loss in frontal cortex biopsies in Alzheimer's disease: Correlation with cognitive severity. *Annals of Neurology*, *27*(5), 457–464.

DeKosky, S. T., Scheff, S. W., & Styren, S. D. (1996). Structural correlates of cognition in dementia: Quantification and assessment of synaptic change. *Neurodegeneration*, *5*, 417–421.

Derkach, V., Barria, A., & Soderling, T. R. (1999). Ca2+/calmodulin-kinase II enhances channel conductance of alpha-amino-3-hydroxy-5-methyl-4-isoxazolepropionate type glutamate receptors. *Proceedings of the National Academy of Sciences of the United States of America*, *96*(6), 3269–3274.

DeSouza, S., Fu, J., States, B. A., & Ziff, E. B. (2002). Differential palmitoylation directs the AMPA receptor-binding protein ABP to spines or to intracellular clusters. *Journal of Neuroscience*, *22*(9), 3493–3503.

Dong, H., O'Brien, R. J., Fung, E. T., Lanahan, A. A., Worley, P. F., & Huganir, R. L. (1997). GRIP: A synaptic PDZ domain-containing protein that interacts with AMPA receptors. *Nature*, *386*(6622), 279–284.

Doyle, C. A., & McDougle, C. J. (2012). Pharmacologic treatments for the behavioral symptoms associated with autism spectrum disorders across the lifespan. *Dialogues in Clinical Neuroscience*, *14*(3), 263–279.

Dragunow, M. (2008). The adult human brain in preclinical drug development. *Nature Reviews Drug Discovery*, 7(8), 659–666.

Drejer, J., & Honoré, T. (1988). New quinoxalinediones show potent antagonism of quisqualate responses in cultured mouse cortical neurons. *Neuroscience Letters*, *87*, 104–108.

Dudek, S. M., & Bear, M. F. (1992). Homosynaptic long-term depression in area CA1 of hippocampus and effects of N-methyl-D-aspartate receptor blockade. *Proceedings of the National Academy of Sciences of the United States of America*, *89*, 4363–4367.

Dunah, A. W., Wang, Y., Yasuda, R. P., Kameyama, K., Huganir, R. L., et al. (2000). Alterations in subunit expression, composition, and phosphorylation of striataln-methyl-D-aspartate glutamate receptors in a rat 6-hydroxydopamine model of Parkinson's disease. *Molecular Pharmacology*, *57*(2), 342–352.

Durand, C. M., Betancur, C., Boeckers, T. M., Bockmann, J., Chaste, P., Fauchereau, F., et al. (2007). Mutations in the gene encoding the synaptic scaffolding protein SHANK3 are associated with autism spectrum disorders. *Nature Genetics*, *39*(1), 25–27.

Durand, C. M., Perroy, J., Loll, F., Perrais, D., Fagni, L., et al. (2012). SHANK3 mutations identified in autism lead to modification of dendritic spine morphology via an actin-dependent mechanism. *Molecular Psychiatry*, *17*(1), 71–84.

Eggert, K., Skogar, O., Amar, K., Luotonen, L., Kuoppamäki, M., et al. (2010). Direct switch from levodopa/benserazide or levodopa/carbidopa to levodopa/carbidopa/entacapone in Parkinson's disease patients with wearing-off: Efficacy, safety and feasibility—An open-label, 6-week study. *Journal of Neural Transmission*, *117*(3), 333–342.

Ehlers, M. D. (2000). Reinsertion or degradation of AMPA receptors determined by activity-dependent endocytic sorting. *Neuron*, *28*(2), 511–525.

Ehrlich, I., & Malinow, R. (2004). Postsynaptic density 95 controls AMPA receptor incorporation during long-term potentiation and experience-driven synaptic plasticity. *Journal of Neuroscience*, *24*(4), 916–927.

El-Husseini Ael, D., Schnell, E., Dakoji, S., Sweeney, N., Zhou, Q., et al. (2002). Synaptic strength regulated by palmitate cycling on PSD-95. *Cell*, *108*(6), 849–863.

Elias, G. M., Funke, L., Stein, V., Grant, S. G., Bredt, D. S., et al. (2006). Synapse-specific and developmentally regulated targeting of AMPA receptors by a family of MAGUK scaffolding proteins. *Neuron, 52*(2), 307–320.

Elias, G. M., & Nicoll, R. A. (2007). Synaptic trafficking of glutamate receptors by MAGUK scaffolding proteins. *Trends in Cell Biology, 17*(7), 343–352.

Elkashef, A., Kahn, R., Yu, E., Iturriaga, E., Li, S. H., Anderson, A., et al. (2012). Topiramate for the treatment of methamphetamine addiction: A multi-center placebo-controlled trial. *Addiction, 107*(7), 1297–1306.

Emond, M. R., Montgomery, J. M., Huggins, M. L., Hanson, J. E., Mao, L., et al. (2010). AMPA receptor subunits define properties of state-dependent synaptic plasticity. *The Journal of Physiology, 588*(11), 1929–1946.

Epis, R., Marcello, E., Gardoni, F., Vastagh, C., Malinverno, M., Balducci, C., et al. (2010). Blocking ADAM10 synaptic trafficking generates a model of sporadic Alzheimer's disease. *Brain, 133*, 3323–3335.

Etherton, M. R., Blaiss, C. A., Powell, C. M., & Südhof, T. C. (2009). Mouse neurexin-1alpha deletion causes correlated electrophysiological and behavioral changes consistent with cognitive impairments. *Proceedings of the National Academy of Sciences of the United States of America, 106*(42), 17998–18003.

Everett, K. V., Chioza, B., Aicardi, J., Aschauer, H., Brouwer, O., Callenbach, P., et al. (2007). Linkage and association analysis of CACNG3 in childhood absence epilepsy. *European Journal of Human Genetics, 15*, 463–472.

Famous, K. R., Kumaresan, V., Sadri-Vakili, G., Schmidt, H. D., Mierke, D. F., Cha, J. H., et al. (2008). Phosphorylation-dependent trafficking of GluR2-containing AMPA receptors in the nucleus accumbens plays a critical role in the reinstatement of cocaine seeking. *Journal of Neuroscience, 28*(43), 11061–11070.

Fan, M. M. Y., Fernandes, H. B., Zhang, L. Y. J., Hayden, M. R., & Raymond, L. A. (2007). Altered NMDA receptor trafficking in a yeast artificial chromosome transgenic mouse model of Huntington's disease. *The Journal of Neuroscience, 27*(14), 3768–3779.

Fan, M. M., & Raymond, L. A. (2007). N-methyl-D-aspartate (NMDA) receptor function and excitotoxicity in Huntington's disease. *Progress in Neurobiology, 81*(5-6), 272–293.

Fassio, A., Patry, L., Congia, S., Onofri, F., Piton, A., et al. (2011). SYN1 loss-of-function mutations in autism and partial epilepsy cause impaired synaptic function. *Human Molecular Genetics, 20*(12), 2297–2307.

Fernandez, F., Morishita, W., Zuniga, E., Nguyen, J., Blank, M., et al. (2007). Pharmacotherapy for cognitive impairment in a mouse model of Down syndrome. *Nature Neuroscience, 10*(4), 411–413. Epub 2007 Feb 25.

Fitzjohn, S. M., Doherty, A. J., & Collingridge, G. L. (2006). Promiscuous interactions between AMPA-Rs and MAGUKs. *Neuron, 52*(2), 222–224.

Fourie, C., Kim, H., Waldvogel, H., Wong, J. M., McGregor, A., et al. (2014). Differential changes in postsynaptic density proteins in postmortem Huntington's disease and Parkinson's disease human brains. *Journal of Neurodegenerative Diseases*. Article ID 938530.

French, J. A., Krauss, G. L., Biton, V., Squillacote, D., Yang, H., Laurenza, A., et al. (2012). Adjunctive perampanel for refractory partial-onset seizures, randomized phase III study 304. *Neurology, 79*, 589–596.

Fukata, Y., Dimitrov, A., Boncompain, G., Vielemeyer, O., Perez, F., & Fukata, M. (2013). Local palmitoylation cycles define activity-regulated postsynaptic subdomains. *Journal of Cell Biology, 202*(1), 145–161.

Fukata, Y., Tzingounis, A. V., Trinidad, J. C., Fukata, M., Burlingame, A. L., Nicoll, R. A., et al. (2005). Molecular constituents of neuronal AMPA receptors. *Journal of Cell Biology, 169*(3), 399–404.

Gabbott, P. L., Warner, T. A., Jays, P. R., Salway, P., & Busby, S. J. (2005). Prefrontal cortex in the rat: Projections to subcortical autonomic, motor, and limbic centers. *Journal of Comparative Neurology*, *492*(2), 145–177.

Gardoni, F. (2008). MAGUK proteins: New targets for pharmacological intervention in the glutamatergic synapse. *European Journal of Pharmacology*, *585*(1), 147–152.

Gardoni, F., Marcello, E., & Di Luca, M. (2009). Postsynaptic density-membrane associated guanylate kinase proteins (PSD-MAGUKs) and their role in CNS disorders. *Neuroscience*, *158*(1), 324–333.

Gardoni, F., Mauceri, D., Fiorentini, C., Bellone, C., Missale, C., Cattabeni, F., et al. (2003). CaMKII-dependent phosphorylation regulates SAP97/NR2A interaction. *Journal of Biological Chemistry*, *278*(45), 44745–44752.

Gardoni, F., Picconi, B., Ghiglieri, V., Polli, F., Bagetta, V., Bernardi, G., et al. (2006). A critical interaction between NR2B and MAGUK in L-DOPA induced dyskinesia. *Neurobiology of Disease*, *26*(11), 2914–2922.

Gardoni, F., Sgobio, C., Pendolino, V., Calabresi, P., & Di Luca, M. (2012). Targeting NR2A-containing NMDA receptors reduces L-DOPA-induced dyskinesias. *Neurobiology of Aging*, *33*(9), 2138–2144.

Gates, M., Ogden, A., & Bleakman, D. (2001). Pharmacological effects of AMPA receptor potentiators LY392098 and LY404187 on rat neuronal AMPA receptors in vitro. *Neuropharmacology*, *40*(8), 984–991.

Geiger, J. R., Melcher, T., Koh, D. S., Sakmann, B., Seeburg, P. H., Jonas, P., et al. (1995). Relative abundance of subunit mRNAs determines gating and Ca2+ permeability of AMPA receptors in principal neurons and interneurons in rat CNS. *Neuron*, *15*(1), 193–204.

Gerges, N. Z., Backos, D. S., Rupasinghe, C. N., Spaller, M. R., & Esteban, J. A. (2006). Dual role of the exocyst in AMPA receptor targeting and insertion into the postsynaptic membrane. *EMBO Journal*, *25*(8), 1623–1634.

Gilman, S. R., Iossifov, I., Levy, D., Ronemus, M., Wigler, M., & Vitkup, D. (2011). Rare de novo variants associated with autism implicate a large functional network of genes involved in formation and function of synapses. *Neuron*, *70*(5), 898–907.

Gladding, C. M., Sepers, M. D., Xu, J., Zhang, L. Y., Milnerwood, A. J., et al. (2012). Calpain and STriatal-Enriched protein tyrosine phosphatase (STEP) activation contribute to extrasynaptic NMDA receptor localization in a Huntington's disease mouse model. *Human Molecular Genetics*, *21*(17), 3739–3752.

Glebov, O. O., Tigaret, C. M., Mellor, J. R., & Henley, J. M. (2015). Clathrin-independent trafficking of AMPA receptors. *Journal of Neuroscience*, *35*(12), 4830–4836.

Gomperts, S. N., Carroll, R. C., Malenka, R. C., & Nicoll, R. A. (2000). Distinct roles for ionotropic and metabotropic glutamate receptors in the maturation of excitatory synapses. *Journal of Neuroscience*, *20*(6), 2229–2237.

Gong, Y., Lippa, C. F., Zhu, J., Lin, Q., & Rosso, A. L. (2009). Disruption of glutamate receptors at Shank-postsynaptic platform in Alzheimer's disease. *Brain Research*, *1292*, 191–198.

Grabrucker, A. M., Schmeisser, M. J., Udvardi, P. T., Arons, M., Schoen, M., et al. (2011). Amyloid beta protein-induced zinc sequestration leads to synaptic loss via dysregulation of the ProSAP2/Shank3 scaffold. *Molecular Neurodegeneration*, *6*, 65.

Graf, E. R., Kang, Y., Hauner, A. M., & Craig, A. M. (2006). Structure function and splice site analysis of the synaptogenic activity of the neurexin-1 beta LNS domain. *The Journal of Neuroscience*, *26*, 4256–4265.

Graham, R. K., Pouladi, M. A., Joshi, P., Lu, G., Deng, Y., Wu, N.-P., et al. (2009). Differential susceptibility to excitotoxic stress in YAC128 mouse models of Huntington disease between initiation and progression of disease. *The Journal of Neuroscience*, *29*(7), 2193–2204.

Greger, I. H., Khatri, L., Kong, X., & Ziff, E. B. (2003). AMPA receptor tetramerization is mediated by Q/R editing. *Neuron*, *40*(4), 763–774.

Groc, L., Heine, M., Cognet, L., Brickley, K., Stephenson, F. A., Lounis, B., et al. (2004). Regulation of the lateral mobilities of AMPA and NMDA receptors. *Nature Neuroscience*, 7(7), 695–697.

Gryder, D. S., & Rogawski, M. A. (2003). Selective antagonism of GluR5 kainate-receptor-mediated synaptic currents by topiramate in rat basolateral amygdala neurons. *Journal of Neuroscience*, *23*(18), 7069–7074.

Guglielmo, R., Martinotti, G., Quatrale, M., Ioime, L., Kadilli, I., Di Nicola, M., et al. (2015). Topiramate in alcohol use disorders: Review and update. *CNS Drugs*, *29*(5), 383–395.

Guilmatre, A., Huguet, G., Delorme, R., & Bourgeron, T. (2014). The emerging role of SHANK genes in neuropsychiatric disorders. *Developmental Neurobiology*, *74*(2), 113–122.

Hamos, J. E., DeGennaro, L. J., & Drachman, D. (1989). Synaptic loss in Alzheimer's Disease and other dementias. *Neurology*, *39*(3), 355–361.

Hampson, R. E., Rogers, G., Lynch, G., & Deadwyler, S. A. (1998a). Facilitative effects of the ampakine CX516 on short-term memory in rats: Correlations with hippocampal neuronal activity. *Journal of Neuroscience*, *18*(7), 2748–2763.

Hampson, R. E., Rogers, G., Lynch, G., & Deadwyler, S. A. (1998b). Facilitative effects of the ampakine CX516 on short-term memory in rats: Enhancement of delayed-nonmatch-to-sample performance. *Journal of Neuroscience*, *18*(7), 2740–2747.

Hanada, T. (2014a). The AMPA receptor as a therapeutic target in epilepsy, preclinical and clinical evidence. *Journal of Receptor Ligand and Channel Research*, 7, 39–50.

Hanada, T. (2014b). The discovery and development of perampanel for the treatment of epilepsy. *Expert Opinion on Drug Discovery*, *9*, 449–458.

Hanada, T., Hashizume, Y., Tokuhara, N., Takenaka, O., Kohmura, N., Ogasawara, A., et al. (2011). Perampanel, a novel., orally active., noncompetitive AMPA-receptor antagonist that reduces seizure activity in rodent models of epilepsy. *Epilepsia*, *52*, 1331–1340.

Hardingham, G. E., Fukunaga, Y., & Bading, H. (2002). Extrasynaptic NMDARs oppose synaptic NMDARs by triggering CREB shut-off and cell death pathways. *Nature Neuroscience*, *5*(5), 405–414.

Hayashi, Y., Shi, S. H., Esteban, J. A., Piccini, A., Poncer, J. C., & Malinow, R. (2000). Driving AMPA receptors into synapses by LTP and CaMKII: Requirement for GluR1 and PDZ domain interaction. *Science*, *287*(5461), 2262–2267.

He, K., Song, L., Cummings, L. W., Goldman, J., Huganir, R. L., & Lee, H. (2009). Stabilization of Ca2+-permeable AMPA receptors at perisynaptic sites by GluR1-S845 phosphorylation. *Proceedings of the National Academy of Sciences*, *106*(47), 20033–20038.

Heine, M., Groc, L., Frischknecht, R., Bé, J., Lounis, B., Rumbaugh, G., et al. (2008). Surface mobility of postsynaptic AMPARs tunes synaptic transmission. *Science*, *320*, 201–205.

Hodgson, J. G., Agopyan, N., Gutekunst, C.-A., Leavitt, B. R., LePiane, F., Singaraja, R., et al. (1999). A YAC mouse model for Huntington's disease with full-length mutant huntingtin, cytoplasmic toxicity, and selective striatal neurodegeneration. *Neuron*, *23*(1), 181–192.

Hollmann, M., O'Shea-Greenfield, A., Rogers, S. W., & Heinemann, S. (1989). Cloning by functional expression of a member of the glutamate receptor family. *Nature*, *342*(6250), 643–648.

Hoover, W. B., & Vertes, R. P. (2007). Anatomical analysis of afferent projections to the medial prefrontal cortex in the rat. *Brain Structure and Function*, *212*(2), 149–179.

Horton, A. C., & Ehlers, M. D. (2004). Secretory trafficking in neuronal dendrites. *Nature Cell Biology*, *6*(7), 585–591.

Hoze, N., Nair, D., Hosy, E., Sieben, C., Manley, S., Herrmann, A., et al. (2012). Heterogeneity of AMPA receptor trafficking and molecular interactions revealed by superresolution analysis of live cell imaging. *Proceedings of the National Academy of Sciences*, *109*(42), 17052–17057.

Hsieh, H., Boehm, J., Sato, C., Iwatsubo, T., Tomita, T., et al. (2006). AMPAR removal underlies Abeta-induced synaptic depression and dendritic spine loss. *Neuron*, *52*(5), 831–843.

Hu, H., Qin, Y., Bochorishvili, G., Zhu, Y., van Aelst, L., et al. (2008). Ras signaling mechanisms underlying impaired GluR1-dependent plasticity associated with fragile X syndrome. *Journal of Neuroscience*, *28*(31), 7847–7862.

Huber, K. M., Gallagher, S. M., Warren, S. T., & Bear, M. F. (2002). Altered synaptic plasticity in a mouse model of fragile X mental retardation. *Proceedings of the National Academy of Sciences of the United States of America*, *99*(11), 7746–7750.

Hume, R. I., Dingledine, R., & Heinemann, S. F. (1991). Identification of a site in glutamate receptor subunits that controls calcium permeability. *Science*, *253*(5023), 1028–1031.

Hung, A. Y., Futai, K., Sala, C., Valtschanoff, J. G., Ryu, J., Woodworth, M. A., et al. (2008). Smaller dendritic spines, weaker synaptic transmission, but enhanced spatial learning in mice lacking Shank1. *The Journal of Neuroscience*, *28*(7), 1697–1708.

Ichtchenko, K., Hata, Y., Nguyen, T., Ullrich, B., Missler, M., Moomaw, C., et al. (1995). Neuroligin 1: A splice site-specific ligand for β-neurexins. *Cell*, *81*(3), 435–443.

Ingvar, M., Ambros-Ingerson, J., Davis, M., Granger, R., Kessler, M., et al. (1997). Enhancement by an ampakine of memory encoding in humans. *Experimental Neurology*, *146*(2), 553–559.

Jakus, R., Graf, M., Ando, R. D., Balogh, B., Gacsalyi, I., Levay, G., et al. (2004). Effect of two noncompetitive AMPA receptor antagonists GYKI 52466 and GYKI 53405 on vigilance, behavior and spike-wave discharges in a genetic rat model of absence epilepsy. *Brain Research*, *1008*, 236–244.

Jamain, S., Betancur, C., Quach, H., Philippe, A., Fellous, M., Giros, B., et al. (2002). Linkage and association of the glutamate receptor 6 gene with autism. *Molecular Psychiatry*, 7(3), 302–310.

Jamain, S., Quach, H., Betancur, C., Rastam, M., Colineaux, C., Gillberg, I. C., et al. (2003). Mutations of the X-linked genes encoding neuroligins NLGN3 and NLGN4 are associated with autism. *Nature Genetics*, *34*(1), 27–29.

Jarabek, B. R., Yasuda, R. P., & Wolfe, B. B. (2004). Regulation of proteins affecting NMDA receptor-induced excitotoxicity in a Huntington's mouse model. *Brain*, *127*(3), 505–516.

Jaskolski, F., Mayo-Martin, B., Jane, D. E., & Henley, J. M. (2009). Dynamin-dependent membrane drift recruits AMPA receptors to dendritic spines. *Journal of Biological Chemistry*, *284*(18), 12491–12503.

Jeyifous, O., Waites, C. L., Specht, C. G., Fujisawa, S., Schubert, M., Lin, E. I., et al. (2009). SAP97 and CASK mediate sorting of NMDA receptors through a previously unknown secretory pathway. *Nature Neuroscience*, *12*(8), 1011–1019.

Johnson, B. A., Roache, J. D., Ait-Daoud, N., Gunderson, E. W., Haughey, H. M., Wang, X. Q., et al. (2013). Topiramate's effects on cocaine-induced subjective mood, craving and preference for money over drug taking. *Addiction Biology*, *18*(3), 405–416.

Jourdi, H., Hsu, Y. T., Zhou, M., Qin, Q., Bi, X., & Baudry, M. (2009). Positive AMPA receptor modulation rapidly stimulates BDNF release and increases dendritic mRNA translation. *Journal of Neuroscience*, *29*(27), 8688–8697.

Julio-Pieper, M., Flor, P. J., Dinan, T. G., & Cryan, J. F. (2011). Exciting times beyond the brain: Metabotropic glutamate receptors in peripheral and non-neural tissues. *Pharmacological Reviews*, *63*(1), 35–58.

Kaila, K., Ruusuvuori, E., Seja, P., Voipio, J., & Puskarjov, M. (2014). GABA actions and ionic plasticity in epilepsy. *Current Opinion in Neurobiology*, *26*, 34–41.

Kalivas, P. W., & Volkow, N. D. (2011). New medications for drug addiction hiding in glutamatergic neuroplasticity. *Molecular Psychiatry*, *16*(10), 974–986.

Kalk, N. J., & Lingford-Hughes, A. R. (2014). The clinical pharmacology of acamprosate. *British Journal of Clinical Pharmacology*, *77*(2), 315–323.

Kaufman, A. M., Milnerwood, A. J., Sepers, M. D., Coquinco, A., She, K., Wang, L., et al. (2012). Opposing roles of synaptic and extrasynaptic NMDA receptor signaling in cocultured striatal and cortical neurons. *The Journal of Neuroscience*, *32*(12), 3992–4003.

Keinanen, K., Wisden, W., Sommer, B., Werner, P., Herb, A., Verdoorn, T. A., et al. (1990). A family of AMPA-selective glutamate receptors. *Science*, *249*(4968), 556–560.

Kennard, J. T., Barmanray, R., Sampurno, S., Ozturk, E., Reid, C. A., Paradiso, L., et al. (2011). Stargazin and AMPA receptor membrane expression is increased in the somatosensory cortex of Genetic Absence Epilepsy Rats from Strasbourg. *Neurobiology of Disease*, *42*, 48–54.

Kennedy, M. J., Davison, I. G., Robinson, C. G., & Ehlers, M. D. (2010). Syntaxin-4 defines a domain for activity-dependent exocytosis in dendritic spines. *Cell*, *141*(3), 524–535.

Kerr, J. M., & Blanpied, T. A. (2012). Subsynaptic AMPA receptor distribution is acutely regulated by actin-driven reorganization of the postsynaptic density. *Journal of Neuroscience*, *32*(2), 658–673.

Kim, J. H., Anwyl, R., Suh, Y. H., Djamgoz, M. B., & Rowan, M. J. (2001). Use-dependent effects of amyloidogenic fragments of (beta)-amyloid precursor protein on synaptic plasticity in rat hippocampus in vivo. *Journal of Neuroscience*, *21*(4), 1327–1333.

Kim, E., Cho, K. O., Rothschild, A., & Sheng, M. (1996). Heteromultimerization and NMDA receptor-clustering activity of Chapsyn-110, a member of the PSD-95 family of proteins. *Neuron*, *17*(1), 103–113.

Kim, C. H., Chung, H. J., Lee, H., & Huganir, R. L. (2001). Interaction of the AMPA receptor subunit GluR2/3 with PDZ domains regulates hippocampal long-term depression. *Proceedings of the National Academy of Sciences*, *98*(20), 11725–11730.

Kim, H. G., Kishikawa, S., Higgins, A. W., Seong, I. S., Donovan, D. J., Shen, Y., et al. (2008). Disruption of neurexin 1 associated with autism spectrum disorder. *American Journal of Human Genetics*, *82*(1), 199–207.

Kim, J. H., & Lawrence, A. J. (2014). Drugs currently in Phase II clinical trials for cocaine addiction. *Expert Opinion on Investigational Drugs*, *23*(8), 1105–1122.

Kim, E., Niethammer, M., Rothschild, A., Jan, Y. N., & Sheng, M. (1995). Clustering of Shaker-type K+ channels by interaction with a family of membrane-associated guanylate kinases. *Nature*, *378*, 85–88.

Kim, E., & Sheng, M. (1996). Differential K+ channel clustering activity of PSD-95 and SAP97, two related membrane-associated putative guanylate kinases. *Neuropharmacology*, *35*(7), 993–1000.

Knight, H. M., Maclean, A., Irfan, M., Naeem, F., Cass, S., Pickard, B. S., et al. (2008). Homozygosity mapping in a family presenting with schizophrenia., epilepsy and hearing impairment. *European Journal of Human Genetics*, *16*, 750–758.

Knobloch, M., & Mansuy, I. M. (2008). Dendritic spine loss and synaptic alterations in Alzheimer's disease. *Molecular Neurobiology*, *37*(1), 73–82.

Kobylecki, C., Cenci, M. A., Crossman, A. R., & Ravenscroft, P. (2010). Calcium-permeable AMPA receptors are involved in the induction and expression of l-DOPA-induced dyskinesia in Parkinson's disease. *Journal of Neurochemistry*, *114*(2), 499–511.

Kolodziejczyk, K., Parsons, M. P., Southwell, A. L., Hayden, M. R., & Raymond, L. A. (2014). Striatal synaptic dysfunction and hippocampal plasticity deficits in the Hu97/18 mouse model of Huntington disease. *PLoS One*, *9*(4), e94562.

Kornau, H. C., Schenker, L. T., Kennedy, M. B., & Seeburg, P. H. (1995). Domain interaction between NMDA receptor subunits and the postsynaptic density protein PSD-95. *Science*, *269*(5231), 1737–1740.

Kramer, L., Satlin, A., Krauss, G., French, J., Perucca, E., Ben-Menachem, E., et al. (2014). Perampanel for adjunctive treatment of partial-onset seizures, a pooled dose response analysis of phase-III trials. *Epilepsia*, *55*, 423–431.

Krauss, G. L., Serratosa, J. M., Villanueva, V., Endziniene, M., Hong, Z., French, J., et al. (2012). Randomized phase III study 306, adjunctive perampanel for refractory partial-onset seizures. *Neurology*, *78*, 1408–1415.

Langan, Y. M., Lucas, R., Jewell, H., Toublanc, N., Schaefer, H., Sander, J. W., et al. (2003). Talampanel., a new antiepileptic drug, single- and multiple-dose pharmacokinetics and initial 1-week experience in patients with chronic intractable epilepsy. *Epilepsia*, *44*, 46–53.

Larson, J., & Lynch, G. (1986). Induction of synaptic potentiation in hippocampus by patterned stimulation involves two events. *Science*, *232*(4753), 985–988.

Larson, J., & Munkácsy, E. (2015). Theta-burst LTP. *Brain Research*, *1621*, 38–50.

Lau, C. G., & Zukin, R. S. (2007). NMDA receptor trafficking in synaptic plasticity and neuropsychiatric disorders. *Nature Review Neuroscience*, *8*(6), 413–426.

Laumonnier, F., Bonnet-Brilhault, F., Gomot, M., Blanc, R., David, A., et al. (2004). X-linked mental retardation and autism are associated with a mutation in the NLGN4 gene, a member of the neuroligin family. *American Journal of Human Genetics*, *74*, 552–557.

Lauterborn, J. C., Lynch, G., Vanderklish, P., Arai, A., & Gall, C. M. (2000). Positive modulation of AMPA receptors increases neurotrophin expression by hippocampal and cortical neurons. *Journal of Neuroscience*, *20*(1), 8–21.

Leblond, C. S., Heinrich, J., Delorme, R., Proepper, C., Betancur, C., Huguet, G., et al. (2012). Genetic and functional analyses of SHANK2 mutations suggest a multiple hit model of autism spectrum disorders. *PLoS Genetics*, *8*(2), e1002521.

Leblond, C. S., Nava, C., Polge, A., Gauthier, J., Huguet, G., Lumbroso, S., et al. (2014). Meta-analysis of SHANK mutations in autism spectrum disorders: A gradient of severity in cognitive impairments. *PLoS Genetics*, *10*, e1004580.

Lee, H. K., Barbarosie, M., Kameyama, K., Bear, M. F., & Huganir, R. L. (2000). Regulation of distinct AMPA receptor phosphorylation sites during bidirectional synaptic plasticity. *Nature*, *405*(6789), 955–959.

Lees, A., Fahn, S., Eggert, K. M., Jankovic, J., Lang, A., et al. (2012). Perampanel, an AMPA antagonist, found to have no benefit in reducing "off" time in Parkinson's disease. *Movement Disorders*, *27*(2), 284–288.

Legutko, B., Li, X., & Skolnick, P. (2001). Regulation of BDNF expression in primary neuron culture by LY392098, a novel AMPA receptor potentiator. *Neuropharmacology*, *40*(8), 1019–1027.

Leonard, A. S., Davare, M. A., Horne, M. C., Garner, C. C., & Hell, J. W. (1998). SAP97 is associated with the alpha-amino-3-hydroxy-5-methylisoxazole-4-propionic acid receptor GluR1 subunit. *Journal of Biological Chemistry*, *273*(31), 19518–19524.

Letts, V. A., Felix, R., Biddlecome, G. H., Arikkath, J., Mahaffey, C. L., Valenzuela, A., et al. (1998). The mouse stargazer gene encodes a neuronal Ca2+-channel gamma subunit. *Nature Genetics*, *19*(4), 340–347.

Letts, V. A., Valenzuela, A., Kirley, J. P., Sweet, H. O., Davisson, M. T., & Frankel, W. N. (1997). Genetic and physical maps of the stargazer locus on mouse chromosome 15. *Genomics*, *43*(1), 62–68.

Li, D., Specht, C. G., Waites, C. L., Butler-Munro, C., Leal-Ortiz, S., et al. (2011). SAP97 directs NMDA receptor spine targeting and synaptic plasticity. *Journal of Physiology*, *589*(Pt. 18), 4491–4510.

Liao, D., Zhang, X., O'Brien, R., Ehlers, M. D., & Huganir, R. L. (1999). Regulation of morphological postsynaptic silent synapses in developing hippocampal neurons. *Nature Neuroscience*, *2*(1), 37–43.

Lim, S., Naisbitt, S., Yoon, J., Hwang, J. I., Suh, P. G., Sheng, M., et al. (1999). Characterization of the Shank family of synaptic proteins. Multiple genes, alternative splicing, and differential expression in brain and development. *Journal of Biological Chemistry*, *274*(41), 29510–29518.

Lin, E. I., Jeyifous, O., & Green, W. N. (2013). CASK regulates SAP97 conformation and its interactions with AMPA and NMDA receptors. *Journal of Neuroscience*, *33*(29), 12067–12076.

Lin, D. T., Makino, Y., Sharma, K., & Hayashi, T. (2009). Regulation of AMPA receptor extrasynaptic insertion by 4.1 N, phosphorylation and palmitoylation. *Nature Neuroscience*, *12*(7), 879–887.

Lionel, A. C., Vaags, A. K., Sato, D., Gazzellone, M. J., Mitchell, E. B., et al. (2013). Rare exonic deletions implicate the synaptic organizer Gephyrin (GPHN) in risk for autism, schizophrenia and seizures. *Human Molecular Genetics*, *22*(10), 2055–2066.

Lisman, J. (1989). A mechanism for the Hebb and the anti-Hebb processes underlying learning and memory. *Proceedings of the National Academy of Sciences of the United States of America*, *86*(23), 9574–9578.

Lisman, J. E., Raghavachari, S., & Tsien, R. W. (2007). The sequence of events that underlie quantal transmission at central glutamatergic synapses. *Nature Reviews. Neuroscience*, *8*(July), 597–609.

Lomeli, H., Mosbacher, J., Melcher, T., Hoger, T., Geiger, J. R., Kuner, T., et al. (1994). Control of kinetic properties of AMPA receptor channels by nuclear RNA editing. *Science*, *266*(5191), 1709–1713.

Lovell, M. A. (2009). A potential role for alterations of zinc and zinc transport proteins in the progression of Alzheimer's disease. *Journal of Alzheimer's Disease*, *16*(3), 471–483.

Lovell, M. A., Robertson, J. D., Teesdale, W. J., Campbell, J. L., & Markesbery, W. R. (1998). Copper, iron and zinc in Alzheimer's disease senile plaques. *Journal of Neurological Sciences*, *158*(1), 47–52.

Lu, J., Helton, T. D., Blanpied, T. A., & Rácz, B. (2007). Postsynaptic positioning of endocytic zones and AMPA receptor cycling by physical coupling of dynamin-3 to Homer. *Neuron*, *55*(6), 874–889.

Lu, W., Man, H., Ju, W., Trimble, W. S., MacDonald, J. F., & Wang, Y. T. (2001). Activation of synaptic NMDA receptors induces membrane insertion of new AMPA receptors and LTP in cultured hippocampal neurons. *Neuron*, *29*(1), 243–254.

Lu, W., & Ziff, E. B. (2005). PICK1 interacts with ABP/GRIP to regulate AMPA receptor trafficking. *Neuron*, *47*(3), 407–421.

Lue, R. A., Marfatia, S. M., Branton, D., & Chishti, A. H. (1994). Cloning and characterization of hdlg: The human homologue of the Drosophila discs large tumor suppressor binds to protein 4.1. *Proceedings of the National Academy of Sciences*, *91*(21), 9818–9822.

Luo, C. X., Lin, Y. H., Qian, X. D., Tang, Y., Zhou, H. H., et al. (2014). Interaction of nNOS with PSD-95 negatively controls regenerative repair after stroke. *Journal of Neuroscience*, *34*(40), 13535–13548.

Lüscher, C., Xia, H., Beattie, E. C., Carroll, R. C., von Zastrow, M., Malenka, R. C., et al. (1999). Role of AMPA receptor cycling in synaptic transmission and plasticity. *Neuron, 24*(3), 649–658.

Luthi-Carter, R., Apostol, B. L., Dunah, A. W., DeJohn, M. M., Farrell, L. A., Bates, G. P., et al. (2003). Complex alteration of NMDA receptors in transgenic Huntington's disease mouse brain: Analysis of mRNA and protein expression, plasma membrane association, interacting proteins, and phosphorylation. *Neurobiology of Disease, 14*(3), 624–636.

Lynch, G., & Gall, C. M. (2006). Ampakines and the threefold path to cognitive enhancement. *Trends in Neurosciences, 29*(10), 554–562.

Lynch, G., Rex, C. S., Chen, L. Y., & Gall, C. M. (2008). The substrates of memory: Defects, treatments, and enhancement. *European Journal of Pharmacology, 585*(1), 2–13.

Ma, J. Z., Johnson, B. A., Yu, E., Weiss, D., McSherry, F., Saadvandi, J., et al. (2013). Fine-grain analysis of the treatment effect of topiramate on methamphetamine addiction with latent variable analysis. *Drug and Alcohol Dependence, 130*(1-3), 45–51.

Ma, D. Q., Whitehead, P. L., Menold, M. M., Martin, E. R., Ashley-Koch, A. E., Mei, H., et al. (2005). Identification of significant association and gene-gene interaction of GABA receptor subunit genes in autism. *American Journal of Human Genetics, 77*(3), 377–388.

MacGillavry, H. D., Song, Y., Raghavachari, S., & Blanpied, T. A. (2013). Nanoscale scaffolding domains within the postsynaptic density concentrate synaptic AMPA receptors. *Neuron, 78*(4), 615–622.

Mack, V., Burnashev, N., Kaiser, K. M., Rozov, A., Jensen, V., Hvalby, O., et al. (2001). Conditional restoration of hippocampal synaptic potentiation in Glur-A-deficient mice. *Science, 292*(5526), 2501–2504.

Mackowiak, M., O'Neill, M. J., Hicks, C. A., Bleakman, D., & Skolnick, P. (2002). An AMPA receptor potentiator modulates hippocampal expression of BDNF: An in vivo study. *Neuropharmacology, 43*(1), 1–10.

Makino, H., & Malinow, R. (2009). AMPA receptor incorporation into synapses during LTP: The role of lateral movement and exocytosis. *Neuron, 64*(3), 381–390.

Malenka, R. C. (2003). Synaptic plasticity and AMPA receptor trafficking. *Annals of the New York Academy of Sciences, 1003*, 1–11.

Malinow, R., & Malenka, R. C. (2002). AMPA receptor trafficking and synaptic plasticity. *Annual Review of Neuroscience, 25*, 103–126.

Mameli, M., Halbout, B., Creton, C., Engblom, D., Parkitna, J. R., Spanagel, R., et al. (2009). Cocaine-evoked synaptic plasticity: Persistence in the VTA triggers adaptations in the NAc. *Nature Neuroscience, 12*(8), 1036–1041.

Man, H. Y., Lin, J. W., Ju, W. H., Ahmadian, G., Liu, L., Becker, L. E., et al. (2000). Regulation of AMPA receptor-mediated synaptic transmission by clathrin-dependent receptor internalization. *Neuron, 25*(3), 649–662.

Man, H., Sekine-aizawa, Y., & Huganir, R. L. (2007). Regulation of alpha-amino-3-hydroxy-5-methyl-4-isoxazolepropionic acid receptor trafficking through PKA phosphorylation of the Glu receptor 1 subunit. *Proceedings of the National Academy of Sciences of the United States of America, 104*(9), 3579–3584.

Man, H. Y., Wang, Q., Lu, W. Y., Ju, W., Ahmadian, G., Liu, L., et al. (2003). Activation of PI3-kinase is required for AMPA receptor insertion during LTP of mEPSCs in cultured hippocampal neurons. *Neuron, 38*(4), 611–624.

Mandal, M., Wei, J., Zhong, P., Cheng, J., Duffney, L. J., et al. (2011). Impaired alpha-amino-3-hydroxy-5-methyl-4-isoxazolepropionic acid (AMPA) receptor trafficking and function by mutant huntingtin. *Journal of Biological Chemistry, 286*(39), 33719–33728.

Mansour, M., Nagarajan, N., Nehring, R. B., Clements, J. D., & Rosenmund, C. (2001). Heteromeric AMPA receptors assemble with a preferred subunit stoichiometry and spatial arrangement. *Neuron, 32*(5), 841–853.

Marcello, E., Epis, R., Saraceno, C., Gardoni, F., Borroni, B., Cattabeni, F., et al. (2010). SAP97-mediated local trafficking is altered in Alzheimer disease patients' hippocampus. *Neurobiology of Aging*, *33*, 1–10.

Marcello, E., Gardoni, F., Di Luca, M., & Pérez-Otaño, I. (2010). An arginine stretch limits ADAM10 exit from the endoplasmic reticulum. *Journal of Biological Chemistry*, *285*(14), 10376–10384.

Marcello, E., Gardoni, F., Mauceri, D., Romorini, S., Jeromin, A., Epis, R., et al. (2007). Synapse-associated protein-97 mediates alpha-secretase ADAM10 trafficking and promotes its activity. *Journal of Neuroscience*, *27*(7), 1682–1691.

Marcello, E., Saraceno, C., Musardo, S., Vara, H., De La Fuente, A. G., Pelucchi, S., et al. (2013). Endocytosis of synaptic ADAM10 in neuronal plasticity and Alzheimer's disease. *Journal of Clinical Investigation*, *123*, 2523–2538.

Martin, J. B., & Gusella, J. F. (1986). Huntington's disease. Pathogenesis and management. *New England Journal of Medicine*, *315*(20), 1267–1276.

Masugi-Tokita, M., Tarusawa, E., Watanabe, M., Molnár, E., Fujimoto, K., & Shigemoto, R. (2007). Number and density of AMPA receptors in individual synapses in the rat cerebellum as revealed by SDS-digested freeze-fracture replica labeling. *Journal of Neuroscience*, *27*(8), 2135–2144.

Matsuda, S., Mikawa, S., & Hirai, H. (1999). Phosphorylation of serine-880 in GluR2 by protein kinase C prevents its C terminus from binding with glutamate receptor-interacting protein. *Journal of Neurochemistry*, *73*(4), 1765–1768.

Matsuzaki, M., Ellis-Davies, G. C., Nemoto, T., Miyashita, Y., Iino, M., & Kasai, H. (2001). Dendritic spine geometry is critical for AMPA receptor expression in hippocampal CA1 pyramidal neurons. *Nature Neuroscience*, *4*(11), 1086–1092.

McCormick, D. A., & Contreras, D. (2001). On the cellular and network bases of epileptic seizures. *Annual Review of Physiology*, *63*, 815–846.

McGlade-McCulloh, E., Yamamoto, H., Tan, S.-E., Brickey, D. A., & Soderling, T. R. (1993). Phosphorylation and regulation of glutamate receptors by calcium/calmodulin-dependent protein kinase II. *Nature*, *362*(6421), 640–642.

McLaughlin, M., Hale, R., Ellston, D., Gaudet, S., Lue, R. A., & Viel, A. (2002). The distribution and function of alternatively spliced insertions in hDlg. *Journal of Biological Chemistry*, *277*(8), 6406–6412.

Mejias, R., Adamczyk, A., Anggono, V., Niranjan, T., Thomas, G. M., et al. (2011). Gain-of-function glutamate receptor interacting protein 1 variants alter GluA2 recycling and surface distribution in patients with autism. *Proceedings of the National Academy of Sciences of the United States of America*, *108*(12), 4920–4925.

Mellone, M., & Gardoni, F. (2013). Modulation of NMDA receptor at the synapse: Promising therapeutic interventions in disorders of the nervous system. *European Journal of Pharmacology*, *719*(1-3), 75–83.

Menuz, K., & Nicoll, R. A. (2008). Loss of inhibitory neuron AMPA receptors contributes to ataxia and epilepsy in stargazer mice. *Journal of Neuroscience*, *28*, 10599–10603.

Meyer, G., Varoqueaux, F., Neeb, A., Oschlies, M., & Brose, N. (2004). The complexity of PDZ domain-mediated interactions at glutamatergic synapses: A case study on neuroligin. *Neuropharmacology*, *47*(5), 724–733.

Migaud, M., Charlesworth, P., Dempster, M., Webster, L. C., Watabe, A. M., et al. (1998). Enhanced long-term potentiation and impaired learning in mice with mutant postsynaptic density-95 protein. *Nature*, *396*(6710), 433–439.

Miller, Y., Ma, B., & Nussinov, R. (2010). Zinc ions promote Alzheimer Abeta aggregation via population shift of polymorphic states. *Proceedings of the National Academy of Sciences of the United States of America*, *107*(21), 9490–9495.

Milnerwood, A. J., Gladding, C. M., Pouladi, M. A., Kaufman, A. M., Hines, R. M., Boyd, J. D., et al. (2010). Early increase in extrasynaptic NMDA receptor signaling and expression contributes to phenotype onset in Huntington's disease mice. *Neuron*, *65*(2), 178–190.

Minnerup, J., Sutherland, B. A., Buchan, A. M., & Kleinschnitz, C. (2012). Neuroprotection for stroke: Current status and future perspectives. *International Journal of Molecular Sciences*, *13*(9), 11753–11772.

Miu, P., Jarvie, K. R., Radhakrishnan, V., Gates, M. R., Ogden, A., et al. (2001). Novel AMPA receptor potentiators LY392098 and LY404187: Effects on recombinant human AMPA receptors in vitro. *Neuropharmacology*, *40*(8), 976–983.

Montgomery, K. E., Kessler, M., & Arai, A. C. (2009). Modulation of agonist binding to AMPA receptors by 1-(1,4-benzodioxan-6-ylcarbonyl)piperidine (CX546): Differential effects across brain regions and GluA1-4/transmembrane AMPA receptor regulatory protein combinations. *Journal of Pharmacology and Experimental Therapeutics*, *331*(3), 965–974.

Montgomery, J. M., Pavlidis, P., & Madison, D. V. (2001). Pair recordings reveal all-silent synaptic connections and the postsynaptic expression of long-term potentiation. *Neuron*, *29*(3), 691–701.

Montgomery, J. M., Zamorano, P. L., & Garner, C. C. (2004). MAGUKs in synapse assembly and function: An emerging view. *Cellular and Molecular Life Sciences CMLS*, *61*(7-8), 911–929.

Mori, K., Iwao, K., Miyoshi, Y., Nakagawara, A., Kofu, K., et al. (1998). Identification of brain-specific splicing variants of the hDLG1 gene and altered splicing in neuroblastoma cell lines. *Journal of Human Genetics*, *43*(2), 123–127.

Mulkey, R. M., Endo, S., Shenolikar, S., & Malenka, R. C. (1994). Involvement of a calcineurin/inhibitor-1 phosphatase cascade in hippocampal long-term depression. *Nature*, *369*(6480), 486–488.

Mulkey, R. M., Herron, C. E., & Malenka, R. C. (1993). An essential role for protein phosphatases in hippocampal long-term depression. *Science*, *261*(5124), 1051–1055.

Mulkey, R. M., & Malenka, R. C. (1992). Mechanisms underlying induction of homosynaptic long-term depression in area CA1 of the hippocampus. *Neuron*, *9*(5), 967–975.

Müller, B., Kistner, U., Veh, R. W., Cases-Langhoff, C., Becker, B., Gundelfinger, E. D., et al. (1995). Molecular characterization and spatial distribution of SAP97, a novel presynaptic protein homologous to SAP90 and the Drosophila discs-large tumor suppressor protein. *Journal of Neuroscience*, *15*(3 Pt. 2), 2354–2366.

Murphy, K. P. S. J., Carter, R. J., Lione, L. A., Mangiarini, L., Mahal, A., Bates, G. P., et al. (2000). Abnormal synaptic plasticity and impaired spatial cognition in mice transgenic for exon 1 of the human Huntington's disease mutation. *The Journal of Neuroscience*, *20*(13), 5115–5123.

Murray, T. K., Whalley, K., Robinson, C. S., Ward, M. A., Hicks, C. A., et al. (2003). LY503430, a novel alpha-amino-3-hydroxy-5-methylisoxazole-4-propionic acid receptor potentiator with functional, neuroprotective and neurotrophic effects in rodent models of Parkinson's disease. *The Journal of Pharmacology and Experimental Therapeutics*, *306*(2), 752–762.

Musardo, S., Marcello, E., Gardoni, F., & Di Luca, M. (2014). ADAM10 in synaptic physiology and pathology. *Neurodegenerative Diseases*, *13*, 72–74.

Na, E. S., Morris, M. J., Nelson, E. D., & Monteggia, L. M. (2014). GABAA receptor antagonism ameliorates behavioral and synaptic impairments associated with MeCP2 overexpression. *Neuropsychopharmacology*, *39*(8), 1946–1954.

Nair, D., Hosy, E., Petersen, J. D., Constals, A., Giannone, G., Choquet, D., et al. (2013). Super-resolution imaging reveals that AMPA receptors inside synapses are dynamically organized in nanodomains regulated by PSD95. *Journal of Neuroscience*, *33*(32), 13204–13224.

Naisbitt, S., Kim, E., Tu, J. C., Xiao, B., Sala, C., Valtschanoff, J. G., et al. (1999). Shank, a novel family of postsynaptic density proteins that binds to the NMDA receptor/PSD-95/GKAP complex and cortactin. *Neuron, 23*, 569–582.

Nakamoto, M., Nalavadi, V., Epstein, M. P., Narayanan, U., Bassell, G. J., & Warren, S. T. (2007). Fragile X mental retardation protein deficiency leads to excessive mGluR5-dependent internalization of AMPA receptors. *Proceedings of the National Academy of Sciences of the United States of America, 104*(39), 15537–15542.

Nakanishi, N., Shneider, N. A., & Axel, R. (1990). A family of glutamate receptor genes: Evidence for the formation of heteromultimeric receptors with distinct channel properties. *Neuron, 5*(5), 569–581.

Namba, T., Morimoto, K., Sato, K., Yamada, N., & Kuroda, S. (1994). Antiepileptogenic and anticonvulsant effects of NBQX, a selective AMPA receptor antagonist, in the rat kindling model of epilepsy. *Brain Research, 1994*(638), 36–44.

Nash, J. E., Johnston, T. H., Collingridge, G. L., Garner, C. C., & Brotchie, J. M. (2005). Subcellular redistribution of the synapse-associated proteins PSD-95 and SAP97 in animal models of Parkinson's disease and L-DOPA-induced dyskinesia. *The Federation of American Societies for Experimental Biology, 19*, 583–585.

Nehring, R. B., Wischmeyer, E., Doring, F., Veh, R. W., Sheng, M., et al. (2000). Neuronal inwardly rectifying K(+) channels differentially couple to PDZ proteins of the PSD-95/SAP90 family. *The Journal of Neuroscience, 20*, 156–162.

Neuhofer, D., Henstridge, C. M., Dudok, B., Sepers, M., Lassalle, O., et al. (2015). Functional and structural deficits at accumbens synapses in a mouse model of Fragile X. *Frontiers in Cellular Neuroscience, 9*, 100.

Newpher, T. M., & Ehlers, M. D. (2008). Glutamate receptor dynamics in dendritic microdomains. *Neuron, 58*, 472–497.

Niethammer, M., Kim, E., & Sheng, M. (1996). Interaction between the C terminus of NMDA receptor subunits and multiple members of the PSD-95 family of membrane-associated guanylate kinases. *The Journal of Neuroscience, 16*(7), 2157–2163.

Nishijima, H., Suzuki, S., Kon, T., Funamizu, Y., Ueno, T., et al. (2014). Morphologic changes of dendritic spines of striatal neurons in the levodopa-induced dyskinesia model. *Movement Disorders, 29*(3), 336–343.

Nishimune, A., Isaac, J. T., Molnar, E., Noel, J., Nash, S. R., Tagaya, M., et al. (1998). NSF binding to GluR2 regulates synaptic transmission. *Neuron, 21*(1), 87–97.

Nusser, Z., Lujan, R., Laube, G., Roberts, J. D. B., Molnár, E., & Somogyi, P. (1998). Cell type and pathway dependence of synaptic AMPA receptor number and variability in the hippocampus. *Neuron, 21*, 545–559.

Oh, M. C., Derkach, V. A., Guire, E. S., & Soderling, T. R. (2006). Extrasynaptic membrane trafficking regulated by GluR1 serine 845 phosphorylation primes AMPA receptors for long-term potentiation. *The Journal of Biological Chemistry, 281*(2), 752–758.

Oliet, S. H., Malenka, R. C., & Nicoll, R. A. (1997). Two distinct forms of long-term depression coexist in CA1 hippocampal pyramidal cells. *Neuron, 18*(6), 969–982.

O'Neill, M. J., Murray, T. K., Clay, M. P., Lindstrom, T., Yang, C. R., et al. (2005). LY503430: Pharmacology, pharmacokinetics, and effects in rodent models of Parkinson's disease. *CNS Drug Reviews, 11*(1), 77–96.

Opazo, P., & Choquet, D. (2011). A three-step model for the synaptic recruitment of AMPA receptors. *Molecular and Cellular Neuroscience, 46*(1), 1–8.

Opazo, P., Labrecque, S., Tigaret, C. M., Frouin, A., Wiseman, P. W., De Koninck, P., et al. (2010). CaMKII Triggers the Diffusional Trapping of Surface AMPARs through Phosphorylation of Stargazin. *Neuron, 67*(2), 239–252.

Ornstein, P. L., Zimmerman, D. M., Arnold, M. B., Bleisch, T. J., Cantrell, B., et al. (2000). Biarylpropylsulfonamides as novel, potent potentiators of 2-amino-3- (5-

methyl-3-hydroxyisoxazol-4-yl)- propanoic acid (AMPA) receptors. *Journal of Medicinal Chemistry, 43*(23), 4354–4358.
Osten, P., Khatri, L., Perez, J. L., Köhr, G., Giese, G., Daly, C., et al. (2000). Mutagenesis reveals a role for ABP/GRIP binding to GluR2 in synaptic surface accumulation of the AMPA receptor. *Neuron, 27*(2), 313–325.
Osten, P., Srivastava, S., Inman, G. J., Vilim, F. S., Khatri, L., Lee, L. M., et al. (1998). The AMPA receptor GluR2 C terminus can mediate a reversible, ATP-dependent interaction with NSF and alpha- and beta-SNAPs. *Neuron, 21*(1), 99–110.
Osterweil, E. K., Chuang, S. C., Chubykin, A. A., Sidorov, M., Bianchi, R., et al. (2013). Lovastatin corrects excess protein synthesis and prevents epileptogenesis in a mouse model of fragile X syndrome. *Neuron, 77*(2), 243–250.
Pagnamenta, A. T., Khan, H., Walker, S., Gerrelli, D., Wing, K., et al. (2011). Rare familial 16q21 microdeletions under a linkage peak implicate cadherin 8 (CDH8) in susceptibility to autism and learning disability. *Journal of Medical Genetics, 48*(1), 48–54.
Palmer, C. L., Lim, W., Hastie, P. G., Toward, M., Korolchuk, V. I., Burbidge, S. A., et al. (2005). Hippocalcin functions as a calcium sensor in hippocampal LTD. *Neuron, 47*(4), 487–494.
Park, M., Penick, E. C., Edwards, J. G., Kauer, J. A., & Ehlers, M. D. (2004). Recycling endosomes supply AMPA receptors for LTP. *Science, 305*(5692), 1972–1975.
Park, M., Salgado, J. M., Ostroff, L., Helton, T. D., Robinson, C. G., Harris, K. M., et al. (2006). Plasticity-induced growth of dendritic spines by exocytic trafficking from recycling endosomes. *Neuron, 52*(5), 817–830.
Park, P., Volianskis, A., Sanderson, T. M., Bortolotto, Z. A., Jane, D. E., et al. (2013). NMDA receptor-dependent long-term potentiation comprises a family of temporally overlapping forms of synaptic plasticity that are induced by different patterns of stimulation. *Philosophical Transactions of the Royal Society of London. Series B, Biological Sciences, 369*(1633), 20130131.
Parsons, M. P., & Raymond, L. A. (2014). Extrasynaptic NMDA receptor involvement in central nervous system disorders. *Neuron, 82*(2), 279–293.
Passafaro, M., Piech, V., & Sheng, M. (2001). Subunit-specific temporal and spatial patterns of AMPA receptor exocytosis in hippocampal neurons. *Nature Neuroscience, 4*(9), 917–926.
Patterson, M. A., Szatmari, E. M., & Yasuda, R. (2010). AMPA receptors are exocytosed in stimulated spines and adjacent dendrites in a Ras-ERK-dependent manner during long-term potentiation. *Proceedings of the National Academy of Sciences, 107*(36), 15951–15956.
Paz, J. T., Bryant, A. S., Peng, K., Fenno, L., Yizhar, O., Frankel, W. N., et al. (2011). A new mode of corticothalamic transmission revealed in the Gria4(-/-). Model of absence epilepsy. *Nature Neuroscience, 14*, 1167–1173.
Peca, J., Feliciano, C., Ting, J. T., Wang, W., Wells, M. F., Venkatraman, T. N., et al. (2011). Shank3 mutant mice display autistic-like behaviours and striatal dysfunction. *Nature, 472*(7344), 437–442.
Peineau, S., Nicolas, C. S., Bortolotto, Z. A., Bhat, R. V., Ryves, W. J., et al. (2009). A systematic investigation of the protein kinases involved in NMDA receptor-dependent LTD: Evidence for a role of GSK-3 but not other serine/threonine kinases. *Molecular Brain, 2*, 22.
Penzes, P., & Vanleeuwen, J. E. (2011). Impaired regulation of synaptic actin cytoskeleton in Alzheimer's disease. *Brain Research Reviews, 67*(1-2), 184–192.
Perez, J. L., Khatri, L., Chang, C., Srivastava, S., Osten, P., & Ziff, E. B. (2001). PICK1 targets activated protein kinase Calpha to AMPA receptor clusters in spines of hippocampal neurons and reduces surface levels of the AMPA-type glutamate receptor subunit 2. *Journal of Neuroscience, 21*(15), 5417–5428.
Petralia, R. S., Wang, Y. X., & Wenthold, R. J. (2003). Internalization at glutamatergic synapses during development. *European Journal of Neuroscience, 18*, 3207–3217.

Petrini, E. M., Lu, J., Cognet, L., Lounis, B., Ehlers, M. D., & Choquet, D. (2009). Endocytic trafficking and recycling maintain a pool of mobile surface AMPA receptors required for synaptic potentiation. *Neuron, 63*(1), 92–105.

Pham, E., Crews, L., Ubhi, K., Hansen, L., Adame, A., et al. (2010). Progressive accumulation of amyloid-beta oligomers in Alzheimer's disease and in amyloid precursor protein transgenic mice is accompanied by selective alterations in synaptic scaffold proteins. *FEBS Journal, 277*(14), 3051–3067.

Picconi, B., Pisani, A., Barone, I., Bonsi, P., Centonze, D., et al. (2005). Pathological synaptic plasticity in the striatum: Implications for Parkinson's disease. *Neurotoxicology, 26*(5), 779–783.

Pickard, L., Noel, J., Duckworth, J. K., Fitzjohn, S. M., Henley, J. M., et al. (2001). Transient synaptic activation of NMDA receptors leads to the insertion of native AMPA receptors at hippocampal neuronal plasma membranes. *Neuropharmacology, 41*(6), 700–713.

Pierce, R. C., & Wolf, M. E. (2013). Psychostimulant-induced neuroadaptations in nucleus accumbens AMPA receptor transmission. *Cold Spring Harbor Perspectives in Medicine, 3*(2), a012021.

Ping, A., Xi, J., Prasad, B. M., Wang, M. H., & Kruzich, P. J. (2008). Contributions of nucleus accumbens core and shell GluR1 containing AMPA receptors in AMPA- and cocaine-primed reinstatement of cocaine-seeking behavior. *Brain Res, 1215*, 173–182.

Poncer, J. C. (2003). Hippocampal long term potentiation: Silent synapses and beyond. *Journal of Physiology, Paris, 97*(4-6), 415–422.

Proctor, D. T., Coulson, E. J., & Dodd, P. R. (2010). Reduction in post-synaptic scaffolding PSD-95 and SAP-102 protein levels in the Alzheimer inferior temporal cortex is correlated with disease pathology. *Journal of Alzheimer's Disease, 21*(3), 795–811.

Purcell, A. E., Jeon, O. H., Zimmerman, A. W., Blue, M. E., & Pevsner, J. (2001). Postmortem brain abnormalities of the glutamate neurotransmitter system in autism. *Neurology, 57*(9), 1618–1628.

Purgianto, A., Scheyer, A. F., Loweth, J. A., Ford, K. A., Tseng, K. Y., & Wolf, M. E. (2013). Different adaptations in AMPA receptor transmission in the nucleus accumbens after short vs long access cocaine self-administration regimens. *Neuropsychopharmacology, 38*(9), 1789–1797.

Quirk, J. C., & Nisenbaum, E. S. (2002). LY404187: A novel positive allosteric modulator of AMPA receptors. *CNS Drug Reviews, 8*(3), 255–282.

Rácz, B., Blanpied, T. A., Ehlers, M. D., & Weinberg, R. J. (2004). Lateral organization of endocytic machinery in dendritic spines. *Nature Neuroscience*, 7(9), 917–918.

Raghavachari, S., & Lisman, J. E. (2004). Properties of quantal transmission at CA1 synapses. *Journal of Neurophysiology, 92*(4), 2456–2467.

Ramanathan, S., Woodroffe, A., Flodman, P. L., Mays, L. Z., Hanouni, M., et al. (2004). A case of autism with an interstitial deletion on 4q leading to hemizygosity for genes encoding for glutamine and glycine neurotransmitter receptor sub-units (AMPA 2, GLRA3, GLRB) and neuropeptide receptors NPY1R, NPY5R. *BMC Medical Genetics, 5*, 10.

Rao, P. S., Bell, R. L., Engleman, E. A., & Sari, Y. (2015). Targeting glutamate uptake to treat alcohol use disorders. *Frontiers in Neuroscience, 9*, 144.

Rex, C. S., Lauterborn, J. C., Lin, C. Y., Kramár, E. A., Rogers, G. A., et al. (2006). Restoration of long-term potentiation in middle-aged hippocampus after induction of brain-derived neurotrophic factor. *Journal of Neurophysiology, 96*(2), 677–685. Epub 2006 May 17.

Robinson, T. E., & Becker, J. B. (1986). Enduring changes in brain and behavior produced by chronic amphetamine administration: A review and evaluation of animal models of amphetamine psychosis. *Brain Research, 396*(2), 157–198.

Rogawski, M. A. (2011). Revisiting AMPA receptors as an antiepileptic drug target. *Epilepsy Currents*, *11*, 56–63.

Rogawski, M. A. (2013). AMPA receptors as a molecular target in epilepsy therapy. *Acta Neurologica Scandinavica. Supplementum*, *197*, 9–18.

Roselli, F., Hutzler, P., Wegerich, Y., Livrea, P., & Almeida, O. F. (2009). Disassembly of shank and homer synaptic clusters is driven by soluble beta-amyloid(1-40) through divergent NMDAR-dependent signalling pathways. *PLoS One*, *4*(6), e6011.

Roussignol, G., Ango, F., Romorini, S., Tu, J. C., Sala, C., et al. (2005). Shank expression is sufficient to induce functional dendritic spine synapses in aspiny neurons. *The Journal of Neuroscience*, *25*, 3560–3570.

Sala, C., Piëch, V., Wilson, N. R., Passafaro, M., Liu, G., & Sheng, M. (2001). Regulation of dendritic spine morphology and synaptic function by Shank and Homer. *Neuron*, *31*(1), 115–130.

Sala, C., Roussignol, G., Meldolesi, J., & Fagni, L. (2005). Key role of the postsynaptic density scaffold proteins shank and homer in the functional architecture of Ca2 + homeostasis at dendritic spines in hippocampal neurons. *The Journal of Neuroscience*, *25*(18), 4587–4592.

Sans, N., Petralia, R. S., Wang, Y.-X., Blahos, J., Hell, J. W., & Wenthold, R. J. (2001). A developmental change in NMDA receptor-associated proteins at hippocampal synapses. *The Journal of Neuroscience*, *20*(3), 1260–1271.

Scheiffele, P., Fan, J., Choih, J., Fetter, R., & Serafini, T. (2000). Neuroligin expressed in nonneuronal cells triggers presynaptic development in contacting axons. *Cell*, *101*(6), 657–669.

Scheyer, A. F., Wolf, M. E., & Tseng, K. Y. (2014). A protein synthesis-dependent mechanism sustains calcium-permeable AMPA receptor transmission in nucleus accumbens synapses during withdrawal from cocaine self-administration. *Journal of Neuroscience*, *34*(8), 3095–3100.

Schlüter, O. M., Xu, W., & Malenka, R. C. (2006). Alternative N-terminal domains of PSD-95 and SAP97 govern activity-dependent regulation of synaptic AMPA receptor function. *Neuron*, *51*(1), 99–111.

Schmeisser, M. J., Ey, E., Wegener, S., Bockmann, J., Stempel, A. V., et al. (2012). Autistic-like behaviours and hyperactivity in mice lacking ProSAP1/Shank2. *Nature*, *486*(7402), 256–260.

Schmidt, H. D., & Pierce, R. C. (2010). Cocaine-induced neuroadaptations in glutamate transmission: Potential therapeutic targets for craving and addiction. *Annals of the New York Academy of Sciences*, *1187*, 35–75.

Schnell, E., Sizemore, M., Karimzadegan, S., Chen, L., Bredt, D. S., & Nicoll, R. A. (2002). Direct interactions between PSD-95 and stargazin control synaptic AMPA receptor number. *Proceedings of the National Academy of Sciences*, *99*(21), 13902–13907.

Selkoe, D. J. (2002). Alzheimer's disease is a synaptic failure. *Science*, *298*(5594), 789–791.

Seo, S., & Leitch, B. (2014). Altered thalamic $GABA_A$-receptor subunit expression in the stargazer mouse model of absence epilepsy. *Epilepsia*, *55*, 224–232.

Seo, S., & Leitch, B. (2015). Synaptic changes in $GABA_A$ receptor expression in the thalamus of the stargazer mouse model of absence epilepsy. *Neuroscience*, *306*, 28–38.

Shao, Y., Cuccaro, M. L., Hauser, E. R., Raiford, K. L., Menold, M. M., et al. (2003). Fine mapping of autistic disorder to chromosome 15q11-q13 by use of phenotypic subtypes. *American Journal of Human Genetics*, *72*(3), 539–548.

Shao, C. Y., Mirra, S. S., Sait, H. B., Sacktor, T. C., & Sigurdsson, E. M. (2011). Postsynaptic degeneration as revealed by PSD-95 reduction occurs after advanced Aβ and tau pathology in transgenic mouse models of Alzheimer's disease. *Acta Neuropathologica*, *122*(3), 285–292.

Shen, L., Liang, F., Walensky, L. D., & Huganir, R. L. (2000). Regulation of AMPA receptor GluR1 subunit surface expression by a 4. 1N-linked actin cytoskeletal association. *Journal of Neuroscience*, *20*(21), 7932–7940.

Sheng, M., & Hoogenraad, C. C. (2007). The postsynaptic architecture of excitatory synapses: A more quantitative view. *Annual Review of Biochemistry*, *76*, 823–847.

Sheng, M., & Kim, E. (2000). The Shank family of scaffold proteins. *Journal of Cell Science*, *113*(Pt. 11), 1851–1856.

Shepherd, J. D., & Huganir, R. L. (2007). The cell biology of synaptic plasticity: AMPA receptor trafficking. *Annual Review of Cell and Developmental Biology*, *23*, 613–643.

Shi, S., Hayashi, Y., Esteban, J. A., & Malinow, R. (2001). Subunit-specific rules governing AMPA receptor trafficking to synapses in hippocampal pyramidal neurons. *Cell*, *105*(3), 331–343.

Shi, S. H., Hayashi, Y., Petralia, R. S., Zaman, S. H., Wenthold, R. J., Svoboda, K., et al. (1999). Rapid spine delivery and redistribution of AMPA receptors after synaptic NMDA receptor activation. *Science*, *284*(5421), 1811–1816.

Shinn, A. K., & Greenfield, S. F. (2010). Topiramate in the treatment of substance-related disorders: A critical review of the literature. *Journal of Clinical Psychiatry*, *71*(5), 634–648.

Silverdale, M. A., Kobylecki, C., Hallett, P. J., Li, Q., Dunah, A. W., et al. (2010). Synaptic recruitment of AMPA glutamate receptor subunits in levodopa-induced dyskinesia in the MPTP-lesioned nonhuman primate. *Synapse*, *64*(2), 177–180.

Silverman, J. L., Oliver, C. F., Karras, M. N., Gastrell, P. T., & Crawley, J. N. (2013). AMPAKINE enhancement of social interaction in the BTBR mouse model of autism. *Neuropharmacology*, *64*, 268–282.

Simmons, D. A., Mehta, R. A., Lauterborn, J. C., Gall, C. M., & Lynch, G. (2011). Brief ampakine treatments slow the progression of Huntington's disease phenotypes in R6/2 mice. *Neurobiology of Disease*, *41*(2), 436–444.

Simmons, D. A., Rex, C. S., Palmer, L., Pandyarajan, V., Fedulov, V., et al. (2009). Up-regulating BDNF with an ampakine rescues synaptic plasticity and memory in Huntington's disease knockin mice. *Proceedings of the National Academy of Sciences of the United States of America*, *106*(12), 4906–4911.

Sommer, B., Kohler, M., Sprengel, R., & Seeburg, P. H. (1991). RNA editing in brain controls a determinant of ion flow in glutamate-gated channels. *Cell*, *67*(1), 11–19.

Srivastava, S., Osten, P., Vilim, F. S., Khatri, L., Inman, G., States, B., et al. (1998). Novel anchorage of GluR2/3 to the postsynaptic density by the AMPA receptor–binding protein ABP. *Neuron*, *21*(3), 581–591.

Staubli, U., Rogers, G., & Lynch, G. (1994). Facilitation of glutamate receptors enhances memory. *Proceedings of the National Academy of Sciences of the United States of America*, *91*(2), 777–781.

Staudinger, J., Lu, J., & Olson, E. N. (1997). Specific interaction of the PDZ domain protein PICK1 with the COOH terminus of protein kinase C-alpha. *Journal of Biological Chemistry*, *272*(51), 32019–32024.

Steinhoff, B. J. (2015). The AMPA receptor antagonist perampanel in the adjunctive treatment of partial-onset seizures, clinical trial evidence and experience. *Therapeutic Advances in Neurological Disorders*, *8*, 137–147.

Suh, S. W., Jensen, K. B., Jensen, M. S., Silva, D. S., Kesslak, P. J., et al. (2000). Histochemically-reactive zinc in amyloid plaques, angiopathy, and degenerating neurons of Alzheimer's diseased brains. *Brain Research*, *852*(2), 274–278.

Sun, Y., Savanenin, A., Reddy, P. H., & Liu, Y. F. (2001). Polyglutamine-expanded Huntingtin promotes sensitization of N-methyl-d-aspartate receptors via post-synaptic density 95—JBC. *The Journal of Biological Chemistry*, *276*, 24713–24718.

Sun, Q., & Turrigiano, G. G. (2011). PSD-95 and PSD-93 play critical but distinct roles in synaptic scaling up and down. *Journal of Neuroscience*, *31*(18), 6800–6808.

Suvrathan, A., Hoeffer, C. A., Wong, H., Klann, E., & Chattarji, S. (2010). Characterization and reversal of synaptic defects in the amygdala in a mouse model of fragile X syndrome. *Proceedings of the National Academy of Sciences of the United States of America*, *107*(25), 11591–11596.

Swanson, G. T., Kamboj, S. K., & Cull-Candy, S. G. (1997). Single-channel properties of recombinant AMPA receptors depend on RNA editing, splice variation, and subunit composition. *Journal of Neuroscience*, *17*(1), 58–69.

Szepesi, Z., Hosy, E., Ruszczycki, B., Bijata, M., Pyskaty, M., Bikbaev, A., et al. (2014). Synaptically released matrix metalloproteinase activity in control of structural plasticity and the cell surface distribution of GluA1-AMPA receptors. *PLoS One*, *9*(5), e98274.

Tallaksen-Greene, S. J., & Albin, R. L. (1994). Localization of ampa-selective excitatory amino acid receptor subunits in identified populations of striatal neurons. *Neuroscience*, *61*(3), 509–519.

Tanaka, H., & Hirano, T. (2012). Visualization of subunit-specific delivery of glutamate receptors to postsynaptic membrane during hippocampal long-term potentiation. *Cell Reports*, *1*(4), 291–298.

Tanaka, J., Matsuzaki, M., Tarusawa, E., Momiyama, A., Molnár, E., Kasai, H., et al. (2005). Number and density of AMPA receptors in single synapses in immature cerebellum. *Journal of Neuroscience*, *25*, 799–807.

Tao-Cheng, J. H., Crocker, V. T., Winters, C. A., Azzam, R., Chludzinski, J., & Reese, T. S. (2011). Trafficking of AMPA receptors at plasma membranes of hippocampal neurons. *Journal of Neuroscience*, *31*(13), 4834–4843.

Tao-Cheng, J. H., Dosemeci, A., Gallant, P. E., Smith, C., & Reese, T. (2010). Activity induced changes in the distribution of Shanks at hippocampal synapses. *Neuroscience*, *168*(1), 11–17.

Tardin, C., Cognet, L., Bats, C., Lounis, B., & Choquet, D. (2003). Direct imaging of lateral movements of AMPA receptors inside synapses. *EMBO Journal*, *22*(18), 4656–4665.

Taylor, C. P., & Vartanian, M. G. (1992). Probenecid pretreatment enhances anticonvulsant action of NBQX in mice. *European Journal of Pharmacology*, *213*, 151–153.

Terry, R. D., Masliah, E., Salmon, D. P., Butters, N., DeTeresa, R., et al. (1991). Physical basis of cognitive alterations in Alzheimer's disease: Synapse loss is the major carrelate of cognitive impairment. *Annals of Neurology*, *30*, 572–580.

Thomas, P., Valton, L., & Genton, P. (2006). Absence and myoclonic status epilepticus precipitated by antiepileptic drugs in idiopathic generalized epilepsy. *Brain*, *129*, 1281–1292.

Tomita, S., Adesnik, H., Sekiguchi, M., Zhang, W., Wada, K., et al. (2005). Stargazin modulates AMPA receptor gating and trafficking by distinct domains. *Nature*, *435*(7045), 1052–1058.

Tomita, S., Chen, L., Kawasaki, Y., Petralia, R. S., Wenthold, R. J., Nicoll, R. A., et al. (2003). Functional studies and distribution define a family of transmembrane AMPA receptor regulatory proteins. *Journal of Cell Biology*, *161*(4), 805–816.

Tomita, S., Fukata, M., Nicoll, R. A., & Bredt, D. S. (2004). Dynamic interaction of stargazin-like TARPs with cycling AMPA receptors at synapses. *Science*, *303*, 1508–1512.

Toro, R., Konyukh, M., Delorme, R., Leblond, C., Chaste, P., et al. (2010). Key role for gene dosage and synaptic homeostasis in autism spectrum disorders. *Trends in Genetics*, *26*(8), 363–372.

Torres-Peraza, J. F., Giralt, A., Garcia-Martinez, J. M., Pedrosa, E., Canals, J. M., & Alberch, J. (2008). Disruption of striatal glutamatergic transmission induced by mutant huntingtin involves remodeling of both postsynaptic density and NMDA receptor signaling. *Neurobiology of Disease, 29*(3), 409–421.

Traynelis, S. F., Wollmuth, L. P., McBain, C. J., Menniti, F. S., Vance, K. M., Ogden, K. K., et al. (2010). Glutamate receptor ion channels: Structure, regulation, and function. *Pharmacological Reviews, 62*(3), 405–496.

Trzepacz, P. T., Cummings, J., Konechnik, T., Forrester, T. D., Chang, C., et al. (2013). Mibampator (LY451395) randomized clinical trial for agitation/aggression in Alzheimer's disease. *International Psychogeriatrics, 25*(5), 707–719.

Uchino, S., Wada, H., Honda, S., Nakamura, Y., Ondo, Y., et al. (2006). Direct interaction of post-synaptic density-95/Dlg/ZO-1 domain-containing synaptic molecule Shank3 with GluR1 alpha-amino-3-hydroxy-5-methyl-4-isoxazole propionic acid receptor. *Journal of Neurochemistry, 97*(4), 1203–1214.

Usdin, M. T., Shelbourne, P. F., Myers, R. M., & Madison, D. V. (1999). Impaired synaptic plasticity in mice carrying the Huntington's disease mutation. *Human Molecular Genetics, 8*(5), 839–846.

Valtschanoff, J. G., & Weinberg, R. J. (2001). Laminar organization of the NMDA receptor complex within the postsynaptic density. *Journal of Neuroscience, 21*(4), 1211–1217.

Vandergriff, J., Huff, K., Bond, A., & Lodge, D. (2001). Potentiation of responses to AMPA on central neurones by LY392098 and LY404187 in vivo. *Neuropharmacology, 40*(8), 1003–1009.

Vogler, E. C., & Busciglio, J. (2014). Disruption of zinc neuromodulation by Aß oligomers: Therapeutic implications. *Current Pharmaceutical Design, 20*(15), 2520–2524.

Von Ossowski, L., Tossavainen, H., Von Ossowski, I., Cai, C., Aitio, O., Fredriksson, K., et al. (2006). Peptide binding and NMR analysis of the interaction between SAP97 PDZ2 and GluR-A: Potential involvement of a disulfide bond. *Biochemistry, 45*, 5567–5575.

Voorn, P., Vanderschuren, L. J., Groenewegen, H. J., Robbins, T. W., & Pennartz, C. M. (2004). Putting a spin on the dorsal-ventral divide of the striatum. *Trends in Neurosciences, 27*(8), 468–474.

Waites, C. L., Specht, C. G., Hartel, K., Leal-Ortiz, S., Genoux, D., Li, D., et al. (2009). Synaptic SAP97 isoforms regulate AMPA receptor dynamics and access to presynaptic glutamate. *Journal of Neuroscience, 29*(14), 4332–4345.

Wakabayashi, K., Narisawa-Saito, M., Iwakura, Y., Arai, T., Ikeda, K., et al. (1999). Phenotypic down-regulation of glutamate receptor subunit GluR1 in Alzheimer's disease. *Neurobiology of Aging, 20*(3), 287–295.

Wang, Z., Edwards, J. G., Riley, N., William Provance, D., Li, X., Davison, I. G., et al. (2009). Mysoin Vb mobilizes recycling endosomes and AMPA receptors for postsynaptic plasticity. *Cell, 135*(3), 535–548.

Wang, H., Kim, S. S., & Zhuo, M. (2010). Roles of fragile X mental retardation protein in dopaminergic stimulation-induced synapse-associated protein synthesis and subsequent alpha-amino-3-hydroxyl-5-methyl-4-isoxazole-4-propionate (AMPA) receptor internalization. *Journal of Biological Chemistry, 285*(28), 21888–21901.

Wang, X., McCoy, P. A., Rodriguiz, R. M., Pan, Y., Je, H., et al. (2011). Synaptic dysfunction and abnormal behaviors in mice lacking major isoforms of Shank3. *Human Molecular Genetics, 20*(15), 3093–3108.

Ward, S. E., Bax, B. D., & Harries, M. (2010). Challenges for and current status of research into positive modulators of AMPA receptors. *British Journal of Pharmacology, 160*(2), 181–190.

Wasterlain, C. G., Naylor, D. E., Liu, H., Niquet, J., & Baldwin, R. (2013). Trafficking of NMDA receptors during status epilepticus: Therapeutic implications. *Epilepsia, 54*(Suppl. 6), 78–80.

Wenthold, R. J., Petralia, R. S., Blahos, J., II, & Niedzielski, A. S. (1996). Evidence for multiple AMPA receptor complexes in hippocampal CA1/CA2 neurons. *Journal of Neuroscience, 16*(6), 1982–1989.

WHO (2015). World Health Organisation Epilepsy Fact Sheet No 999. http//www.who.int/mediacentre/factsheets/fs999/en/.

Williams, K., Brignell, A., Randall, M., Silove, N., & Hazell, P. (2013). Selective serotonin reuptake inhibitors (SSRIs) for autism spectrum disorders (ASD). *Cochrane Database of Systematic Reviews, 8*, CD004677.

Wolf, M. E., & Ferrario, C. R. (2010). AMPA receptor plasticity in the nucleus accumbens after repeated exposure to cocaine. *Neuroscience and Biobehavioral Reviews, 35*(2), 185–211.

Won, H., Lee, H., Gee, H., Yung, M., Won, K., et al. (2012). Autistic-like social behaviour in Shank2-mutant mice improved by restoring NMDA receptor function. *Nature, 486*(7402), 261–265.

Won, H., Mah, W., & Kim, E. (2013). Autism spectrum disorder causes, mechanisms, and treatments: Focus on neuronal synapses. *Frontiers in Molecular Neuroscience, 5*, 6–19.

Wu, H., Nash, J. E., Zamorano, P., & Garner, C. C. (2002). Interaction of SAP97 with minus-end-directed actin motor myosin VI. Implications for AMPA receptor trafficking. *Journal of Biological Chemistry, 277*(34), 30928–30934.

Wyszynski, M., Valtschanoff, J. G., Naisbitt, S., Dunah, A. W., Kim, E., Standaert, D. G., et al. (1999). Association of AMPA receptors with a subset of glutamate receptor-interacting protein in vivo. *Journal of Neuroscience, 19*(15), 6528–6537.

Xia, J., Chung, H. J., Wihler, C., Huganir, R. L., & Linden, D. J. (2000). Cerebellar long-term depression requires PKC-regulated interactions between GluR2/3 and PDZ domain-containing proteins. *Neuron, 28*(2), 499–510.

Xia, Y. F., Kessler, M., & Arai, A. C. (2005). Positive alpha-amino-3-hydroxy-5-methyl-4-isoxazolepropionic acid (AMPA) receptor modulators have different impact on synaptic transmission in the thalamus and hippocampus. *Journal of Pharmacology and Experimental Therapeutics, 313*(1), 277–285.

Xia, J., Zhang, X., Staudinger, J., & Huganir, R. L. (1999). Clustering of AMPA receptors by the synaptic PDZ domain-containing protein PICK1. *Neuron, 22*(1), 179–187.

Xu, J., Kurup, P., Zhang, Y., Goebel-Goody, S. M., Wu, P. H., et al. (2009). Extrasynaptic NMDA receptors couple preferentially to excitotoxicity via calpain-mediated cleavage of STEP. *Journal of Neuroscience, 29*, 9330–9343.

Yamaguchi, S., Donevan, S. D., & Rogawski, M. A. (1993). Anticonvulsant activity of AMPA/kainate antagonists, comparison of GYKI 52466 and NBOX in maximal electroshock and chemoconvulsant seizure models. *Epilepsy Research, 15*, 179–184.

Yao, I., Hata, Y., Hirao, K., Deguchi, M., Ide, N., Takeuchi, M., et al. (1999). Synamon, a novel neuronal protein interacting with synapse-associated protein 90/postsynaptic density-95-associated protein. *Journal of Biological Chemistry, 274*(39), 27463–27466.

Yoo, J., Bakes, J., Bradley, C., Collingridge, G. L., & Kaang, B. K. (2013). Shank mutant mice as an animal model of autism. *Philosophical Transactions of the Royal Society of London, Series B: Biological Sciences, 369*, 20130143.

Yudowski, G. A. (2007). Real-time imaging of discrete exocytic events mediating surface delivery of AMPA receptors. *Journal of Neuroscience, 27*(41), 11112–11121.

Yuki, D., Sugiura, Y., Zaima, N., Akatsu, H., Takei, S., Yao, I., et al. (2014). DHA-PC and PSD-95 decrease after loss of synaptophysin and before neuronal loss in patients with Alzheimer's disease. *Scientific Reprots, 4*, 7130.

Zamanillo, D., Sprengel, R., Hvalby, O., Jensen, V., Burnashev, N., Rozov, A., et al. (1999). Importance of AMPA receptors for hippocampal synaptic plasticity but not for spatial learning. *Science, 284*(5421), 1805–1811.

Zhao, M. G., Ko, S. W., Wu, L. J., Toyoda, H., Xu, H., et al. (2006). Enhanced presynaptic neurotransmitter release in the anterior cingulate cortex of mice with chronic pain. *Journal of Neuroscience*, *26*(35), 8923–8930.

Zheng, N., Jeyifous, O., Munro, C., Montgomery, J. M., & Green, W. N. (2015). Synaptic activity regulates AMPA receptor trafficking through different recycling pathways. *eLife*, *13*, 4.

Zheng, C. Y., Seabold, G. K., Horak, M., & Petralia, R. S. (2011). MAGUKs, synaptic development, and synaptic plasticity. *The Neuroscientist*, *17*(5), 493–512.

Zhou, L., Li, F., Xu, H. B., Luo, C. X., Wu, H. Y., et al. (2010). Treatment of cerebral ischemia by disrupting ischemia-induced interaction of nNOS with PSD-95. *Nature Medicine*, *16*(12), 1439–1443.

Zhou, H. H., Tang, Y., Zhang, X. Y., Luo, C. X., Gao, L. Y., et al. (2015). Delayed administration of Tat-HA-NR2B9c promotes recovery after stroke in rats. *Stroke*, *46*(5), 1352–1358.

Zhou, W., Zhang, L., Guoxiang, X., Mojsilovic-Petrovic, J., Takamiya, K., Sattler, R., et al. (2008). GluR1 controls dendrite growth through its binding partner, SAP97. *Journal of Neuroscience*, *28*(41), 10220–10233.

Zhou, H. H., Zhang, A. X., Zhang, Y., & Zhu, D. Y. (2012). Cloning, expression, and purification of a recombinant Tat-HA-NR2B9c peptide. *Protein Expression and Purification*, *85*(2), 239–245.

Zitzer, H., Honck, H. H., Bachner, D., Richter, D., & Kreienkamp, H. J. (1999). Somatostatin receptor interacting protein defines a novel family of multidomain proteins present in human and rodent brain. *The Journal of Biological Chemistry*, *274*, 32997–33001.

CHAPTER SEVEN

pH-Sensitive K^+ Currents and Properties of K_2P Channels in Murine Hippocampal Astrocytes

Johannes Weller, Christian Steinhäuser[1], Gerald Seifert
Institute of Cellular Neurosciences, Medical Faculty, University of Bonn, Bonn, Germany
[1]Corresponding author: e-mail address: christian.steinhaeuser@ukb.uni-bonn.de

Contents

Advances in Protein Chemistry and Structural Biology, Volume 103
ISSN 1876-1623
http://dx.doi.org/10.1016/bs.apcsb.2015.10.005

Abstract

Based on their intimate spatial association with synapses and the capillary, astrocytes are critically involved in the control of ion, transmitter, and energy homeostasis as well as regulation of the cerebral blood flow. Under pathophysiological conditions, dysfunctional astrocytes can no longer assure homeostatic control although the underlying mechanisms are poorly understood. Specifically, neurological diseases are often accompanied by acidification of the extracellular space, but the properties of astrocytes in such an acidic environment are still a matter of debate. To meet the homeostatic requirements, astrocytes are equipped with intercellular gap junctions, inwardly rectifying K^+ (Kir) channels, and two-pore domain K^+ (K_2P) channels. One goal of the present study was to overview current knowledge about astrocyte K^+ channel function during acidosis. In addition, we combined functional and molecular analyses to clarify how low pH affects K^+ channel function in astrocytes freshly isolated from the developing mouse hippocampus. Extracellular acidification led to a decrease of K^+ currents in astrocytes, probably due to modulation of Kir4.1 channels. After blocking Kir4.1 channels, low pH enhanced K^+ current amplitudes. This current activation was mimicked by modulators of TREK-1 channels, which belong to the K_2P channels family. We found no evidence for the presence of acid-sensitive ion channels and transient receptor potential vanilloid receptors in hippocampal astrocytes. In conclusion, the assembly of astrocytic K^+ channels allows tolerating short, transient acidification, and glial Kir4.1 and K_2P channels can be considered promising new targets in brain diseases accompanied by pH shifts.

1. INTRODUCTION

Astrocytes play a key role in brain ion homeostasis and neurotransmitter uptake, thereby modulating synaptic and neurovascular signaling. Of particular interest is the alteration of astrocyte function under pathophysiological conditions. Pathological conditions such as tissue inflammation, cerebral ischemia, trauma, and epileptic seizures lead to acidification of the extracellular space via accumulation of lactic acid and H^+, which might render pH regulation of neurons and glial cells difficult (Nedergaard, Goldman, Desai, & Pulsinelli, 1991; reviewed by Siesjo et al., 1993). For example, severe ischemia can reduce the pH of brain parenchyma to 6.0 (Rehncrona, 1985) and under hyperglycemic conditions astrocytic acidosis may reach 1–2 orders of magnitude more acidic values (Kraig & Chesler, 1990). However, astrocytic pathophysiology under acidic conditions remains elusive.

Regulation of intracellular pH is mediated by intracellular pH buffers and the activity of ion transporters and exchangers that may extrude protons. Major pH regulating transport systems expressed in astrocytes are the

Na^+/H^+ exchanger NHE1, the electrogenic Na^+/HCO_3^- cotransporter NBCe1, and probably the HCO_3^-/Cl^- anion exchangers (Chesler, 2005; Steinhäuser, Seifert, & Deitmer, 2013). Under ischemic conditions, the fall of extracellular pH and Na^+ diminish the driving force for Na^+-coupled transporters and exchangers leading to profound intracellular acidification (Chesler, 2005; Rose, Waxman, & Ransom, 1998). In cultured astrocytes, long-lasting exposure to low pH entails irreversible decrease of intracellular pH and cell death (Nedergaard et al., 1991). Extracellular acidification is also accompanied by an almost complete block of gap junctional coupling of astrocytes (Wallraff et al., 2006), rendering K^+ buffering insufficient. Recent cell culture work claimed that astrocytes express proton-activated acid-sensitive ion channels (ASICa) and transient receptor potential vanilloid receptors (TRPV1) (Huang et al., 2010). Astrocytes located in the ventral part of the brainstem respond to a decrease of extracellular pH with Ca^{2+} elevation and vesicular ATP release (Kasymov et al., 2013). It was suggested that the inhibition of pH-sensitive inwardly rectifying K^+ (Kir) channels Kir4.1/Kir5.1 by acidification or CO_2 administration leads to depolarization (Wenker, Kreneisz, Nishiyama, & Mulkey, 2010).

Among the K^+ channels identified in the plasma membrane of astrocytes *in situ* after fresh isolation are Kir4.1 channels and the two-pore domain K^+ (K_2P) channels, Tandem of P-domains in a weakly inward rectifying K^+ channel 1 (TWIK-1), and TWIK-related K^+ channel 1 (TREK-1) (Seifert et al., 2009; Zhou et al., 2009). Notably, K_2P channels confer new properties to astrocytes as they are regulated by neurotransmitters, pH changes, pressure, polyunsaturated fatty acids (PUFAs), or lipids. Recently, K_2P channels were identified as a release pore for glutamate in astrocytes, and they were shown to form heteromeric channels within the K_2P channel family (Hwang et al., 2014; Woo et al., 2012). Kir4.1 K^+ channels are potentially inhibited by acidic intracellular pH (Pessia, Imbrici, D'Adamo, Salvatore, & Tucker, 2001), while TREK-1 channels are inhibited by decrease of extracellular pH (Sandoz, Douguet, Chatelain, Lazdunski, & Lesage, 2009).

In the present work, we wanted to clarify how pH affects ion channel function in the membrane of astrocytes freshly isolated from the hippocampus. Specifically, we asked whether members of the ASIC family, TRPV receptors and K_2P channels are activated through decreasing pH. Freshly isolated astrocytes were employed to improve clamp control, avoid potential culture artifacts, and avoid indirect effects through neighboring cells as inevitable in the slice preparation. Our pharmacological and single-cell transcript analyses hinted at a prominent expression of TREK-1 while we found no evidence for the presence of acid-sensitive ASICs and TRPV channels. We conclude that the composition of K^+ channels in the astrocytic plasma membrane is assembled to tolerate short-time acidification and maintain K^+ membrane currents, keeping the negative resting membrane potential relatively unchanged.

2. MATERIALS AND METHODS

Mice were kept under standard housing conditions and all experiments were carried out in accordance with local, state, and European regulations.

2.1 Acute Slice Preparation of Murine Hippocampi

Transgenic human glial fibrillary acidic protein/enhanced green fluorescent protein (Tg(hGFAP/EGFP)) mice (Nolte et al., 2001) of postnatal day (P) 8 to P12 were anesthetized, sacrificed by decapitation and their brains were dissected, washed, and cut into 300 μm thick coronal slices using a vibratome (VT 1000S; Leica Microsystems). Slice preparation was performed at 6 °C in Ca^{2+}-free HEPES-buffered external solution containing (in mM) 150 NaCl, 5 KCl, 2 $MgSO_4$, 10 HEPES, 10 glucose, 1 sodium pyruvate, pH adjusted to 7.38, bubbled with oxygen. Subsequently, slices were stored for at least 30 min in artificial cerebrospinal fluid (ACSF) at 4 °C, containing (in mM) 126 NaCl, 3 KCl, 2 $CaCl_2$, 2 $MgSO_4$, 1.25 NaH_2PO_4, 37 $NaHCO_3$, 10 glucose, and 40 sucrose. The solution was gassed with carbogen (95% O_2, 5% CO_2, pH 7.4).

2.2 Preparation of Freshly Isolated Cells

Freshly isolated cells were obtained as described previously (Matthias et al., 2003). Briefly, slices obtained as above were incubated in oxygenated ACSF containing papain (24 U/ml) and L-cysteine (0.24 mg/ml) (8–12 min, room temperature, bubbled with carbogen). After washing, the CA1 region of the hippocampus (strata radiatum and lacunosum-moleculare) was dissected under a stereomicroscope (STEMI 2000; Zeiss), and cells were isolated using tungsten needles and Pasteur pipettes.

2.3 HEK-293 Cell Culture and Transfection

HEK-293 cells were maintained in plastic dishes in DMEM with Glutamax and 25 mM glucose (Life Technologies) supplemented with 10% fetal calf serum (PAA Laboratories) and penicillin/streptomycin (PAA Laboratories). For transfection, we incubated 0.4 μg DNA with 2 μl transfection medium (TurboFect; Fermentas) in 100 μl DMEM without supplements in well plates for 20 min, and then HEK cells were transferred on top of the transfection reagent/DNA mixture. Cells were held for 2 days at 37 °C in an incubator (5% CO_2). For transient expression, we used the whole coding sequence of the Kir4.1 channel (kindly provided by L. Wang-Eckardt and V. Gieselmann).

2.4 Electrophysiological Recordings

For recordings *in situ*, slices were transferred to a recording chamber and continuously perfused with carbogenated standard ACSF containing (in mM): 126 NaCl, 3 KCl, 2 $CaCl_2$, 2 $MgSO_4$, 1.25 NaH_2PO_4, 26 $NaHCO_3$, and 10 glucose. Cells were visualized using an upright microscope equipped with an infrared DIC system (Eclipse E600FN, Nikon, Germany) at 60-fold magnification. The infrared image was captured with a CCD camera (COHU; model 4912-5100M; Poway; USA). Whole-cell voltage-clamp experiments were performed on cells identified by endogenous fluorescence. Pipettes fabricated from borosilicate capillaries (Science Products, Germany) had a resistance of 3–6 MΩ when filled with internal solution containing (in mM) 130 KCl, 0.5 $CaCl_2$, 2 $MgCl_2$, 5 BAPTA, 10 HEPES, and 3 Na_2-ATP, pH 7.25. Pipette capacitance was compensated for. Currents were recorded using the patch-clamp technique in the whole-cell configuration with an EPC-7 patch-clamp amplifier (HEKA Elektronik, Lambrecht, Germany), filtered at 3 and 10 kHz, digitized with an ITC 16 (NPI electronic), and sampled at 6 and 30 kHz by an interface connected to a computer system, which also served as a stimulus generator. Recordings were monitored with TIDA software (HEKA). Holding potential for experiments in slices was −70 mV.

Experiments on isolated astrocytes and HEK-293 cells employed a customized concentration clamp device connected to an EPC-7 amplifier and TIDA software (Seifert & Steinhäuser, 1995). Cells were recorded in HEPES-buffered external solution containing (in mM) 150 NaCl, 5 KCl, 2 $CaCl_2$, 2 $MgCl_2$, 10 glucose, 10 HEPES, pH adjusted to 7.38. Visual control was achieved with an inverted microscope (Zeiss Axiovert 135) equipped with conventional bright-field DIC. Holding potential for experiments with isolated cells was −60 mV. All recordings were performed at 23–25 °C. Ion channel modulators and blockers were applied by transferring the isolated cell with a tube electrode to the different solutions. Concentrations of ion channel modulators and blockers were as follows: 100 μM amiloride (AMI), 4 mM 4-aminopyridine (4-AP), 10 μM arachidonic acid (AA), 100 μM $BaCl_2$, 100 μM capsaicin (CAP), 10 μM α-linolenic acid (LA), 10 μM prostaglandin E_2 (PGE_2), 100 μM riluzole (RIL), 10 mM tetraethylammonium chloride (TEA), 0.4 mM trinitrophenol (TNP). All reagents were purchased form Sigma unless stated otherwise. AMI, AA, CAP, LA, PGE_2, RIL, and TNP were suspended in DMSO (final concentration 0.05–0.2%), aliquoted, and added to bath solution immediately before use. For experiments in elevated K^+ (20 mM), HEPES-buffered external solution consisted of (in mM) 121 NaCl, 20 KCl, 2 $CaCl_2$, 2 $MgCl_2$, 10 Glucose, 10 HEPES, 10 TEA, 4 4-AP, 0.1 Ba^{2+}.

2.5 Cell Harvesting and Single-Cell RT-PCR

After recording of astrocytes in the stratum radiatum (sr) of the hippocampal CA1 region *in situ*, the cell content was carefully sucked into the recording pipette under optical control with a CCD camera (COHU) using IR-DIC optics. The pipette content was transferred to a tube filled with 3 μl of DEPC-treated water, frozen in liquid nitrogen, and stored at −20 °C. Single-cell transcript analysis was performed as reported (Matthias et al., 2003; Seifert, Rehn, Weber, & Steinhäuser, 1997). Briefly, reverse transcription (RT, 37 °C, 1 h) was performed by adding first-strand buffer (Life Technologies), dithiothreitol (DTT; 10 mM), dNTPs (4 × 250 μM; Life Technologies), RNasin (20 U; Promega), random hexamer primers (50 μM; Roche), and reverse transcriptase (SuperscriptIII; 100 U; Life Technologies). For detection of TASK1 and TASK3 mRNA, DNaseI digestion was performed prior to the RT reaction (Seifert et al., 2009).

A multiplex two-round single-cell PCR was performed with primers for ASICs, TRPVs, Kir4.1, and subunits of the K_2P channel family (see Table 1). The first PCR was performed after adding PCR buffer, $MgCl_2$ (2.5 mM), primers (200 nM each), and 5 U of Taq polymerase (Life Technologies) to the RT product (final volume, 50 μl). Forty-five cycles were performed (denaturation at 94 °C, 25 s; annealing at 49 °C, 2 min for the first five cycles, and 45 s for the remaining cycles; extension at 72 °C,

Table 1 Primers for Single-Cell RT-PCR

Gene	Sequence	Position	Product Length (bp)	GeneBank Accession No.
ASIC1a/2	se 5′-AACGGGCTGGAGATCATGCTGGAC as 5′-AGGGCTGGGTCTGCACACTCCTT	796, 646 1171, 1021	389	NM_007384, NM_009597
ASIC1a/2 (nested)	se 5′-GTGGCTCCAGGSTTCCAGACSTT as 5′-CCATGCGGCAGTTGCAGTTYTCCA	943, 793 1103, 953	184	
TRPV1	se 5′-TGCCCTATCATCACCGTCAGC as 5′-AGTGCCATGTTCCGCCTTTCAAT	217 628	434	NM_001001445
TRPV1 (nested)	se 5′-CGCCCCCAAGGCTCTATGAT as 5′-GCCTTTCAATGGCAATGTGTA	323 617	316	
TRPV2	se 5′-TTTCCCGAAAGTTCACCGAGTGGT as 5′-ACCCCCAAGCAGGATCAGAATGTG	962 1297	359	NM_011706
TRPV2 (nested)	se 5′-ACCGCATGGTGGTTTTAGAGC as 5′-AAGTCGCTTTTGATGAGGGAATGG	1094 1247	177	
TASK1	se 5′-TTCTACTTCGCCATCACCGTCATC as 5′-AGAAAGTCCAGCGCTCGTAGTAGG	250 536	310	NM_010608

Continued

Table 1 Primers for Single-Cell RT-PCR—cont'd

Gene	Sequence	Position	Product Length (bp)	GeneBank Accession No.
TASK1 (nested)	se 5′-CTTCGCCATCACCGTCATCACCAC as 5′-CGCATGCCCAGCCCCCTCTT	255 449	195	
TASK3	se 5′-AGCCGAAGAAGTCCGTCTCAGAGG as 5′-CAGCCGCCCCAAGGCACAGG	126 510	404	NM_001033876
TASK3 (nested)	se 5′-AAATTCGCCGGTTCCTTCTAC as 5′-GCCACAGCACTTCTTGATACG	235 424	210	
Synaptophysin	se 5′-AGGTGCTGCAGTGGGTCTTT as 5′-TCTGGCGGCACATAGGCATCT	89 539	471	NM_009305
Synaptophysin (nested)	se 5′-CGGCTGAGCGTGGAGTGTGC as 5′-AGGGCCCCCATGGAGTAGAGGAA	154 346	215	
S100 β	se 5′-AGGCCATGGTTGCCCTCATTGAT as 5′-ACTCATGGCAGGCCGTGGTCA	17 242	246	NM_009115
S100β (nested)	se 5′-TACTCCGGGCGAGAGGGTGACAA as 5′-GGCGACGAAGGCCATGAACTCC	52 216	186	

"Se" and "as" mark sense and antisense primers. Position 1 is the first nucleotide of the initiation codon. All sense and antisense primers except those for TASK1 and TASK3 are located on different exons, respectively.

25 s; final elongation at 72 °C, 7 min). An aliquot (2 μl) of the PCR product was used as a template for the second PCR (35 cycles; annealing at 54 °C, 2 min for the first five cycles, and 45 s for the remaining cycles) using nested primers (Table 1). The conditions were the same as described for the first round, but dNTPs (4 × 50 μM) and Platinum Taq polymerase (2.5 U; Invitrogen) were added. Products were identified with gel electrophoresis using a molecular weight marker (low-molecular-weight marker; New England Biolabs). As a positive control, RT-PCR for single cells and for total RNA from mouse brain were run in parallel. Negative controls were performed using distilled water or bath solution for RT-PCR.

2.6 Preparation of Brain Tissue for Semiquantitative RT-PCR

Tissue samples were isolated from 200 μm thick coronal brain slices. By using a stereo microscope, the strata radiatum and lacunosum-moleculare of the CA1 region were isolated with a small scalpel by removing strata pyramidale and oriens. For regional comparison, in some cases, the radiatum of the CA1 region was further separated from the lacunosum-moleculare and underwent sqRT-PCR analysis. At each developmental stage, at least three mice from different litters were used. Total RNA was isolated from tissue samples with Trizol (Invitrogen) and dissolved in 10 μl DEPC-treated water. Genomic DNA was removed with DNase treatment in a mixture containing PCR buffer, 2.5 mM $MgCl_2$, 10 mM DTT (all Life Technologies), 20 U DNaseI (Roche), and 40 U RNase inhibitor (Promega; final volume 20 μl; incubation at 37 °C for 30 min). Subsequently, mRNA was isolated using oligo$(dT)_{25}$-linked Dynabeads (Life Technologies) and the adherent mRNA was suspended in DEPC-treated water (20 μl). RT (37 °C, 1 h) was performed by adding a mastermix containing first-strand buffer (Life Technologies), dithiothreitol (DTT; 10 mM), dNTPs (4 × 250 μM; Life Technologies), RNasin (40 U; Promega), random hexamer primers (50 μM; Roche), and reverse transcriptase (SuperscriptIII; 200 U; Invitrogen, final volume 40 μl). Semiquantitative PCR was performed in a reaction volume of 12.5 μl. The reaction mixture contained TaqMan™ mastermix and the primers/probe mix (Life Technologies). One-microliter cDNA from the RT product was added to each well. PCRs for K_2P channel subunits and β-actin were run in separate wells. Negative controls (water) were also performed in each run. Samples were incubated at 50 °C (2 min), and after denaturation (95 °C, 10 min), 50 cycles were performed (denaturation at 94 °C, 15 s; primer annealing and extension at 60 °C, 60 s). Fluorescence intensity was readout during each annealing/extension step.

2.7 Data Analysis

Membrane capacitance and series resistance were determined offline using IGOR software. Membrane conductance was calculated by dividing membrane current through the corresponding driving force at −130 and +20 mV. The rectification index (RI) was determined by comparing the chord conductances of blocker-sensitive and modulator-evoked currents at −40 and −130 mV (in standard solution):

$$\mathrm{RI} = \frac{I_{(-40\ \mathrm{mV})}}{(-40\ \mathrm{mV} - E_{\mathrm{rev}})} \Big/ \frac{I_{(-130\ \mathrm{mV})}}{(-130\ \mathrm{mV} - E_{\mathrm{rev}})} \quad (1)$$

or at 0 and −90 mV (20 mM K^+ solution):

$$\mathrm{RI} = \frac{I_{(0\ \mathrm{mV})}}{-E_{\mathrm{rev}}} \Big/ \frac{I_{(-90\ \mathrm{mV})}}{(-90\ \mathrm{mV} - E_{\mathrm{rev}})} \quad (2)$$

where I is the amplitude of the respective currents, and E_{rev} its extrapolated reversal potential.

The KCNK/β-actin gene expression ratio was determined by comparing C_{T} values of the target gene with those of the reference gene, β-actin. The relative quantification of different genes was determined according to the following equation:

$$R_n = X \times E^{\mathrm{CT}} \text{ or, in the logarithmic form,} \\ \log Y = \log X + C_{\mathrm{T}} \times \log E \quad (3)$$

with R_n being the fluorescence intensity, X the input copy number, E the efficiency of amplification, and C_{T} the cycle number at threshold. By quantification of KCNK gene expression against that of β-actin, C_{T} was determined for each gene at the same R_n and the difference in input copy number was estimated according to Eq. (4):

$$\frac{X_{\mathrm{KCNK}}}{X_{\beta\text{-actin}}} = \frac{E_{\beta\text{-actin}}{}^{\mathrm{CT}_{\beta\text{-actin}}}}{E_{\mathrm{KCNK}}{}^{\mathrm{CT}_{\mathrm{KCNK}}}} \quad (4)$$

The amplification efficiency was 1.73, 1.70, 1.74, and 1.76 for KCNK1, KCNK2, KCNK10, and β-actin, respectively.

Statistical significance ($p < 0.05$) was calculated using Student's *t*-test for paired data or ANOVA followed by TUKEY's post-hoc test.

3. RESULTS

3.1 Effect of Acidic Extracellular pH on Astrocytic Membrane Currents

Recent work on cultured astrocytes demonstrated large inward currents upon application of bath solution with a pH of 6.0. These currents were thought to be mediated by acid-sensitive channels of the ASIC family and transient receptor potential channel vanilloid type I (TRPV1) (Huang et al., 2010). To test whether low pH inhibits K^+ channels and elicits channel/receptor-mediated inward currents *in situ*, we performed patch-clamp experiments on astrocytes isolated from the hippocampus of Tg(hGFAP/EGFP) mice. Astrocytes were identified through their EGFP fluorescence, morphology, and passive current pattern (Matthias et al., 2003). Upon acidification of bath solution to pH 6.0, we observed a significant reduction of outward (to 85.0 ± 9.2%, $V = +20$ mV) and inward currents (to 84.4 ± 8.1%, $V = -130$ mV) compared to the control obtained at pH 7.4 ($n = 15$). Notably, we did not observe an increase in membrane conductance, which would have indicated expression of acid-sensitive channels or receptors (Fig. 1A). The resting potential (−63.2 ± 7.8 mV; $n = 10$) remained unchanged at pH 6.0 (−62.5 ± 9.3 mV), but the input resistance (46 ± 19 MΩ; pH 7.4) increased (57 ± 26 MΩ; pH 6.0). In the majority of cells, the pH-sensitive currents reversed close to the K^+ equilibrium potential (−69 ± 10 mV, $n = 10$; Fig. 1B) indicating that they were mainly carried by K^+, possibly through Kir4.1 or K_2P channels, which are strongly expressed by astrocytes in the hippocampus (Seifert et al., 2009). A significant reduction of in- and outward conductance was already obvious at pH 7.0 ($n = 4$) and at pH 6.5 ($n = 3$; not shown).

The pH sensitivity of Kir4.1 channels was tested after stable expression in HEK-293 cells. Similar to astrocytes, switching to bath solution with pH 6.0 did not affect the resting potential of Kir4.1-expressing HEK-293 cells (−68.1 ± 5.7 mV, pH 7.4; −69.5 ± 4.4 mV, pH 6.0) but led to an increase in input resistance (by 8.7 ± 5.2%, $n = 5$) due to a reduction of the membrane conductance at low pH (to 92.9 ± 5.3%, $V = -130$ mV and to 92.5 ± 3.6%, $V = +20$ mV; $n = 5$; not shown). Accordingly, the acidification-induced reduction of astrocytic membrane conductance might have been due to modulation of Kir4.1 channels.

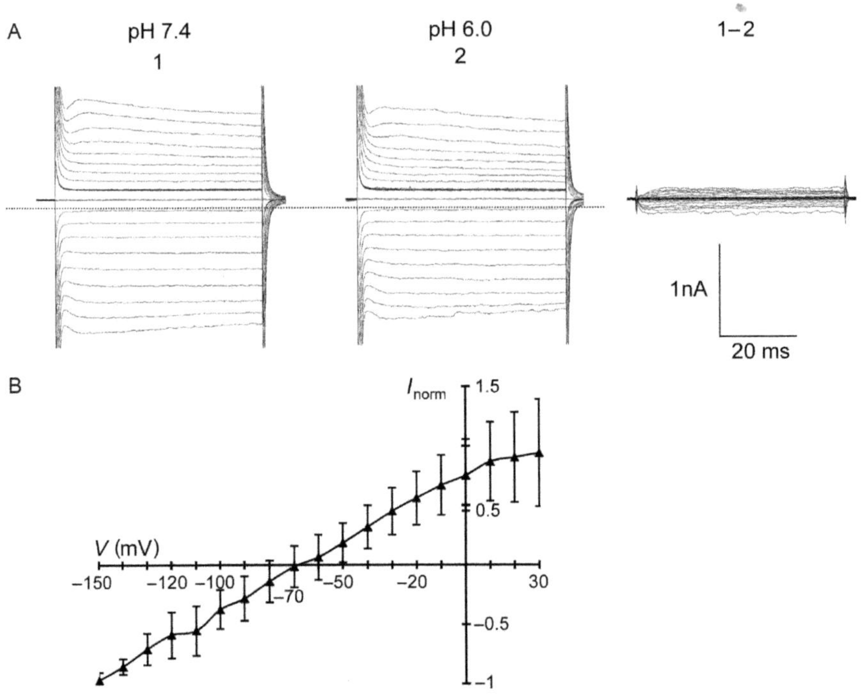

Figure 1 pH sensitivity of freshly isolated astrocytes. (A) Whole-cell membrane currents were activated in HEPES-buffered solution at pH 7.4 (left panel) and pH 6.0 (middle) by de- and hyperpolarization between −150 and +30 mV (10 mV increment, holding potential −60 mV). Dotted lines indicate zero current. pH-sensitive currents were determined by subtraction of currents at pH 6.0 from those at pH 7.4 (right). (B) *I*/*V*-relation of the pH-sensitive currents as indicated in (A, right panel) ($n = 10$, normalized to maximum inward current in each cell).

3.2 Low pH Increases Membrane Currents When Kir4.1 Channels Were Blocked

To test whether acidic pH selectively inhibits Kir4.1 channel-mediated currents in astrocytes, they were blocked with 100 μM Ba^{2+} prior to exposing the cells to bath solution with pH 6.0 and Ba^{2+}. After Ba^{2+} application (pH 7.4), the cells' resting potential depolarized (from -64 ± 9 to -47 ± 10 mV, $n = 23$), which was accompanied by an increase in input resistance (from 63 ± 38 to 220 ± 144 MΩ) and a reduction of membrane conductance (to $63 \pm 26\%$ and $25 \pm 13\%$ at +20 and −130 mV, respectively; Fig. 2A). In the presence of Ba^{2+}, acidification (from pH 7.4 to 6.0) did not change resting potential. However, in most cells the membrane conductance significantly increased under these conditions (to $119 \pm 19\%$ and $114 \pm 10\%$ at −130 and +20 mV, respectively; $n = 15/23$). The pH-induced conductance

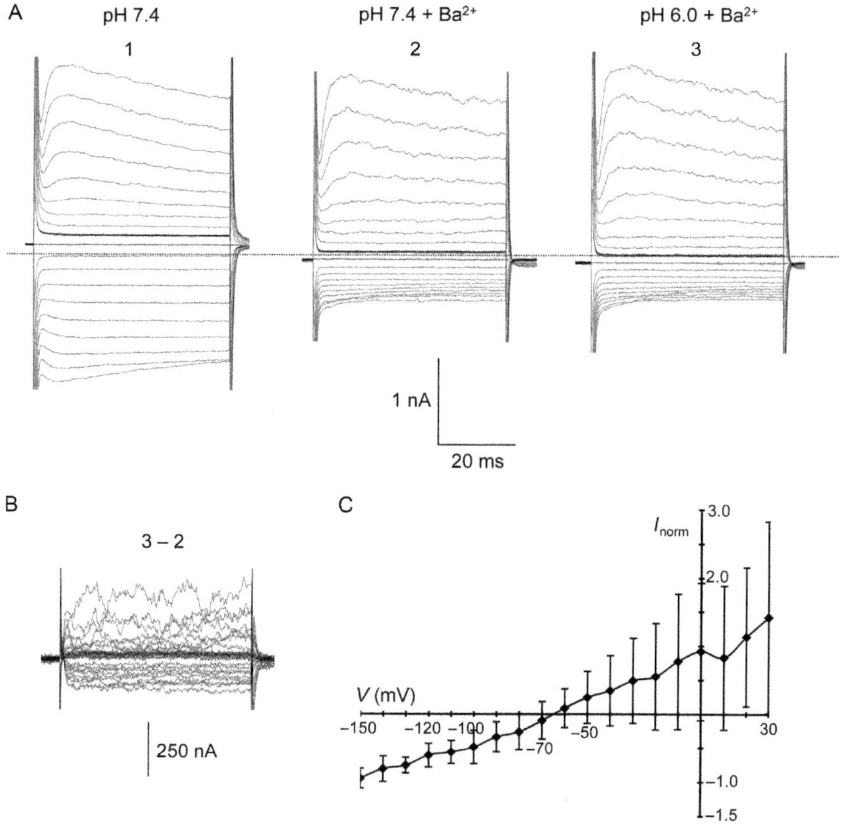

Figure 2 Low pH elicited membrane currents in the presence of Ba^{2+}. (A) Membrane currents were activated as described in Fig. 1A. After evoking control currents (left), HEPES-buffered bath solution supplemented with Ba^{2+} (100 μM) was applied at pH 7.4 (middle) and pH 6.0 (right). (B) The low pH-induced currents were obtained by subtraction of the respective current families. (C) *I*/*V*-relation of the pH 6.0-induced currents as it indicated in (B) ($n = 11$, normalized to maximum inward current in each cell).

increase was reversible when switching back to pH 7.4. This pH-dependent current recorded in the absence of Kir channels reversed at −65 mV and showed weak outward rectification (Fig. 2C).

3.3 pH-Induced Membrane Currents Are Insensitive to Blockers of Voltage-Activated K^+ Channels

To increase pH-evoked currents, the extracellular K^+ concentration was iso-osmolarily raised to 20 mM, which shifted the K^+ equilibrium potential to −47 mV. We examined if the low pH-induced conductance is sensitive to blockers of voltage-activated K^+ currents. In presence of tetraethylammonium chloride (TEA, 10 mM), 4-aminopyridine (4-AP, 4 mM), and Ba^{2+} (100 μM) in 20 mM $[K^+]_o$, acidification to pH 6.0 lead to a reversible conductance increase (at +20 mV: 132±24%; at −130 mV: 121±17%; $n=18$) compared to the control at pH 7.4 (Fig. 3A). Low pH did not affect the resting potential ($V=-34.0\pm5.6$ mV and -33.9 ± 6.5 mV at pH 7.4 and 6.0, respectively; $n=18$) but led to a decrease in input resistance (from 154±108 to 119±75 MΩ). The I/V curve of low pH-activated currents in high K^+ solution showed outward rectification (RI: 1.43) and reversed direction at −40 mV ($n=13$), close to the K^+ equilibrium potential (Fig. 3B).

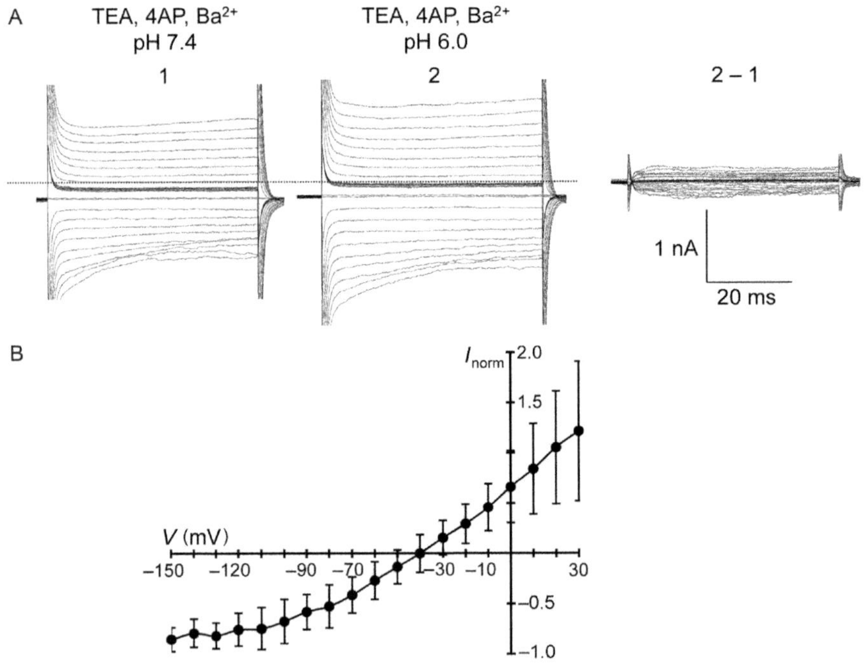

Figure 3 Low pH extracellular solution evoked outward rectifying membrane currents in the presence of K^+ channel blockers. (A) Freshly isolated astrocytes were incubated in bath solution with 20 mM $[K^+]_o$ supplemented with TEA (10 mM), 4-AP (4 mM), and Ba^{2+} (100 μM) and currents were evoked as described in Fig. 1A. After obtaining control currents (left), the pH was switched to pH 6.0 (middle) and low pH-evoked currents were isolated (right). (B) I/V-relation of low pH-induced currents ($n=13$, normalized to the maximum inward current in each cell).

3.4 Does Acidic Bath Solution Activate ASIC- or TRPV1-Mediated Currents?

To further evaluate whether ASIC contributed to the conductance activated at acid pH in the presence of Ba^{2+}, the ASIC blocker amiloride was added. Amiloride (100 μM) did not affect astrocytic membrane currents (+20 mV: 105±9%; −130 mV: 102±6%; $n=7$; Fig. 4A). Because ASIC1a-mediated currents have been reported to be transient and inhibited by extracellular Ca^{2+} (Waldmann, Champigny, Bassilana, Heurteaux, & Lazdunski, 1997), astrocytes were rapidly (<100 ms) exposed to pH 6.0 in normal ($n=4$) and nominally Ca^{2+}-free extracellular solution ($n=6$). No responses were observed under either condition (Fig. 4B).

Acid-sensitive TRPV1 channels are sensitive to various stimuli, expressed by polymodal nociceptive neurons and activated by noxious heat (>43 °C), the pungent vanilloid capsaicin, and acidic pH (4.5–5.0) (Julius, 2013). To test for activation of TRPV1-mediated currents in astrocytes, voltage-activated and Kir currents were blocked as before and the cells were exposed to capsaicin (100 μM; pH 7.4). Capsaicin did not affect the membrane conductance (+20 mV: 92±9%; −130 mV: 100±6%; $n=7$; Fig. 4C).

To examine expression of ASIC and TRPV1 on the transcript level, we performed single-cell RT-PCR by harvesting the cytoplasm from individual astrocytes and subsequent functional analysis. A two-round PCR was performed, with primers common to ASIC1a and ASIC2 transcripts in the first PCR and gene-specific primers in the second round. None of the astrocytes tested contained mRNAs encoding ASICs ($n=6$; Fig. 4D). In contrast, ASIC1a and ASIC2 were frequently found in CA1 pyramidal neurons (in 60% and 33% of cells analyzed, $n=15$; Fig. 4D and E). Similarly, TRPV channels were not or only very rarely expressed by hippocampal astrocytes (TRPV1: $n=0/7$; TRPV2: $n=1/7$; Fig. 4D and E). Astrocytes were identified through coexpression of S100β ($n=13$).

3.5 Does Low Extracellular pH Activate K_2P Channels in Astrocytes?

The *I*/*V* relation of the low pH-activated responses in the presence of TEA, 4-AP, and Ba^{2+} resembled arachidonic acid-induced currents through TREK-1 channels in astrocytes (Seifert et al., 2009). The latter may open upon acidification of the cytoplasm (Honore, 2007). To test whether low pH and PUFAs activate the same K^+ channel subunits, we applied linolenic

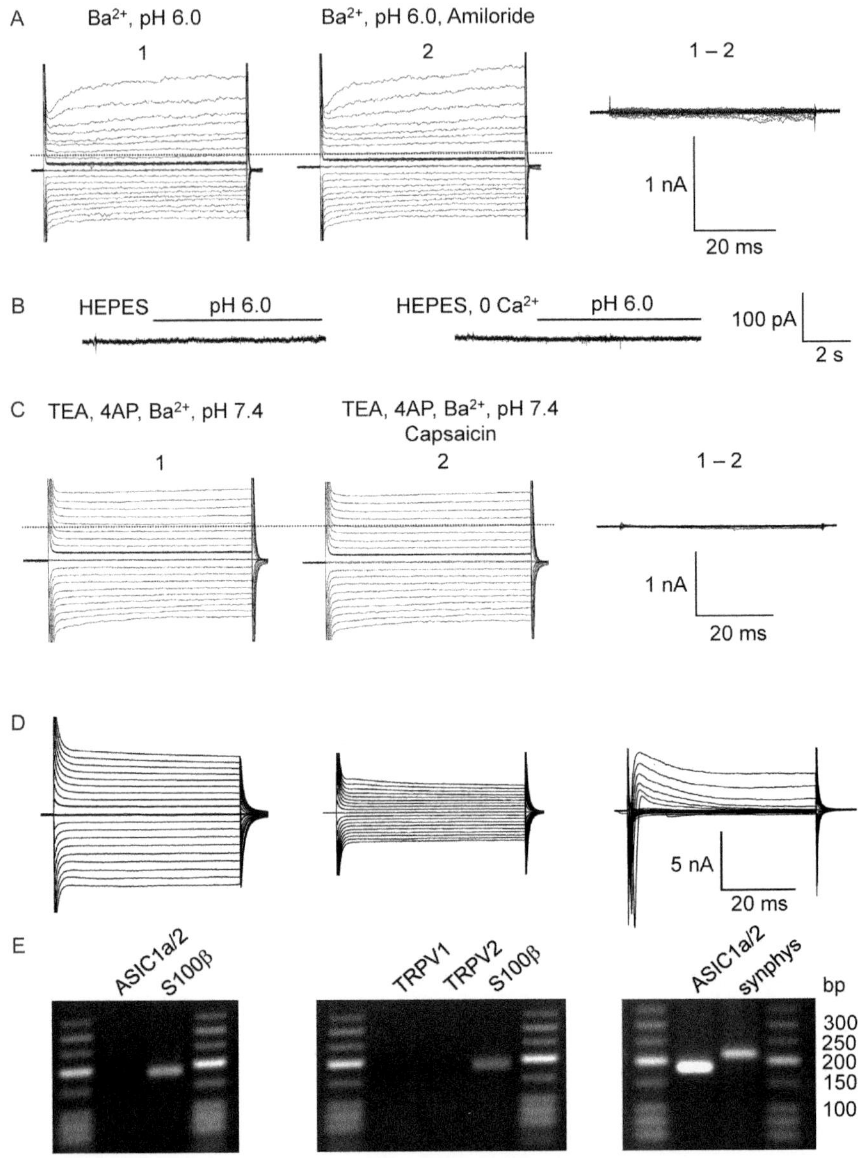

Figure 4 Freshly isolated astrocytes lack ASICs and TRPV1. (A) Membrane currents were evoked as described in Fig. 1A at pH 6.0 in the presence of Ba^{2+} (100 μM, left) and Ba^{2+} plus amiloride (100 μM each, middle). The responses were insensitive to the ASIC blocker amiloride (right). (B) At a holding potential of −60 mV, bath solution with 2 mM Ca^{2+} (left) and without Ca^{2+} (middle) was fast switched from pH 7.4 to 6.0. (C) Currents were recorded in the presence of Ba^{2+} (100 μM), 4-AP (4 mM), and TEA (10 mM) at pH 7.4, before (left) and after incubation with the TRPV1 agonist capsaicin (100 μM, middle). Subtracting both current families revealed lack of TRPV1.

and arachidonic acid in high extracellular K^+ solution (20 mM) in presence of the above K^+ channel blockers, which do not inhibit TREK-1 channels. α-Linolenic acid (10 μM) increased the membrane conductance (+20 mV: 172 ± 50%; −130 mV: 131 ± 26%; $n = 8$). The corresponding *I*/*V* relation reversed at −30 mV and showed outward rectification (RI: 1.7; $n = 7$; Fig. 5A and D). A similar conductance increase was observed after application of arachidonic acid (10 μM; +20 mV: 158 ± 44%; −130 mV: 149 ± 41%; $n = 10$). The *I*/*V* relation reversed at −35 mV and showed outward rectification (RI: 1.7, $n = 10$; Fig. 5B and D). In contrast, prostaglandin E2 (PGE_2, 10 μM), a product of cyclooxygenase-mediated conversion of arachidonic acid that rather inactivates TREK-1 currents (Maingret et al., 2000), left the membrane currents unaffected ($n = 7$; not shown). The membrane crenator trinitrophenol (TNP), another TREK-1-opener (Miller et al., 2003), also activated a conductance (400 μM; +20 mV: 130 ± 17%; −130 mV: 118 ± 8%; $n = 8$). These currents reversed at −26 mV and showed outward rectification (RI: 2.4; $n = 8$; Fig. 5C and D). Eventually, the TREK-1-activator riluzole (100 μM) also enhanced astrocytic membrane currents (+20 mV: 133 ± 22%; −130 mV: 114 ± 20%; $n = 9$; Fig. 5D). The activation of putative K_2P channel-mediated currents was always reversible. The currents evoked by low extracellular pH, PUFAs, TNP, and riluzole were all sensitive to quinine (200 μM), a K_2P channel blocker (Lesage & Lazdunski, 2000) (at +20 mV: pH 6.0 to 71 ± 24%, $n = 7$; α-linolenic acid to 59 ± 15%, $n = 3$; TNP to 30 ± 14%, $n = 3$; riluzole to 43 ± 1%, $n = 4$). A similar decrease of membrane conductance was observed at −130 mV. Together, these data strongly suggest that low pH activates TREK channels in hippocampal astrocytes.

3.6 Developmental Upregulation of K_2P Channel mRNA in the Hippocampus

To assess the expression of the acid-sensitive K_2P subunits TASK1 and TASK3, we performed single-cell RT-PCR subsequent to functional

(D) Representative current patterns of astrocytes (left, middle) and a CA1 pyramidal neuron (right). Cells were de- and hyperpolarized *in situ* between −160 and +20 mV (10 mV increment, holding potential −70 mV). (E) After recording, the cytoplasm was harvested and single-cell RT-PCR performed. Astrocytes expressed transcripts for the marker S100β (186 bp) but lacked ASICs and TRPVs (left, middle). In contrast, mRNA-encoding ASIC1/2 (184 bp) was found in synaptophysin-expressing CA1 pyramidal neurons (right).

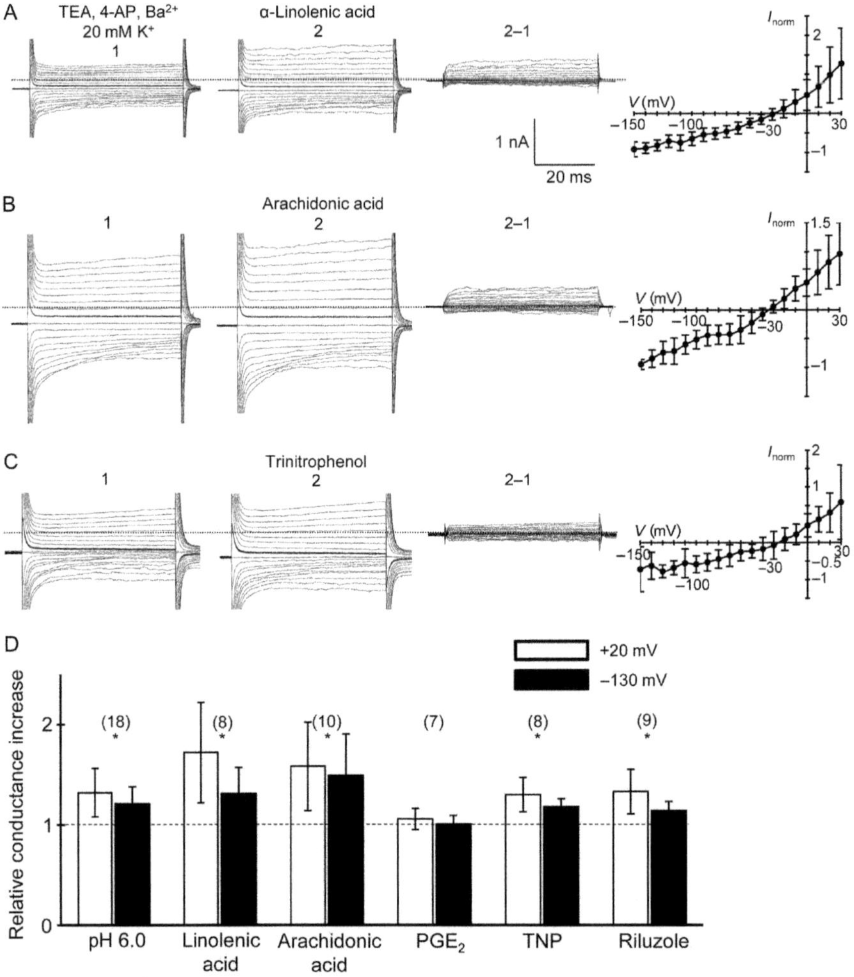

Figure 5 TREK-1 channel agonists mimic the effect of low pH. (A–C) Currents were activated as described in Fig. 1A in a solution containing 20 mM K^+ and the K^+ channel blocker TEA (10 mM), 4 AP (4 mM), and Ba^{2+} (100 μM) (left panels). α-Linolenic acid (10 μM), arachidonic acid (10 μM), and trinitrophenol (400 μM) enhanced the responses (second row), which were isolated by subtracting currents at respective voltages (third row). The right panel shows the corresponding *I*/*V*-relations, normalized to maximum inward currents. (D) summarizes the effect of TREK-1 channel modulators at +20 mV (open bars) and −130 mV (filled). The concentrations of PGE_2 and riluzole were 10 and 100 μM, respectively. Asterisks indicate significant changes through modulator application at pH 7.4, or after switching the bath solution to pH 6.0. Cell numbers are given in parentheses.

analysis. Both subunits were coexpressed by CA1 pyramidal neurons ($n=5$). In contrast, astrocytes consistently lacked TASK1 ($n=12$) and only rarely expressed TASK3 ($n=1/7$; not shown). For cell type identification, primers for Kir4.1 (astrocytes) and synaptophysin (neurons) were used.

Next, we investigated the developmental regulation of the K2P subunits TWIK-1 (KCNK1), TREK-1 (KCNK2), and TREK2 (KCNK10), which are expressed by hippocampal astrocytes (Seifert et al., 2009). mRNA was isolated from small tissue samples of the hippocampal CA1 region and semi-quantitative real-time RT-PCR was performed. β-actin served as a reference gene. An increase of KCNK1 and KCNK2 expression was observed during the first three postnatal weeks, reaching a maximum at p21 (Fig. 6A left and middle panel, 6B). In contrast, the abundance of KCNK10 was lower by a factor of 10 at p21 and remained constant by p60 (Fig. 6A right panel and 6B).

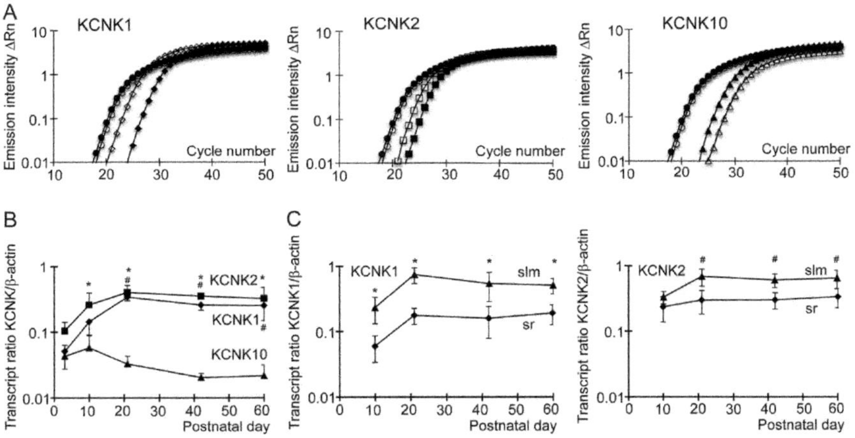

Figure 6 Developmental regulation of KCNK1, KCNK2, KCNK10 mRNA expression in the CA1 region of hippocampus. (A) Amplification plots for the respective KCNK subunits (diamonds, squares, triangles) and β-actin (circles) from mice at p3 (filled symbols) and p21 (open) using real-time sqRT-PCR. Note that for KNCK1 and KCNK2, the difference in threshold cycles (taken at $\Delta Rn=0.1$) at p3 exceeded the p21 value. (B) The gene ratios, KCNK/β-actin, were determined according to Eq. (4) and plotted against age. Mean and SD were obtained at p3 to p60 (eight samples each). Differences between p3 and later ages (*) and p10 and later ages (#) were statistically significant. KCNK10 mRNA expression showed no developmental regulation. (C) The ratios KCNK1/β-actin and KCNK2/β-actin were determined and plotted against postnatal age. Higher ratios were observed for stratum lacunosum-moleculare (triangles) versus stratum radiatum (diamonds). Data were obtained from six samples each.

By comparing mRNA expression in CA1 subregions, higher KCNK1 expression was found in the stratum lacunosum-moleculare (slm) versus sr (Fig. 6C left). Similarly, between p21 and p60 KCNK2 mRNA was more abundant in the slm (Fig. 6C right). KCNK1 was upregulated in both subregions between p10 and p20, while KCNK2 expression in the sr remained unchanged (Fig. 6C right).

4. DISCUSSION

4.1 ASICs and TRPVs in Astrocytes

ASICs comprise a subfamily of the ENaC/degenerin channel family, which are gated by low pH and permeable to Na^+ and to a lower extent also to Ca^{2+} (Kellenberger & Schild, 2015). Their expression in the peripheral nervous system is well established, where the channels transmit nociceptive stimuli produced by acidification. ASICs are also expressed in CNS neurons, e.g., in the hippocampus. In cultured hippocampal neurons, ASIC1a was expressed in soma, dendrites, and axons, although synaptic stimulation did not elicit postsynaptic currents mediated by ASICs (Alvarez de la Rosa et al., 2003). ASIC1a is a proton receptor localized on postsynaptic spines and their density is increased after Ca^{2+} elevation and activation of CaMKinase II (Zha, Wemmie, Green, & Welsh, 2006). It is currently thought that ASICs regulate neuronal activity pH-dependently. A role for neuronal ASICs in acidosis-mediated brain injury is now well established, and these channels represent promising novel pharmacological targets for counteracting ischemic brain injury (see Xiong, Pignataro, Li, Chang, & Simon, 2008 review). ASIC expression has also been described in the oligodendrocyte lineage (Feldman et al., 2008) and in NG2 glia (Lin et al., 2015; Lin, Liu, Huang, & Lien, 2010). Moreover, cell culture work suggested expression of ASIC1 in astrocytes (Huang et al., 2010). However, employing freshly isolated astrocytes we did not find ASIC1a, neither on the functional nor on the transcript level, although we confirmed robust expression of ASIC1a/2 in CA1 neurons.

The multisensory transient receptor potential vanilloid type 1 (TRPV1) channel forms an unspecific cationic pore that is also permeable to protons (Hellwig et al., 2004). TRPV1 is mainly located in primary nociceptive neurons and activated by noxious heat, acids, pungent chemicals as well as basic intracellular pH (Dhaka et al., 2009). In addition, it was found in central neurons. In the hippocampus, TRPV1 is located at presynaptic sites in hippocampal CA1 neurons and contributes to synaptic depression

(Gibson, Edwards, Page, Van Hook, & Kauer, 2008), while in the dentate gyrus, TRPV1 receptors mediate synaptic depression of GABAergic neurotransmission onto somata of granule cells (Chavez, Hernandez, Rodenas-Ruano, Chan, & Castillo, 2014). Recent work has implied the presence of TRPV1 in cultured cortical (Huang et al., 2010) and retinal astrocytes, where the channels were activated by mechanical stress and involved in the regulation of migration (Ho, Lambert, & Calkins, 2014). Immunohistochemistry detected TRPV1-positive astrocytic endfeet and pericytes in the hippocampus (Toth et al., 2005), in astrocytic processes in the circumventricular organs (Mannari, Morita, Furube, Tominaga, & Miyata, 2013), and in spinal dorsal horn astrocytes (Doly, Fischer, Salio, & Conrath, 2004). Transcripts encoding TRPV1 were also found in cultured human astrocytes (Amantini et al., 2007). In our study, astrocytes from the hippocampus did not respond to capsaicin, neither did single-cell RT-PCR detect TRPV1 mRNA. Whether these opposing results are due to use of different preparations (culture vs. fresh isolation) or indicative of region-specific astrocytic heterogeneity remains to be determined.

4.2 pH Sensitivity of Astrocytes and Its Functional Implications

Astrocytes express different transporters and exchangers which regulate the intracellular pH (Steinhäuser et al., 2013). The electrogenic Na^+/HCO_3^- cotransporter, NBCe1, transports 1 Na^+ together with 2 HCO_3^- and can operate in both directions dependent on the actual membrane potential (Deitmer & Schlue, 1989). Knockout studies demonstrated that in astrocytes, extent, and kinetics of recovery from intracellular pH changes largely depend on NBCe1 (Theparambil, Naoshin, Thyssen, & Deitmer, 2015; Theparambil, Ruminot, Schneider, Shull, & Deitmer, 2014). In our experiments, we used nominally HCO_3^- – free HEPES-buffered solutions, but the residual HCO_3^- content might have attenuated intracellular acidification (Theparambil et al., 2014).

In the ventral brainstem, astrocytes may function as pH/CO_2-sensors, leading to an elevation of intracellular Ca^{2+} and release of ATP, which activates neurons of the retrotrapezoid nucleus (RTN) and modulates respiratory activity (Gourine et al., 2010). This pH sensitivity of Ca^{2+}-dependent astrocytic exocytosis of ATP-filled vesicles is region-specific because it was not observed in cortical astrocytes (Kasymov et al., 2013). The mechanism(s) linking extracellular acidification to elevation of intracellular Ca^{2+} in astrocytes is still under debate. It was suggested that Kir4.1/Kir5.1

heteromeric channels in RTN astrocytes convey pH sensitivity to astrocytes, by depolarizing the cells through block of Kir currents (Wenker et al., 2010). In astrocytes of the nucleus of the solitary tract, depolarization upon inhibition of Kir or background K^+ currents compromises glutamate uptake and enhances neuronal excitation (Huda, McCrimmon, & Martina, 2013). Future work has to clarify the molecular basis of astrocytic regulation in respiration and possibly identify new targets for pharmacological intervention in pathological conditions such as decreased respiratory drive or chronic hypercapnia, e.g., chronic obstructive pulmonary disease.

Exposition of cultured astrocytes to low pH, evoked by HCl or lactate, entails a slow drop of intracellular pH and cell death in case of prolonged (hours) application. Both extent and duration of extracellular acidification determine cellular vulnerability. Low intracellular pH affects glycolysis, Ca^{2+} homeostasis, enzymatic activity, gap junction communication, and eventually may induce delayed cell death (Nedergaard et al., 1991; Siesjo et al., 1993). Prolonged oxygen and glucose deprivation are accompanied by altered transmembrane ion gradients can result in reverse operation of the Na^+–H^+-exchanger and increase the Ca^{2+} load of astrocytes, particularly upon reperfusion (Chesler, 2005), while transient acidification leaves the Na^+ gradient intact (Rose et al., 1998). Glial acidosis, induced by oxygen and glucose deprivation or by optogenetic channelrhodopsin-2 activation in astrocytes, activated Ca^{2+}-independent glutamate release and triggered excitotoxicity. Accordingly, preventing glial acidosis might serve as a strategy to prevent ischemic brain damage (Beppu et al., 2014).

4.3 Kir4.1 as a Therapeutic Target in Neurological Disorders?

Kir channels belong to the superfamily of two-membrane domain channels, lack an intrinsic voltage sensor, and have a high open probability at rest. Inward rectification results from positively charged polyamines and Mg^{2+}, which plug the channel from the intracellular site in a voltage-dependent manner. Among the Kir channel family Kir4.1 is expressed in different types of glial cells in the CNS, i.e., astrocytes, oligodendrocytes, NG2 cells, and Müller glial cells (Djukic, Casper, Philpot, Chin, & McCarthy, 2007; Ishii et al., 1997; Seifert et al., 2009; Tang, Taniguchi, & Kofuji, 2009). In astrocytes, Kir4.1 is mainly localized to processes enwrapping synapses and in endfeet around blood vessels (Higashi et al., 2001; Nagelhus, Mathiisen, & Ottersen, 2004). Mice with global or conditional knockout of Kir4.1 developed seizures and died early after birth (Djukic et al., 2007;

Kofuji et al., 2000). Kir4.1-deficient astrocytes have an increased membrane resistance and a depolarized membrane potential (Djukic et al., 2007; Seifert et al., 2009). As a consequence, their ability for K^+ buffering and membrane potential-dependent transport, e.g., glutamate uptake, is severely impaired (Djukic et al., 2007; Kucheryavykh et al., 2007). The significance of Kir4.1 in K^+ uptake and spatial buffering has been confirmed by experiments showing that its deletion impairs K^+ buffering and generates an epileptic phenotype (Chever, Djukic, McCarthy, & Amzica, 2010; Haj-Yasein et al., 2011; Steinhäuser, Seifert, & Bedner, 2012). Under physiological conditions, Kir4.1 regulates synaptic potentiation (Sibille, Pannasch, & Rouach, 2014), extracellular K^+ clearance, and membrane potential (Sibille, Dao, Holcman, & Rouach, 2015), particularly during neuronal activity (Larsen et al., 2014; Larsen & MacAulay, 2014).

In humans, mutations in the KCNJ10 gene, which encodes Kir4.1, cause a multiorgan disorder with clinical features of epilepsy, sensorineural deafness, ataxia, and electrolyte imbalance (EAST/SeSAME syndrome; Bockenhauer et al., 2009; Scholl et al., 2009). Single-nucleotide mutations in KCNJ10 cause missense or nonsense mutations on the protein level, and heterologous expression demonstrated that these mutations result in decreased Kir currents and depolarization (Bockenhauer et al., 2009; Reichold et al., 2010; Williams et al., 2010). Furthermore, they increased the pH sensitivity of Kir4.1 and hampered its surface expression (Sala-Rabanal, Kucheryavykh, Skatchkov, Eaton, & Nichols, 2010; Williams et al., 2010). Single-nucleotide variations in the KCNJ10 gene were identified in DNA from patients with temporal lobe epilepsy and hippocampal sclerosis (TLE-HS) or antecedent febrile seizures (Heuser et al., 2010).

Impaired K^+ buffering and Kir4.1 expression was observed in hippocampal specimens neurosurgically resected from TLE-HS patients (Steinhäuser et al., 2012). Kir4.1 immunoreactivity was profoundly reduced in perivascular endfeet of astrocytes, and this reduction was accompanied by lower dystrophin labeling suggesting a disruption in the dystrophin-associated protein complex (Heuser et al., 2012). It is important to note that besides impaired Kir4.1 expression, dysfunctional astroglial gap junction coupling is crucial for the impaired K^+ clearance as observed in human and experimental TLE-HS (Bedner et al., 2015).

In a mouse model of Huntington's disease, mutant huntingtin protein accumulated in striatal astrocytes and led to decreased Kir4.1 protein and Ba^{2+}-sensitive Kir currents, membrane depolarization, and elevated extracellular K^+ concentration. In this model, viral delivery of Kir4.1-GFP rescued Kir4.1 expression in astrocytes and attenuated the phenotype (Tong et al., 2014).

In patients suffering from multiple sclerosis (MS), an inflammatory autoimmune disease, increased titers of serum IgG antibodies against Kir4.1 were observed. Similarly, injection of MS-specific Kir4.1 serum IgG in mice resulted in altered GFAP expression, loss of Kir4.1 and complement deposit in regions of Kir4.1 loss, indicating functional and structural damage in glial cells (Srivastava et al., 2012). In subcortical white matter tissue of autopsies from MS patients, Kir4.1 loss was observed in astrocytic endfeet and oligodendrocytes of demyelinating lesions, while in the periplaque areas Kir4.1 immunoreactivity was retained or even increased (Schirmer et al., 2014). Glial Kir4.1 channels are a primary target of immune responses in MS and display a decreased functionality after injury or in the diseased CNS (Olsen & Sontheimer, 2008; Seifert, Schilling, & Steinhäuser, 2006). They are sensitive to tricyclic antidepressants and to the serotonin reuptake inhibitor fluoxetine. By searching for specific Kir4.1 blockers with functional assays, some amine derivatives were identified that might exert diuretic effects (Raphemot et al., 2013). Neurological disorders would rather benefit from enhanced Kir4.1 function or expression but a drug increasing Kir4.1 function has not yet been identified (Steinhäuser, Grunnet, & Carmignoto, 2015). However, the steroidal hormone β-estradiol turned out to be beneficial after spinal cord injury, by enhancing Kir4.1 expression, glutamate transporters, and improving K^+ homeostasis (Olsen, Campbell, McFerrin, Floyd, & Sontheimer, 2010).

4.4 Influence of Low Extracellular pH and Ischemia on K_2P Channels in Astrocytes

So far, 15 isoforms of K_2P channels have been identified which are regulated by various stimuli and critically influence the resting membrane potential (Enyedi & Czirjak, 2010). It was suggested that besides Kir4.1, K_2P channels formed by TWIK-1 and TREK-1 contribute to the large K^+ conductance of astrocytes and may form heteromeric channels (Hwang et al., 2014; Seifert et al., 2009; Zhou et al., 2009). Our study provides strong evidence for the functional expression of TREK-1 and reveales a parallel developmental regulation of TWIK-1- and TREK-1-encoding transcripts in the hippocampus. Maximal expression was observed 3 weeks after birth, which is in line with previous transcriptome and protein analyses (Cahoy et al., 2008; Wang et al., 2013). Accordingly, Kir4.1 precedes expression of TWIK-1/TREK-1 by about 7 days (Seifert et al., 2009).

TREK-1 expression increases after cerebral ischemia and it was suggested that quinine-sensitive K^+ channels, among them TREK-1, stimulate proliferation (Wang et al., 2011). In cell culture, inhibition of TREK-1 channels impairs the clearance of glutamate, both under control and hypoxic conditions. The damage-associated pattern protein (DAMP), S100β, is abundantly expressed by astrocytes and released under hypoxic conditions. Additional block of TREK-1 channels exacerbated S100β secretion (Wu et al., 2013). These results indicate that under hypoxic conditions TREK-1 is activated and, through hyperpolarization, might maintain homeostasis and attenuate DAMP secretion. In cultured astrocytes, TREK-1-mediated currents are sensitive to arachidonic acid, low pH_i, and negative pressure (Lu et al., 2014), and our analyses in freshly isolated astrocytes revealed a similar pharmacological profile, assuming that the pH_o changes in our experiments also affected pH_i (Chesler, 2005; Nedergaard et al., 1991). The pH_i regulation *in vivo* is more complex because hypoxia provokes massive changes in transmembrane ionic gradients. Accordingly, results from *in vitro* acidosis may recapitulate only some aspects of systemic ischemia.

In our study, transient application of low pH_o-induced inward and outward currents in isolated hippocampal astrocytes, which were insensitive to amiloride. Moreover, capsaicin did not activate responses, indicating that the low pH_o-mediated currents were not mediated by TRPV1 or ASIC channels. Another study reported quinine-insensitive pH_o-induced currents and suggested that the responses were mediated by yet unknown K^+ channels (Chu et al., 2010). However, quinine block of K_2P channels might be impaired at pH 5 due to di-protonation of the tertiary amine nitrogen atoms. Indeed, we observed a loss of quinine sensitivity of K^+ channels at pH <6 and conclude that the findings of Chu et al. may not exclude functional expression of TREK-1 in astrocytes.

4.5 K_2P Channels in Astrocytes as Therapeutic Targets?

Dysfunctional astrocytes are increasingly recognized to contribute to brain pathologies, e.g., to stroke, trauma, inflammation, epilepsy, Alzheimer's disease, and other neurodegenerative diseases, which has raised the question whether these cells might represent new targets for drug development (Lee & MacLean, 2015; Pekny, Wilhelmsson, & Pekna, 2014). Ion channels represent the second largest group of currently used pharmacological targets in neurologic diseases (Overington, Al-Lazikani, & Hopkins, 2006). Their complex regulation and impact on cellular activity render K_2P channels promising drug targets. Neuroprotective agents such as PUFAs (Blondeau, Widmann, Lazdunski, & Heurteaux, 2002; Fink et al., 1998), riluzole (Duprat et al., 2000), and volatile anesthetics (Patel et al., 1999) are potent activators of TREK-1 and inhibit neuronal excitability. On the other hand, mice with a constitutive knockout of TREK-1 show a higher sensitivity to ischemia, epilepsy, and pain, and the neuroprotective effects of α-linolenic acid or lysophosphatidylcholine are abolished (Alloui et al., 2006; Heurteaux et al., 2004; Noel et al., 2009).

Our present findings confirm and extend previous data on the functional expression of TREK-1 in astrocytes (Hwang et al., 2014; Seifert et al., 2009; Zhou et al., 2009). We show that these glial channels are activated by arachidonic acid, acidification, and membrane crenators mimicking intracellular swelling, i.e., conditions occurring during cerebral ischemia. TREK-1 might be involved in stroke-induced astrogliosis, and hypoxia or acute cerebral ischemia led to upregulation of TREK-1 and astrocytic proliferation (Wang et al., 2011). Systemic administration of riluzole and α-linolenic acid after middle cerebral artery occlusion reduced infarct volume, DNA fragmentation, and cell injury (Heurteaux, Laigle, Blondeau, Jarretou, & Lazdunski, 2006), although the underlying mechanisms and the question whether astrocytes were the primary target of this treatment remained unclear. Similar to Kir channels, TREK-1 contributes to the passive astrocytic membrane conductance and negative resting potential, upon which a number of astrocytic functions depend, including K^+ buffering, neurotransmitter uptake, and pH regulation (Olsen & Sontheimer, 2008; Pasler, Gabriel, & Heinemann, 2007). Impaired astrocytic K^+ siphoning might be involved in cortical spreading depression observed after traumatic brain injury, migraine, or malignant stroke (reviewed in Torrente et al., 2014). The neuroprotective effect of PUFAs that open TREK-1 channels was demonstrated in an animal model of ischemia and kainate-induced seizures. Intracerebroventricular injection of linolenic acid before induction of ischemia and seizures had a pronounced neuroprotective effect (Lauritzen et al., 2000; reviewed by Mathie & Veale, 2007). However, different problems impede the translation of PUFA for neuroprotection in ischemia and epilepsy. A relatively slow dietary uptake might explain some discrepancies between promising basic research and sobering clinical studies, and ongoing research is looking for increased bioavailability to the brain, e.g., via intravenous administration (Bazinet & Laye, 2014). K_2P channels are ubiquitously expressed by both neurons and glial cells, and their detailed involvement in the pathogenic events is not well understood. Moreover, the cell types responsible for the observed neuroprotective effects are mostly unknown. Accordingly, additional work is needed to assess the potential of K_2P channels as therapeutic targets.

ACKNOWLEDGMENT

This work was supported by DFG (STE 552/3), the EU (NSF, EuroEPINOMICS), and BONFOR.

REFERENCES

Alloui, A., Zimmermann, K., Mamet, J., Duprat, F., Noel, J., Chemin, J., et al. (2006). TREK-1, a K+ channel involved in polymodal pain perception. *EMBO Journal, 25*, 2368–2376.

Alvarez de la Rosa, D., Krueger, S. R., Kolar, A., Shao, D., Fitzsimonds, R. M., & Canessa, C. M. (2003). Distribution, subcellular localization and ontogeny of ASIC1 in the mammalian central nervous system. *Journal of Physiology, 546*, 77–87.

Amantini, C., Mosca, M., Nabissi, M., Lucciarini, R., Caprodossi, S., Arcella, A., et al. (2007). Capsaicin-induced apoptosis of glioma cells is mediated by TRPV1 vanilloid receptor and requires p38 MAPK activation. *Journal of Neurochemistry, 102*, 977–990.

Bazinet, R. P., & Laye, S. (2014). Polyunsaturated fatty acids and their metabolites in brain function and disease. *Nature Reviews Neuroscience, 15*, 771–785.

Bedner, P., Dupper, A., Huttmann, K., Muller, J., Herde, M. K., Dublin, P., et al. (2015). Astrocyte uncoupling as a cause of human temporal lobe epilepsy. *Brain, 138*, 1208–1222.

Beppu, K., Sasaki, T., Tanaka, K. F., Yamanaka, A., Fukazawa, Y., Shigemoto, R., et al. (2014). Optogenetic countering of glial acidosis suppresses glial glutamate release and ischemic brain damage. *Neuron, 81*, 314–320.

Blondeau, N., Widmann, C., Lazdunski, M., & Heurteaux, C. (2002). Polyunsaturated fatty acids induce ischemic and epileptic tolerance. *Neuroscience, 109*, 231–241.

Bockenhauer, D., Feather, S., Stanescu, H. C., Bandulik, S., Zdebik, A. A., Reichold, M., et al. (2009). Epilepsy, ataxia, sensorineural deafness, tubulopathy, and KCNJ10 mutations. *New England Journal of Medicine, 360*, 1960–1970.

Cahoy, J. D., Emery, B., Kaushal, A., Foo, L. C., Zamanian, J. L., Christopherson, K. S., et al. (2008). A transcriptome database for astrocytes, neurons, and oligodendrocytes: A new resource for understanding brain development and function. *Journal of Neuroscience, 28*, 264–278.

Chavez, A. E., Hernandez, V. M., Rodenas-Ruano, A., Chan, C. S., & Castillo, P. E. (2014). Compartment-specific modulation of GABAergic synaptic transmission by TRPV1 channels in the dentate gyrus. *Journal of Neuroscience, 34*, 16621–16629.

Chesler, M. (2005). Failure and function of intracellular pH regulation in acute hypoxic-ischemic injury of astrocytes. *Glia, 50*, 398–406.

Chever, O., Djukic, B., McCarthy, K. D., & Amzica, F. (2010). Implication of kir4.1 channel in excess potassium clearance: An in vivo study on anesthetized glial-conditional kir4.1 knock-out mice. *Journal of Neuroscience, 30*, 15769–15777.

Chu, K. C., Chiu, C. D., Hsu, T. T., Hsieh, Y. M., Huang, Y. Y., & Lien, C. C. (2010). Functional identification of an outwardly rectifying pH- and anesthetic-sensitive leak K(+) conductance in hippocampal astrocytes. *European Journal of Neuroscience, 32*, 725–735.

Deitmer, J. W., & Schlue, W.-R. (1989). An inwardly directed electrogenic sodium-bicarbonate co-transport in leech glial cells. *Journal of Physiology*, *411*, 179–194.

Dhaka, A., Uzzell, V., Dubin, A. E., Mathur, J., Petrus, M., Bandell, M., et al. (2009). TRPV1 is activated by both acidic and basic pH. *Journal of Neuroscience*, *29*, 153–158.

Djukic, B., Casper, K. B., Philpot, B. D., Chin, L. S., & McCarthy, K. D. (2007). Conditional knock-out of Kir4.1 leads to glial membrane depolarization, inhibition of potassium and glutamate uptake, and enhanced short-term synaptic potentiation. *Journal of Neuroscience*, *27*, 11354–11365.

Doly, S., Fischer, J., Salio, C., & Conrath, M. (2004). The vanilloid receptor-1 is expressed in rat spinal dorsal horn astrocytes. *Neuroscience Letters*, *357*, 123–126.

Duprat, F., Lesage, F., Patel, A. J., Fink, M., Romey, G., & Lazdunski, M. (2000). The neuroprotective agent riluzole activates the two P domain K^+ channels TREK-1 and TRAAK. *Molecular Pharmacology*, *57*, 906–912.

Enyedi, P., & Czirjak, G. (2010). Molecular background of leak K+ currents: Two-pore domain potassium channels. *Physiological Reviews*, *90*, 559–605.

Feldman, D. H., Horiuchi, M., Keachie, K., Mccauley, E., Bannerman, P., Itoh, A., et al. (2008). Characterization of acid-sensing ion channel expression in oligodendrocyte-lineage cells. *Glia*, *56*, 1238–1249.

Fink, M., Lesage, F., Duprat, F., Heurteaux, C., Reyes, R., Fosset, M., et al. (1998). A neuronal two P domain K+ channel stimulated by arachidonic acid and polyunsaturated fatty acids. *EMBO Journal*, *17*, 3297–3308.

Gibson, H. E., Edwards, J. G., Page, R. S., Van Hook, M. J., & Kauer, J. A. (2008). TRPV1 channels mediate long-term depression at synapses on hippocampal interneurons. *Neuron*, *57*, 746–759.

Gourine, A. V., Kasymov, V., Marina, N., Tang, F., Figueiredo, M. F., Lane, S., et al. (2010). Astrocytes control breathing through pH-dependent release of ATP. *Science*, *329*, 571–575.

Haj-Yasein, N. N., Jensen, V., Vindedal, G. F., Gundersen, G. A., Klungland, A., Ottersen, O. P., et al. (2011). Evidence that compromised K(+) spatial buffering contributes to the epileptogenic effect of mutations in the human kir4.1 gene (KCNJ10). *Glia*, *59*, 1635–1642.

Hellwig, N., Plant, T. D., Janson, W., Schafer, M., Schultz, G., & Schaefer, M. (2004). TRPV1 acts as proton channel to induce acidification in nociceptive neurons. *Journal of Biological Chemistry*, *279*, 34553–34561.

Heurteaux, C., Guy, N., Laigle, C., Blondeau, N., Duprat, F., Mazzuca, M., et al. (2004). TREK-1, a K+ channel involved in neuroprotection and general anesthesia. *EMBO Journal*, *23*, 2684–2695.

Heurteaux, C., Laigle, C., Blondeau, N., Jarretou, G., & Lazdunski, M. (2006). Alpha-linolenic acid and riluzole treatment confer cerebral protection and improve survival after focal brain ischemia. *Neuroscience*, *137*, 241–251.

Heuser, K., Eid, T., Lauritzen, F., Thoren, A. E., Vindedal, G. F., Tauboll, E., et al. (2012). Loss of perivascular Kir4.1 potassium channels in the sclerotic hippocampus of patients with mesial temporal lobe epilepsy. *Journal of Neuropathology and Experimental Neurology*, *71*, 814–825.

Heuser, K., Nagelhus, E. A., Tauboll, E., Indahl, U., Berg, P. R., Lien, S., et al. (2010). Variants of the genes encoding AQP4 and Kir4.1 are associated with subgroups of patients with temporal lobe epilepsy. *Epilepsy Research*, *88*, 55–64.

Higashi, K., Fujita, A., Inanobe, A., Tanemoto, M., Doi, K., Kubo, T., et al. (2001). An inwardly rectifying K(+) channel, Kir4.1, expressed in astrocytes surrounds synapses and blood vessels in brain. *American Journal of Physiology, Cell Physiology*, *281*, C922–C931.

Ho, K. W., Lambert, W. S., & Calkins, D. J. (2014). Activation of the TRPV1 cation channel contributes to stress-induced astrocyte migration. *Glia*, *62*, 1435–1451.
Honore, E. (2007). The neuronal background K2P channels: Focus on TREK1. *Nature Reviews Neuroscience*, *8*, 251–261.
Huang, C., Hu, Z. L., Wu, W. N., Yu, D. F., Xiong, Q. J., Song, J. R., et al. (2010). Existence and distinction of acid-evoked currents in rat astrocytes. *Glia*, *58*, 1415–1424.
Huda, R., McCrimmon, D. R., & Martina, M. (2013). pH modulation of glial glutamate transporters regulates synaptic transmission in the nucleus of the solitary tract. *Journal of Neurophysiology*, *110*, 368–377.
Hwang, E. M., Kim, E., Yarishkin, O., Woo, D. H., Han, K. S., Park, N., et al. (2014). A disulphide-linked heterodimer of TWIK-1 and TREK-1 mediates passive conductance in astrocytes. *Nature Communications*, *5*, 3227.
Ishii, M., Horio, Y., Tada, Y., Hibino, H., Inanobe, A., Ito, M., et al. (1997). Expression and clustered distribution of an inwardly rectifying potassium channel, KAB-2/Kir4.1, on mammalian retinal Muller cell membrane: Their regulation by insulin and laminin signals. *Journal of Neuroscience*, *17*, 7725–7735.
Julius, D. (2013). TRP channels and pain. *Annual Reviews in Cellular and Developmental Biology*, *29*, 355–384.
Kasymov, V., Larina, O., Castaldo, C., Marina, N., Patrushev, M., Kasparov, S., et al. (2013). Differential sensitivity of brainstem versus cortical astrocytes to changes in pH reveals functional regional specialization of astroglia. *Journal of Neuroscience*, *33*, 435–441.
Kellenberger, S., & Schild, L. (2015). International Union of Basic and Clinical Pharmacology. XCI. Structure, function, and pharmacology of acid-sensing ion channels and the epithelial Na+ channel. *Pharmacological Reviews*, *67*, 1–35.
Kofuji, P., Ceelen, P., Zahs, K. R., Surbeck, L. W., Lester, H. A., & Newman, E. A. (2000). Genetic inactivation of an inwardly rectifying potassium channel (Kir4.1 subunit) in mice: Phenotypic impact in retina. *Journal of Neuroscience*, *20*, 5733–5740.
Kraig, R. P., & Chesler, M. (1990). Astrocytic acidosis in hyperglycemic and complete ischemia. *Journal of Cerebral Blood Flow Metabolism*, *10*, 104–114.
Kucheryavykh, Y. V., Kucheryavykh, L. Y., Nichols, C. G., Maldonado, H. M., Baksi, K., Reichenbach, A., et al. (2007). Downregulation of Kir4.1 inward rectifying potassium channel subunits by RNAi impairs potassium transfer and glutamate uptake by cultured cortical astrocytes. *Glia*, *55*, 274–281.
Larsen, B. R., Assentoft, M., Cotrina, M. L., Hua, S. Z., Nedergaard, M., Kaila, K., et al. (2014). Contributions of the Na(+)/K(+)-ATPase, NKCC1, and Kir4.1 to hippocampal K(+) clearance and volume responses. *Glia*, *62*, 608–622.
Larsen, B. R., & MacAulay, N. (2014). Kir4.1-mediated spatial buffering of K(+): Experimental challenges in determination of its temporal and quantitative contribution to K(+) clearance in the brain. *Channels (Austin.)*, *8*, 544–550.
Lauritzen, I., Blondeau, N., Heurteaux, C., Widmann, C., Romey, G., & Lazdunski, M. (2000). Polyunsaturated fatty acids are potent neuroprotectors. *EMBO Journal*, *19*, 1784–1793.
Lee, K. M., & MacLean, A. G. (2015). New advances on glial activation in health and disease. *World Journal of Virology*, *4*, 42–55.
Lesage, F., & Lazdunski, M. (2000). Molecular and functional properties of two-pore-domain potassium channels. *American Journal of Physiology Renal Physiology*, *279*, F793–F801.
Lin, S. H., Chien, Y. C., Chiang, W. W., Liu, Y. Z., Lien, C. C., & Chen, C. C. (2015). Genetic mapping of ASIC4 and contrasting phenotype to ASIC1a in modulating innate fear and anxiety. *European Journal of Neuroscience*, *41*, 1553–1568.
Lin, Y. C., Liu, Y. C., Huang, Y. Y., & Lien, C. C. (2010). High-density expression of Ca^{2+}-permeable ASIC1a channels in NG2 glia of rat hippocampus. *PLoS ONE*, *5*e12665.

Lu, L., Wang, W., Peng, Y., Li, J., Wang, L., & Wang, X. (2014). Electrophysiology and pharmacology of tandem domain potassium channel TREK-1 related BDNF synthesis in rat astrocytes. *Naunyn Schmiedeberg's Archives Pharmacology*, *387*, 303–312.

Maingret, F., Lauritzen, I., Patel, A. J., Heurteaux, C., Reyes, R., Lesage, F., et al. (2000). TREK-1 is a heat-activated background K(+) channel. *EMBO Journal*, *19*, 2483–2491.

Mannari, T., Morita, S., Furube, E., Tominaga, M., & Miyata, S. (2013). Astrocytic TRPV1 ion channels detect blood-borne signals in the sensory circumventricular organs of adult mouse brains. *Glia*, *61*, 957–971.

Mathie, A., & Veale, E. L. (2007). Therapeutic potential of neuronal two-pore domain potassium-channel modulators. *Current Opinion in Investigational Drugs*, *8*, 555–562.

Matthias, K., Kirchhoff, F., Seifert, G., Hüttmann, K., Matyash, M., Kettenmann, H., et al. (2003). Segregated expression of AMPA-type glutamate receptors and glutamate transporters defines distinct astrocyte populations in the mouse hippocampus. *Journal of Neuroscience*, *23*, 1750–1758.

Miller, P., Kemp, P. J., Lewis, A., Chapman, C. G., Meadows, H. J., & Peers, C. (2003). Acute hypoxia occludes hTREK-1 modulation: Re-evaluation of the potential role of tandem P domain K+ channels in central neuroprotection. *Journal of Physiology*, *548*, 31–37.

Nagelhus, E. A., Mathiisen, T. M., & Ottersen, O. P. (2004). Aquaporin-4 in the central nervous system: Cellular and subcellular distribution and coexpression with KIR4.1. *Neuroscience*, *129*, 905–913.

Nedergaard, M., Goldman, S. A., Desai, S., & Pulsinelli, W. A. (1991). Acid-induced death in neurons and glia. *Journal of Neuroscience*, *11*, 2489–2497.

Noel, J., Zimmermann, K., Busserolles, J., Deval, E., Alloui, A., Diochot, S., et al. (2009). The mechano-activated K+ channels TRAAK and TREK-1 control both warm and cold perception. *EMBO Journal*, *28*, 1308–1318.

Nolte, C., Matyash, M., Pivneva, T., Schipke, C. G., Ohlemeyer, C., Hanisch, U. K., et al. (2001). GFAP promoter-controlled EGFP-expressing transgenic mice: A tool to visualize astrocytes and astrogliosis in living brain tissue. *Glia*, *33*, 72–86.

Olsen, M. L., Campbell, S. C., McFerrin, M. B., Floyd, C. L., & Sontheimer, H. (2010). Spinal cord injury causes a wide-spread, persistent loss of Kir4.1 and glutamate transporter 1: Benefit of 17 beta-oestradiol treatment. *Brain*, *133*, 1013–1025.

Olsen, M., & Sontheimer, H. (2008). Functional implications for Kir4.1 channels in glial biology: From K+ buffering to cell differentiation. *Journal of Neurochemistry*, *107*, 589–601.

Overington, J. P., Al-Lazikani, B., & Hopkins, A. L. (2006). How many drug targets are there? *Nature Reviews Drug Discovery*, *5*, 993–996.

Pasler, D., Gabriel, S., & Heinemann, U. (2007). Two-pore-domain potassium channels contribute to neuronal potassium release and glial potassium buffering in the rat hippocampus. *Brain Research*, *1173*, 14–26.

Patel, A. J., Honoré, E., Lesage, F., Fink, M., Romey, G., & Lazdunski, M. (1999). Inhalational anesthetics activate two-pore-domain background K^+ channels. *Nature Neuroscience*, *2*, 422–426.

Pekny, M., Wilhelmsson, U., & Pekna, M. (2014). The dual role of astrocyte activation and reactive gliosis. *Neuroscience Letters*, *565*, 30–38.

Pessia, M., Imbrici, P., D'Adamo, M. C., Salvatore, L., & Tucker, S. J. (2001). Differential pH sensitivity of Kir4.1 and Kir4.2 potassium channels and their modulation by heteropolymerisation with Kir5.1. *Journal of Physiology*, *532*, 359–367.

Raphemot, R., Kadakia, R. J., Olsen, M. L., Banerjee, S., Days, E., Smith, S. S., et al. (2013). Development and validation of fluorescence-based and automated patch clamp-based functional assays for the inward rectifier potassium channel Kir4.1. *Assay and Drug Development Technology*, *11*, 532–543.

Rehncrona, S. (1985). Brain acidosis. *Annals of Emergency Medicine, 14*, 770–776.
Reichold, M., Zdebik, A. A., Lieberer, E., Rapedius, M., Schmidt, K., Bandulik, S., et al. (2010). KCNJ10 gene mutations causing EAST syndrome (epilepsy, ataxia, sensorineural deafness, and tubulopathy) disrupt channel function. *Proceedings of the National Academy of Sciences USA, 107*, 14490–14495.
Rose, C. R., Waxman, S. G., & Ransom, B. R. (1998). Effects of glucose deprivation, chemical hypoxia, and simulated ischemia on Na^+ homeostasis in rat spinal cord astrocytes. *Journal of Neuroscience, 18*, 3554–3562.
Sala-Rabanal, M., Kucheryavykh, L. Y., Skatchkov, S. N., Eaton, M. J., & Nichols, C. G. (2010). Molecular mechanisms of EAST/SeSAME syndrome mutations in Kir4.1 (KCNJ10). *Journal of Biological Chemistry, 285*, 36040–36048.
Sandoz, G., Douguet, D., Chatelain, F., Lazdunski, M., & Lesage, F. (2009). Extracellular acidification exerts opposite actions on TREK1 and TREK2 potassium channels via a single conserved histidine residue. *Proceedings of the National Academy of Sciences USA, 106*, 14628–14633.
Schirmer, L., Srivastava, R., Kalluri, S. R., Bottinger, S., Herwerth, M., Carassiti, D., et al. (2014). Differential loss of KIR4.1 immunoreactivity in multiple sclerosis lesions. *Annals of Neurology, 75*, 810–828.
Scholl, U. I., Choi, M., Liu, T., Ramaekers, V. T., Hausler, M. G., Grimmer, J., et al. (2009). Seizures, sensorineural deafness, ataxia, mental retardation, and electrolyte imbalance (SeSAME syndrome) caused by mutations in KCNJ10. *Proceedings of the National Academy of Science USA, 106*, 5842–5847.
Seifert, G., Hüttmann, K., Binder, D. K., Hartmann, C., Wyczynski, A., Neusch, C., et al. (2009). Analysis of astroglial K^+ channel expression in the developing hippocampus reveals a predominant role of the Kir4.1 subunit. *Journal of Neuroscience, 29*, 7474–7488.
Seifert, G., Rehn, L., Weber, M., & Steinhäuser, C. (1997). AMPA receptor subunits expressed by single astrocytes in the juvenile mouse hippocampus. *Molecular Brain Research, 47*, 286–294.
Seifert, G., Schilling, K., & Steinhäuser, C. (2006). Astrocyte dysfunction in neurological disorders: A molecular perspective. *Nature Reviews Neuroscience*, 7, 194–206.
Seifert, G., & Steinhäuser, C. (1995). Glial cells in the mouse hippocampus express AMPA receptors with an intermediate Ca^{2+} permeability. *European Journal of Neuroscience*, 7, 1872–1881.
Sibille, J., Dao, D. K., Holcman, D., & Rouach, N. (2015). The neuroglial potassium cycle during neurotransmission: Role of Kir4.1 channels. *PLoS Computational Biology, 11*, e1004137.
Sibille, J., Pannasch, U., & Rouach, N. (2014). Astroglial potassium clearance contributes to short-term plasticity of synaptically evoked currents at the tripartite synapse. *Journal of Physiology, 592*, 87–102.
Siesjo, B. K., Katsura, K., Mellergard, P., Ekholm, A., Lundgren, J., & Smith, M. L. (1993). Acidosis-related brain damage. *Progress in Brain Research, 96*, 23–48.
Srivastava, R., Aslam, M., Kalluri, S. R., Schirmer, L., Buck, D., Tackenberg, B., et al. (2012). Potassium channel KIR4.1 as an immune target in multiple sclerosis. *New England Journal of Medicine, 367*, 115–123.
Steinhäuser, C., Grunnet, M., & Carmignoto, G. (2015). Crucial role of astrocytes in temporal lobe epilepsy. *Neuroscience*. http://dx.doi.org/10.1016/j.neuroscience.2014.12.047.
Steinhäuser, C., Seifert, G., & Bedner, P. (2012). Astrocyte dysfunction in temporal lobe epilepsy: K+ channels and gap junction coupling. *Glia, 60*, 1192–1202.
Steinhäuser, C., Seifert, G., & Deitmer, J. W. (2013). Physiology of astrocytes: Ion channels and transporters. In H. Kettenmann & B. Ransom (Eds.), *Neuroglia*. (3rd ed.). New York, USA: Oxford University Press.

Tang, X., Taniguchi, K., & Kofuji, P. (2009). Heterogeneity of Kir4.1 channel expression in glia revealed by mouse transgenesis. *Glia*, *57*, 1706–1715.

Theparambil, S. M., Naoshin, Z., Thyssen, A., & Deitmer, J. W. (2015). Reversed electrogenic sodium bicarbonate cotransporter 1 is the major acid loader during recovery from cytosolic alkalosis in mouse cortical astrocytes. *Journal of Physiology*, *593*, 3533–3547.

Theparambil, S. M., Ruminot, I., Schneider, H. P., Shull, G. E., & Deitmer, J. W. (2014). The electrogenic sodium bicarbonate cotransporter NBCe1 is a high-affinity bicarbonate carrier in cortical astrocytes. *Journal of Neuroscience*, *34*, 1148–1157.

Tong, X., Ao, Y., Faas, G. C., Nwaobi, S. E., Xu, J., Haustein, M. D., et al. (2014). Astrocyte Kir4.1 ion channel deficits contribute to neuronal dysfunction in Huntington's disease model mice. *Nature Neuroscience*, *17*, 694–703.

Torrente, D., Cabezas, R., Avila, M. F., Garcia-Segura, L. M., Barreto, G. E., & Guedes, R. C. (2014). Cortical spreading depression in traumatic brain injuries: Is there a role for astrocytes? *Neuroscience Letters*, *565*, 2–6.

Toth, A., Boczan, J., Kedei, N., Lizanecz, E., Bagi, Z., Papp, Z., et al. (2005). Expression and distribution of vanilloid receptor 1 (TRPV1) in the adult rat brain. *Molecular Brain Research*, *135*, 162–168.

Waldmann, R., Champigny, G., Bassilana, F., Heurteaux, C., & Lazdunski, M. (1997). A proton-gated cation channel involved in acid-sensing. *Nature*, *386*, 173–177.

Wallraff, A., Kohling, R., Heinemann, U., Theis, M., Willecke, K., & Steinhäuser, C. (2006). The impact of astrocytic gap junctional coupling on potassium buffering in the hippocampus. *Journal of Neuroscience*, *26*, 5438–5447.

Wang, W., Putra, A., Schools, G. P., Ma, B., Chen, H., Kaczmarek, L. K., et al. (2013). The contribution of TWIK-1 channels to astrocyte K(+) current is limited by retention in intracellular compartments. *Frontiers in Cellular Neuroscience*, 7, 246.

Wang, M., Song, J., Xiao, W., Yang, L., Yuan, J., Wang, W., et al. (2011). Changes in lipid-sensitive two-pore domain potassium channel TREK-1 expression and its involvement in astrogliosis following cerebral ischemia in rats. *Journal of Molecular Neuroscience*, *46*, 384–392.

Wenker, I. C., Kreneisz, O., Nishiyama, A., & Mulkey, D. K. (2010). Astrocytes in the retrotrapezoid nucleus sense H+ by inhibition of a Kir4.1-Kir5.1-like current and may contribute to chemoreception by a purinergic mechanism. *Journal of Neurophysiology*, *104*, 3042–3052.

Williams, D. M., Lopes, C. M., Rosenhouse-Dantsker, A., Connelly, H. L., Matavel, A., Uchi, J., et al. (2010). Molecular basis of decreased Kir4.1 function in SeSAME/EAST syndrome. *Journal of the American Society of Nephrology*, *21*, 2117–2129.

Woo, D. H., Han, K. S., Shim, J. W., Yoon, B. E., Kim, E., Bae, J. Y., et al. (2012). TREK-1 and Best1 channels mediate fast and slow glutamate release in astrocytes upon GPCR activation. *Cell*, *151*, 25–40.

Wu, X., Liu, Y., Chen, X., Sun, Q., Tang, R., Wang, W., et al. (2013). Involvement of TREK-1 activity in astrocyte function and neuroprotection under simulated ischemia conditions. *Journal of Molecular Neuroscience*, *49*, 499–506.

Xiong, Z. G., Pignataro, G., Li, M., Chang, S. Y., & Simon, R. P. (2008). Acid-sensing ion channels (ASICs) as pharmacological targets for neurodegenerative diseases. *Current Opinion in Pharmacology*, *8*, 25–32.

Zha, X. M., Wemmie, J. A., Green, S. H., & Welsh, M. J. (2006). Acid-sensing ion channel 1a is a postsynaptic proton receptor that affects the density of dendritic spines. *Proceedings of the National Academy of Sciences USA*, *103*, 16556–16561.

Zhou, M., Xu, G., Xie, M., Zhang, X., Schools, G. P., Ma, L., et al. (2009). TWIK-1 and TREK-1 are potassium channels contributing significantly to astrocyte passive conductance in rat hippocampal slices. *Journal of Neuroscience*, *29*, 8551–8564.

CHAPTER EIGHT

SUMOylation and Potassium Channels: Links to Epilepsy and Sudden Death

Hongmei Wu*, Xu Chen*, Jinke Cheng†, Yitao Qi*,[1]
*College of Life Sciences, Shaanxi Normal University, Xi'an, Shaanxi, PR China
†Department of Biochemistry and Molecular Cell Biology, Shanghai Jiao Tong University School of Medicine, Shanghai, PR China
[1]Corresponding author: e-mail address: qiyitao@snnu.edu.cn

Contents

Abstract

Neuronal potassium ion channels play an essential role in the generation of the action potential and excitability of neurons. The dysfunction of ion channel subunits can cause channelopathies, which are associated in some cases with sudden unexplained death in epilepsy SUDEP. The physiological roles of neuronal ion channels have been largely determined, but little is known about the molecular mechanisms underlying neurological channelopathies, especially the determinants of the channels' regulation. SUMO (small ubiquitin-like modifier) proteins covalently conjugate lysine residues in a large

Advances in Protein Chemistry and Structural Biology, Volume 103
ISSN 1876-1623
http://dx.doi.org/10.1016/bs.apcsb.2015.11.009

number of target proteins and modify their functions. SUMO modification (SUMOylation) has emerged as an important regulatory mechanism for protein stability, function, subcellular localization, and protein–protein interactions. Since SUMO was discovered almost 20 years ago, the biological contribution of SUMOylation has not fully understood. It is until recently that the physiological impacts of SUMOylation on the regulation of neuronal potassium ion channels have been investigated. It is well established that SUMOylation controls many aspects of nuclear function, but it is now clear that it is also a key determinant in the function of potassium channels, and SUMOylation has also been implicated in a wide range of channelopathies, including epilepsy and sudden death.

1. INTRODUCTION

Sudden unexplained death in epilepsy (SUDEP) is the most common cause of premature mortality in patients with epilepsy and has been linked to mutations in neuronal ion channel genes. Genes within the channel protein interactome may also contribute to SUDEP pathogenesis. Recent researches have shown that mutation-induced subtle alterations in ion channel properties, regulation, or trafficking may be responsible for seizures, leading to SUDEP, such as those associate with the cognitive deficit and autism-spectrum behaviors (Han et al., 2012) and sudden unexpected death (Kalume et al., 2013) in Dravet syndrome. Apart from the mutation-induced dysfunctions of the neuronal ion channels, the posttranslational modifications of ion channels, including phosphorylation, methylation, acetylation, and ubiquitination have also been implicated in regulation of ion channel function (Hershko, Ciechanover, Heller, Haas, & Rose, 1980; Hershko, Ciechanover, & Rose, 1979; Hochstrasser, 2000; Pickart, 2001).

Many posttranslational modifications have been implicated in regulation of neuronal ion channel function, and SUMOylation has been recently linked to neuronal ion channel degradation and regulation (Zhao, 2007). Over the past several years, modification by small ubiquitin-like modifier (SUMO), a ubiquitin-like protein (UBL), has emerged as a critical factor in multiple biological processes, such as DNA transcription and replication, and the events following DNA damage (Geiss-Friedlander & Melchior, 2007; Johnson, 2004; Ulrich, 2009; Yeh, 2009). SUMOylation of multiple plasma and cytosolic membrane proteins has been implicated in a variety of neuronal functions, and the perturbations in neuronal SUMOylation status contribute to numerous pathological conditions (Anderson, Wilkinson, & Henley, 2009; Berndt, Wilkinson, & Henley, 2012; Henley,

Craig, & Wilkinson, 2014). Through modification of ion channels in neurons, SUMOylation exerts an essential role in controlling membrane potential and cell excitability, which are biophysical markers relevant to epilepsy and sudden death. SUMOylation of a number of neuronal ion channels has been demonstrated, however, the role of the SUMOylation of specific ion channel in neuronal disorders *in vivo* had not been documented until our recent report (Qi, Wang, et al., 2014). Here, we provide an overview of current understanding of the roles of SUMOylation in potassium channels and SUDEP.

2. SUMOylation AND De-SUMOylation MACHINERY

2.1 SUMOylation Pathway

There are more than a dozen UBLs in mammals (Hochstrasser, 2009; Schulman & Harper, 2009). The conjugation of ubiquitin or UBLs to lysine side chains in target proteins is a protein-based posttranslational modification, which is responsible for the functional diversities of modified proteins. Our laboratory and several others reported a novel UBL protein called SUMO1 (also known as sentrin1, GMP1, PIC1, SMT3, and UBL1) about 20 years ago (Boddy, Howe, Etkin, Solomon, & Freemont, 1996; Mahajan, Delphin, Guan, Gerace, & Melchior, 1997; Matunis, Coutavas, & Blobel, 1996; Okura et al., 1996; Shen, Pardington-Purtymun, Comeaux, Moyzis, & Chen, 1996). SUMO can covalently modify a large number of proteins and plays a wide range of roles in different biological processes (Geiss-Friedlander & Melchior, 2007; Johnson, 2004; Ulrich, 2009; Yeh, 2009).

Mammalian cells express four SUMO isoforms: SUMO1, SUMO2, SUMO3, and SUMO4 (Johnson, 2004). SUMO1, SUMO2, and SUMO3 consist of 101, 103, and 95 amino acids, respectively (Kamitani, Kito, Nguyen, Fukuda-Kamitani, & Yeh, 1998). SUMO2 and SUMO3 are more closely related to each other (93.6% identity in a 94 residue overlap) than to SUMO1 (52.4% identity in an 84 residue overlap) and hence are referred to as SUMO2/3. Once conjugated to their cellular substrate, SUMO2 and SUMO3 efficiently form polymeric chains, while in contrast, SUMO1 polychains are rarely observed (Ulrich, 2008). Posttranslational modification of cellular substrates by SUMO4 is questionable. SUMO4 is probably not conjugated under physiological conditions, as it remains in the inactive form *in vivo*, so its biological role is unclear (Bohren, Nadkarni, Song, Gabbay, & Owerbach, 2004; Owerbach, McKay, Yeh, Gabbay, & Bohren, 2005). SUMO homologs have been reported in various species ranging from

Arabidopsis thaliana to *Homo sapiens*, suggesting that SUMO is an evolutionarily conserved protein that may have unique functions in biological processes.

SUMO is readily conjugated to numerous cellular protein substrates via SUMOylation, which has emerged as an important regulatory mechanism in many signaling pathways that alters the function of target proteins (Rytinki & Palvimo, 2009; Wang & Schwartz, 2010). There is a remarkable conservation of mechanisms between ubiquitination and SUMOylation pathways. SUMO conjugation is regulated through an enzyme-controlled cycle, and it is mediated via a sequential process that closely mimics ubiquitination. However, the activating (E1), conjugating (E2), and ligating (E3) enzymes involved in SUMOylation are entirely distinct from ubiquitin E1, E2, and E3.

SUMO paralogs are synthesized as precursors that have to be processed to reveal the C-terminal diglycine motif to become matured forms. The matured SUMO is activated by the formation of a thioester bond between the carboxy-G residue and the C residue on the SUMO-activating enzyme (E1), a heterodimer containing SAE1 (SUMO-activating enzyme subunit 1) and SAE2. Next, SUMO is transferred by a SUMO-conjugating enzyme (E2), Ubc9, to a substrate to form an isopeptide bond between the C-terminal carboxy group of SUMO and the ε-amino group of lysine in the substrate protein, accompanied with a SUMO-ligasing enzyme (E3) (Gareau & Lima, 2010; Gong, Li, Millas, & Yeh, 1999; Johnson, 2004). Our laboratory discovered Ubc9 in a yeast two-hybrid screen using SUMO1 as bait and revealed that Ubc9 was able to form a thiolester adduct with SUMO1 but not with ubiquitin (Gong, Kamitani, Fujise, Caskey, & Yeh, 1997). Thus, in contrast to the large number of E2 used in the ubiquitination pathway, Ubc9 is the only known conjugating enzyme for the SUMOylation pathway (Gong et al., 1997; Johnson & Blobel, 1997).

Three types of SUMO E3 have been discovered: protein inhibitor of the activator of STAT family of proteins (Johnson & Gupta, 2001; Kahyo, Nishida, & Yasuda, 2001); RanBP2, which is localized in the nuclear pore complex (Pichler, Gast, Seeler, Dejean, & Melchior, 2002); and Pc2, which is a component of polycomb protein complexes (Kagey, Melhuish, & Wotton, 2003), and also can be SUMOylated (Kang et al., 2010). Typically, lysine residues modified by SUMO are found within a SUMO modification consensus motif ψKXE (where ψ is a large hydrophobic residue and X is any amino acid), although modification at nonconsensus sites has been reported (Kamitani, Kito, Nguyen, Wada, et al., 1998; Rodriguez,

Dargemont, & Hay, 2001; Yeh, Gong, & Kamitani, 2000). Hence, SUMO conjugation of the target protein requires recognition of this sequence by the E2-SUMO (or E3-SUMO) complex. SUMO2 and SUMO3, like Smt3p, can form chains *in vitro* and *in vivo*, primarily through a conserved lysine (Lys15 in Smt3p, Lys11 in SUMO2 or SUMO3) (Bencsath, Podgorski, Pagala, Slaughter, & Schulman, 2002; Bylebyl, Belichenko, & Johnson, 2003; Tatham et al., 2001).

RanGAP1 and promyelocytic leukemia protein were the first two SUMOylated substrates identified (Kamitani, Nguyen, Kito, Fukuda-Kamitani, & Yeh, 1998; Mahajan et al., 1997; Matunis et al., 1996). Transcription factors are the most common targets for SUMO conjugation, which suggests that SUMOylation plays an important role in regulation of transcription factor activity (Gill, 2003; Ouyang, Valin, & Gill, 2009). The list of SUMOylated proteins has expanded significantly to include nuclear, cytosolic, and plasma membrane proteins (Buschmann, Fuchs, Lee, Pan, & Ronai, 2000; Chauchereau, Amazit, Quesne, Guiochon-Mantel, & Milgrom, 2003; David, Neptune, & DePinho, 2002; Girdwood et al., 2003; Gostissa et al., 1999; Kotaja, Karvonen, Janne, & Palvimo, 2002; Mao, Desai, & Liu, 2000; Mao, Sun, Desai, & Liu, 2000; Poukka, Karvonen, Janne, & Palvimo, 2000). SUMOylation competes with ubiquitination on the same lysine residues, thus functioning almost like an antiubiquitin (Desterro, Rodriguez, & Hay, 1998). SUMOylated RanGAP1 is localized in the nuclear envelope, whereas unmodified RanGAP1 is localized in the cytosol (Hay, Vuillard, Desterro, & Rodriguez, 1999), so SUMOylation can alter a protein's cellular localization. Overall, SUMO covalently modifies a large number of cellular proteins and regulates the localization, function, and protein–protein interaction of proteins (Geiss-Friedlander & Melchior, 2007; Hay, 2005; Yeh, 2009).

2.2 Reversal of SUMOylation by SUMO-Specific Proteases

SUMO conjugation is a dynamic process that is usually short lived, and SUMOylated proteins are rapidly deconjugated by SUMO-specific proteases (SENPs) (Mukhopadhyay & Dasso, 2007; Yeh, 2009).The initial step of SUMOylation requires the inactive SUMO to be cleaved at the carboxy terminus by the hydrolase activity of SENP. SENP exhibits isopeptidase activity to cleave the isopeptide bond between the glycine residue of SUMO and the lysine side chain of substrate (Yeh et al., 2000). The catalytic activity is maintained within a highly conserved 200 amino acid region in the

C-terminus of the proteases. This enzymatic activity of the SENP exposes two glycine residues and generates active SUMO. Studies have shown that SENPs are important determinants of SUMO modification status in cells, and the deconjugation by SENPs plays a crucial role in determining a protein's SUMOylation status and activity (Cheng, Kang, Zhang, & Yeh, 2007; Jiang et al., 2012; Kang et al., 2010; Van Nguyen et al., 2012; Xu et al., 2010).

There are six mammalian SENPs with different subcellular locations and substrate specificities, and they exhibit different endopeptidase activities (Best et al., 2002; Gong, Millas, Maul, & Yeh, 2000; Hang & Dasso, 2002; Kim et al., 2000; Nishida, Kaneko, Kitagawa, & Yasuda, 2001; Nishida, Tanaka, & Yasuda, 2000; Yeh et al., 2000). These SENPs can deconjugate mono-SUMOylated proteins or disassemble polymeric SUMO side chains. SENP1, SENP2, SENP3, and SENP5 are more closely related to the yeast Ulp1, whereas SENP6 and SENP7 are related to Ulp2 (Yeh, 2009). SENP1 and SENP2 can deSUMOylate cellular substrates modified by any of the three SUMO isoforms, while the remaining four SENPs are more efficient at deconjugating SUMO2 and SUMO3 than SUMO1. SENP1, SENP6, and SENP7 are distributed in the nucleoplasm. SENP2 is more compartmentalized in the nuclear pore complex and SENP3 and SENP5 in the nucleolus. In addition, SENP1 and SENP2 express nuclear export sequences that allow these enzymes to shuttle in and out of the nucleus. Differences in the subcellular localization of the SENPs contribute to the selectivity of deSUMOylation for specific cellular proteins. Six SENPs can be divided into three independent subfamilies based on their sequence homology, substrate specificity, and cellular localization: family 1, SENP1 and SENP2; family 2, SENP3 and SENP5; and family 3, SENP6 and SENP7. SENP1 (Cheng et al., 2007), SENP2 (Kang et al., 2010), and SENP6 (unpublished data) knockout mouse embryos do not survive to birth, suggesting that these SENPs are not redundant and must have substrate specificity during development.

SENP1 and SENP2 process both C-terminal hydrolysis and isopeptidase activity in mammalian cells (Mukhopadhyay & Dasso, 2007; Yeh, 2009). SENP1 is a nuclear protease that deconjugates a large number of SUMOylated proteins (Gong et al., 2000) and is more efficient at processing both SUMO1 and SUMO2 than SUMO3 (Shen, Tatham, et al., 2006). There are two residues at the C-terminal side of the cleavage site that make significant contribution to SUMO process by SENP1 (Shen, Tatham, et al., 2006) and SENP2 (Reverter & Lima, 2006). SENP1-(415–643) deconjugates SUMO1- and SUMO2-conjugated RanGAP1 in the same manner *in vitro* (Shen, Tatham, et al., 2006), but it can discriminate between SUMO1-conjugated RanGAP1 and Sp100 (Shen, Dong, Liu, Naismith, & Hay, 2006), suggesting that the structure of its target proteins also impacts its isopeptidase activity. SENP1 plays a key role in the hypoxic response by regulating HIF1α (hypoxia-inducible factor 1α) stability (Cheng et al., 2007) and mitochondrial biogenesis (Cai et al., 2012).

SENP2 has been reported to be tethered to the nuclear pore through binding to Nup153 nucleoporin (Hang & Dasso, 2002; Zhang, Saitoh, & Matunis, 2002). SENP2 is also localized in a yet undefined nuclear speckle that is distinct from the nuclear body (Best et al., 2002). Our laboratory has shown previously that SENP2 contains both nuclear import and export signals that can shuttle between the nucleus and cytoplasm (Itahana, Yeh, & Zhang, 2006). SENP2 deconjugates SUMO1-conjugated RanGAP1 more efficiently than SUMO2-conjugated RanGAP1 *in vitro*, in agreement with its higher hydrolytic activity in processing SUMO2 (Reverter & Lima, 2006). The isopeptidase activities of SENP1 and SENP2 seem to be related to the nature of enzyme's interface with SUMOs. A spliced isoform of mouse SENP2, called SuPr1, can alter the distribution of nuclear POD (promyelocytic leukemia protein oncogenic domain)-associated proteins, such as CBP (CREB-binding protein) and Daxx, and convert Sp3 to a strong activator with diffuse nuclear localization (Best et al., 2002; Ross, Best, Zon, & Gill, 2002). SENP2 is the key regulator of Pc2/CBX4 function, which is critical for embryonic heart development, through regulation of the protein's SUMOylation status and is involved in the binding of polycomb complex to H3K27me3 (Kang et al., 2010). SENP2 also regulates myostatin expression and myogenesis (Qi, Zuo, Yeh, & Cheng, 2014) and has been implicated in Kv7 potassium channel-related seizures and sudden death (Qi, Wang, et al., 2014).

Our laboratory has characterized SENP3 and SENP5 as members constituting a family of nucleolar SENPs with preference for SUMO2/3 (Gong & Yeh, 2006). SENP3 and SENP5 are more active in deconjugating SUMO2/3-conjugated targets than SUMO1-containing ones (Di Bacco et al., 2006; Gong & Yeh, 2006). SENP3 can shuttle from the nucleolus to the nucleus under mild oxidative stress (Huang et al., 2009). It also regulates the transcriptional activity of HIF1α via deSUMOylation of the coregulatory protein p300 (Huang et al., 2009). SENP5 processes the precursor of SUMO3 only when compared with SENP1 and SENP2 (Di Bacco et al., 2006; Gong & Yeh, 2006).

SENP6 and SENP7, which are members of Ulp2 family, have an additional insertion in the catalytic domain, are mainly localized in the nucleus (Shen, Geoffroy, Jaffray, & Hay, 2009), and show very low processing activity (Lima & Reverter, 2008; Mukhopadhyay et al., 2006). In yeast, Ulp2 deletion causes an accumulation of high-molecular-weight conjugated species of Smt3p. In mammalian cells, SENP6 and SENP7 show stronger activity on poly-SUMO2/3 chains than on single SUMO moieties and SUMO1 chains (Lima & Reverter, 2008; Mukhopadhyay et al., 2006). SENP6 regulation of RPA70 plays a critical role in the regulation of DNA repair through homologous recombination (Dou, Huang, Singh, Carpenter, & Yeh, 2010). The structure of SENP7 is unique among well-characterized Ulp family members such as SENP1, SENP2, and Ulp1. The deletion of a distinct SENP7 region, whose location appears to explain its specificity in interactions with an extended SUMO chain, had no effect on the activities of SENP7 (Lima & Reverter, 2008). Differential expression of SENP7 variants regulates epithelial–mesenchymal transition (Bawa-Khalfe et al., 2012).

3. SUMOylation AND POTASSIUM ION CHANNELS

3.1 K2P1

The K2P1 channel (also known as TWIK), the first mammalian member of the K2P superfamily, is background potassium-selective leak channel expressed widely with abundant transcription in many organs and has been implicated in background membrane potassium conductance. Although the channel is expressed at the cell surface in heterologous expression systems, no native currents have been recorded from surface, possibly because the channel is silenced in the basal state (Plant, Rajan, & Goldstein, 2005). Similarly, no K2P1 activity was recorded, and this was attributed to basal SUMOylation of the channels at residue K274 blocking the pore (Rajan, Plant, Rabin, Butler, & Goldstein, 2005). K2P1 channels were expressed at the plasma membrane; however, K2P1 current could only be recorded when lysine-274 was mutated to glutamic acid (K274E), demonstrating that the motif $LK^{274}KF$ is a potential target of SUMOylation via the catalytic activity of Ubc9 enzyme. Expression of a SUMOylation-deficient lysine point mutant (K274E) resulted in the appearance of an active, pH-sensitive, openly rectifying potassium current in COS7 cells (Rajan et al., 2005). K2P1 is inactive in the plasma membrane when covalently modified by SUMO. Consistent with this, expression of wild-type K2P1 together with SENP1 resulted in activation of the channels (Rajan et al., 2005), suggesting an inhibitory role for SUMOylation. Interestingly, another study demonstrated that the mutant K2P1-K274E produces the expected outward potassium current (Feliciangeli et al., 2007). However, the authors of that study did not detect such a current from the positively charged variant K2P1-K274R, suggesting that this effect was not dependent on SUMOylation but was most likely due to the exchange of a positively charged lysine for a negatively charged glutamate. The later study confirmed that the SUMO pathway operates at the plasma membrane of living mammalian cells to regulate K2P1 channels (Table 1). One SUMO1 monomer is sufficient to silence the K2P1 channel, and SUMO2 and SUMO3 together can silence the channel as well (Plant, Dementieva, et al., 2010).

3.2 Kv1.1 and Kv1.2

Kv1.1 and Kv1.2, which are encoded by the two most closely related genes *Kcna1* and *Kcna2*, are the most abundant Kv1 channels in nervous system. These two channels are coexpressed in large axons and commonly occur within the same tetramers. Kv1.1 subunits are widely expressed in the brain, where they regulate action potential propagation and shape, neuronal repetitive firing properties, and neurotransmitter release (Dodson & Forsythe, 2004). Mice-bearing Kv1.1 mutations exhibit hippocampal and peripheral nerve hyperexcitability, severe epilepsy, and premature death (Glasscock, Qian, Yoo, & Noebels, 2007). *Kcna1* loss-of-function mutations also cause neuronal disorders in human, including partial epilepsy, episodic ataxia, and myokymia (Zuberi et al., 1999). Kv1.1 deficiency associated brain-driven cardiac dysfunction has been characterized as a candidate mechanism for SUDEP (Glasscock et al., 2010). Kv1.2-deficient mice manifested more severe seizures characterized by a sudden explosive onset of wild running and jumping, followed by a tonic–clonic seizure, which led to early death less than 3 weeks after birth (Table 1; Brew et al., 2007). Our laboratory has reported that Kv1.1 channel colocalizes with SENP2 in hippocampal neurons and was modified by both SUMO1 and SUMO2 (Qi, Wang, et al., 2014). Kv1.1 and Kv1.2 contain similarly located consensus SUMOylation sites, suggesting that similar SUMO regulation of these channels may also occur in the central nervous system.

Table 1 Effects of SUMOylation on Potassium (K^+) Channels in the Nervous and Cardiac Systems

SUMO/SENP	K^+ Channels	Biophysical Function	Test Models	Biophysical and Biological Effects
SUMO1 and SENP1	K2P1	Determine resting membrane potential and excitability	Oocytes, COS7, and CHO cells	SUMO1 silences the channel, increasing the excitability of neurons; SENP1 reverses this effect (Feliciangeli et al., 2007; Plant, Dementieva, et al., 2010; Plant, Zuniga, Olikara, & Goldstein, 2010; Rajan et al., 2005)
SUMO1, SUMO2/3, and SENP2	Kv1.1 Kv1.2 Kv1.5	Contribute to APD[a] in atrial myocardium	COS7 and Sf-9 cells, knockout mice	SUMOylation inactivates the channel, altering the excitability of both atrial myocytes and vascular smooth muscle cells, thereby increasing susceptibility to AF[b]; also leads to loss of the 4-aminopyridine-sensitive component of IK_{slow}[c] and resistance to drug-induced QT prolongation (Benson et al., 2007)
SUMO1 and SENP1	Kv2.1	Determine intrinsic excitability and APD	CHO cells and hippocampal neurons	SUMOylation increases the excitability of hippocampal neurons and decreases channel current in cardiomyocytes, increasing excitability (Dai, Kolic, Marchi, Sipione, & Macdonald, 2009; Plant, Dowdell, Dementieva, Marks, & Goldstein, 2011)
SUMO1, SUMO2/3, and SENP2	Kv1.1	Control excitability in neurons (link with epilepsy and cardiac arrhythmias)	Knockout mice	SUDEP, heart block due to increased parasympathetic activity in vagus nerves (Glasscock, Yoo, Chen, Klassen, & Noebels, 2010; Zhou, Zhang, Messing, & Chiu, 1998)

Table 1 Effects of SUMOylation on Potassium (K^+) Channels in the Nervous and Cardiac Systems—cont'd

SUMO/SENP	K^+ Channels	Biophysical Function	Test Models	Biophysical and Biological Effects
SUMO1, SUMO2/3, and SENP2	Kv1.2	Control excitability in neurons (link with epilepsy and cardiac arrhythmias)	Knockout mice	Seizures and reduced life span (death within 3 weeks), hypoexcitability, and augmented Kv1 currents in auditory neurons (Brew et al., 2007)
SUMO2/3 and SENP2	Kv7	Control neuronal excitability and spiking behavior (link with epilepsy)	Mice with mutations or knockout	Kv7 mutations causes benign familial neonatal convulsions with seizures; SENP2 deficiency causes SUDEP (Otto, Yang, Frankel, White, & Wilcox, 2006; Qi, Wang, et al., 2014)

[a]APD, action potential duration.
[b]AF, atrial fibrillation.
[c]IK_{slow}, slow potassium ion current.

3.3 Kv1.5

In the central nervous system, the voltage-gated Kv1.5 channel mediates the ultrarapid potassium current, and is involved in microglial cell proliferation and the production and release of nitric oxide following brain injury (Pannasch et al., 2006). Regulation and trafficking of the Kv1.5 channel via posttranslational modification has been suggested (Olson et al., 2006). Kv1.5 has a conserved SUMOylation motif, which is required to accept SUMO molecules transferred by Ubc9, and can be SUMOylated *in vivo* with all three SUMO paralogs (Benson et al., 2007). Inhibition of Kv1.5 SUMOylation by disruption of the conjugation sites or expression of SENP2 led to a selective hyperpolarizing shift in the voltage dependence of steady-state inactivation, indicating that SUMOylation has an important role in fine-tuning channel function (Benson et al., 2007). SUMOylation of Kv1.5 induces biophysical alterations of steady-state inactivation without decreasing current density (Meinecke et al., 2007). Although not demonstrated *in vivo*, SUMOylation of Kv1.5 has been proposed to regulate the excitability of atrial cells. By altering channel inactivation and the excitability of both atrial myocytes and vascular smooth muscle cells, SUMOylation of Kv1.5 increases susceptibility of atrial fibrillation (Table 1).

3.4 Kv2.1

Kv2.1, a voltage-gated potassium channel expressed in the nervous system, heart, skeletal muscle, pulmonary arteries, and pancreas, is important for determining activity-dependent excitability in the brain (Bou-Abboud, Li, & Nerbonne, 2000; Brunet et al., 2004; Li, Guo, Yamada, & Nerbonne, 2004). Elimination of Kv2 channels in mouse ventricular cardiomyocytes leads to prolongation of both the action potential duration and the QT interval (Kodirov et al., 2004; Zhou et al., 2003). The Kv2.1 expressed in HEK293 cells was able to SUMOylate/deSUMOylate by the complex Ubc9/SENP1 (Dai et al., 2009). SUMOylation of Kv2.1 has been shown to inhibit the channel current and alter its activation and recovery properties. This regulatory mechanism has been proposed to exert a modulatory effect on cellular excitability and action potential firing (Dai et al., 2009). Kv2.1 is the major Kv channel responsible for action potential repolarization in β-cells (Jacobson et al., 2007). SUMOylation exerts a strong inhibitory effect on Kv2.1 and regulates cellular excitability in native β-cells (Dai et al., 2009). SUMOylation of surface Kv2.1 in hippocampal neurons alters the biophysical parameters of channel function, leading to modified cellular activity, decreased channel current, and increased excitability (Table 1; Plant et al., 2011). However, the physiological relevance of this mechanism remains to be delineated.

3.5 Kv7.1, Kv7.2, and Kv7.3

Kv7 family of voltage-dependent potassium ion channels consists of five members Kv7.1–7.5, encoded by the *Kcnq*1–5 genes, respectively (Brown & Passmore, 2009). The five members of the KCNQ family play important roles in neuronal physiology and have emerged as critical regulators of pyramidal neuron excitability and spiking behavior. KCNQ channels play important physiological roles in nervous (Delmas & Brown, 2005) and cardiac (Wang et al., 1996) systems. *Kcnq*1 is expressed abundantly in the heart and epithelia, where it has well-established roles. Two mouse models, each carrying a dominant *Kcnq1* point mutation found in patients with long QT syndrome (LQTS) type 1, reproduce the human long QT phenotype (Casimiro et al., 2004). *Kcnq*2 and *Kcnq*3 are highly expressed in neurons and form the M-current (Cooper, Harrington, Jan, & Jan, 2001).

Overexpression of a *Kcnq2* dominant negative subunit in transgenic mice significantly reduced the M-current and caused elevated excitability and seizures (Peters, Hu, Pongs, Storm, & Isbrandt, 2005). Similarly, mice-carrying missense *Kcnq2* or *Kcnq3* mutations identified in patients with benign familial neonatal convulsions exhibited seizures (Table 1; Singh et al., 2008). Our laboratory has reported that Kv7.2 colocalizes with SENP2 in hippocampal neurons, and Kv7.1, Kv7.2, and Kv7.3 channels were modified by SUMO2, but not SUMO1 (Qi, Wang, et al., 2014), indicating the important roles of SUMOylation/deSUMOylation on Kv7/KCNQ channels in neuronal function.

4. ROLE OF SUMOylation IN EPILEPSY AND SUDDEN DEATH

4.1 SUMOylation Pathway Links Epilepsy and Sudden Death

Epilepsy is a serious medical condition with inherited and acquired forms. Inherited disorders of ion channels are a major cause of disease in excitable tissues in humans. It has been estimated that the rate of sudden death is 20-fold higher in epilepsy patients than in the general population, and SUDEP is the most common epilepsy-related cause of sudden death (Sillanpaa & Shinnar, 2010). Although SUDEP is clearly linked to mutations in ion channels (Glasscock et al., 2010; Goldman et al., 2009; Massey, Sowers, Dlouhy, & Richerson, 2014; Surges & Sander, 2012), the mechanism of SUDEP in humans has not been determined, and no biomarkers are available. Mutations in potassium channel subunits represent the largest category of mutations in epileptic conditions linked to channelopathies (Brenner & Wilcox, 2012; Cooper, 2012; Noebels, 2003). Dysfunction of potassium channel genes has been linked to SUDEP in animal models (Glasscock et al., 2010; Goldman et al., 2009). Mutation of individual subunits of the heteromeric potassium channels gives rise to a wide variety of neural and cardiac excitability disorders, yet genes involved in the posttranslational modification within the membrane may also produce disease (Herren, Bers, & Grandi, 2013). For example, SUMOylation of K2P1 or Kv1.5 inactivates these potassium channels in Xenopus oocytes and myocytes, and SUMOylation of Kv2.1 increases the excitability of cultured hippocampal neurons (Benson et al., 2007; Plant, Dementieva, et al., 2010; Plant et al., 2011; Rajan et al., 2005). However, it had not been known whether SUMOylation of potassium channels *in vitro* may be implicated in common disease phenotypes, and whether it is translatable to animal models until we generated SENP2 knockout as well as conditional knockout mouse models.

To explore the role of SENP2 *in vivo*, we generated SENP2 knockout mice and determined that SENP2 is essential for embryonic heart development and suppression of polycomb group protein-mediated gene silencing during embryonic development (Kang et al., 2010). SENP2 may have different biological roles in mouse embryonic development and adult mice. To further investigate the role of SENP2 in adult mice, we generated SENP2-deficient mice models. However, we found that $SENP2^{fxN/fxN}$ mice, in which the neomycin cassette was retained within the floxed SENP2 allele, developed spontaneous convulsive seizures, and died after 6–8 weeks of birth. Interestingly, $SENP2^{fx/fx}$ mice, generated by removing the neomycin cassette from the floxed allele by flippase recognition target recombination, lived normal lifespans (Qi, Wang, et al., 2014). SENP2 is essential for embryonic heart development while it is also important in adult mice.

4.2 Neuronal Hyper-SUMOylation Causes SUDEP

To further determine the *in vivo* effect of SUMOylation on neuronal excitability, we did video electroencephalogram (EEG) recordings in the cortex and hippocampus and revealed high frequency of spontaneous behavioral and featured tonic–clonic seizures before sudden death. Whole-cell patch clamp recordings of hippocampal CA3 pyramidal neurons in brain slices from $SENP2^{fxN/fxN}$ mice showed characteristic hyperexcitability, including spontaneous firing, increased evoked firing frequency, and repetitive bursting upon mossy fiber stimulation.

The causes of sudden death in epilepsy have not been determined. Animal model and clinical studies show that, in addition to neuronal synchrony and hyperexcitability, respiratory and cardiovascular abnormalities during and after seizures contribute to SUDEP, suggesting important interactions between the brain and heart. The $SENP2^{fxN/fxN}$ mice display prolonged sinus bradycardia, a high-degree atrioventricular (AV) conduction block, and AV dissociation prior to lethal asystole following seizures, suggesting that SENP2 deficiency affected the heart as well as the brain. To determine whether the SUDEP in $SENP2^{fxN/fxN}$ mice was neurogenic or cardiogenic in origin, we crossed *SENP2*$^{fx/fx}$ mice with mice expressing either a tamoxifen-inducible Thy1 Cre recombinase or a myosin heavy-chain (MHC) promoter Cre recombinase to generate neuron- or cardiomyocyte-specific *SENP2* deletion, respectively. Mice with the cardiomyocyte-specific *SENP2* deletion (*SENP2*$^{fx/fx}$-MHC) had a normal lifespan and no phenotype. In contrast, neuron-specific *SENP2* deletion (*SENP2*$^{fx/fx}$-Thy1) resulted in seizures and SUDEP. Simultaneous EEG and electrocardiogram (ECG) recordings showed severe cortical seizure activity, bradycardia, a high-degree AV block, and lethal asystole in *SENP2*$^{fx/fx}$-Thy1 mice, which recapitulated the global hypomorphic phenotype, suggesting that the SUDEP phenotype due to SENP2 deficiency is most likely neurogenic, and the change in neuronal excitability with hyper-SUMOylation of potassium channels is responsible for the SUDEP (Qi, Wang, et al., 2014).

4.3 Hyper-SUMOylation of Kv7 Channels Contributes to Epilepsy and Sudden Death

We revealed that the localizations of Kv1.1 and Kv7.2/Kv7.3 channels were similar in cultured *SENP2*$^{+/+}$ and *SENP2*$^{fxN/fxN}$ hippocampal neurons; however, there was an enrichment of SUMOylated Kv1.1 and Kv7.2/Kv7.3 in the brain of *SENP2*$^{fxN/fxN}$ mice. More intriguingly, by using whole-cell patch clamp recording, we found that M-current, but not Kv1.1 current, was significantly diminished in *SENP2*$^{fxN/fxN}$ hippocampal CA3 neurons, suggesting that the increased SUMOylation in *SENP2*$^{fxN/fxN}$ neurons did not affect the localization and express levels but was critical for modulating M-current (Qi, Wang, et al., 2014). SUMOylation has different roles in Kv1 and Kv7, and hyper-SUMOylation of Kv7 channels underlies the increased excitability of the *SENP2*$^{fxN/fxN}$ hippocampal circuitry.

Kv7 channels are best known for their role in mediating the classic M-current, a potassium conductance that increases as neurons approach the action potential threshold with distinct physiological and pharmacological characteristics (Jentsch, 2000). M-current is mediated by Kv7 channel assembled with a hetero-tetrameric Kv7.2 and Kv7.3 subunits (Scott & Holmes, 2012; Surti & Jan, 2005). M-current has been recorded in many types of neurons in the peripheral and central nervous systems, including hippocampal neurons. The M-current plays a critical role in controlling the subthreshold electrical excitability of neurons as well as their responsiveness to synaptic inputs (Gu, Hu, Vervaeke, & Storm, 2008; Gu, Vervaeke, Hu, & Storm, 2005). Blockade or closure of Kv7 channels facilitates membrane depolarization and fast excitatory transmission, which in turn allows neurons to more readily fire action potentials (Otto et al., 2006). Accordingly, pediatric patients with loss-of-function and missense mutations in Kv7 channels develop mild to very severe epilepsy (Charlier et al., 1998; Weckhuysen et al., 2012). SENP2 deficiency leads to reduced M-current and subsequent hyperexcitability in *SENP2*$^{fxN/fxN}$ hippocampal CA3 neurons, and restoration of SENP2 polypeptide levels rescues the M-current and suppress neuronal hyperexcitability. It was concluded that M-current reduction due to SENP2 deficiency was responsible for hippocampal neuronal hyperexcitability in *SENP2*$^{fxN/fxN}$ mice (Qi, Wang, et al., 2014).

Mutation in the *Kcnq1* gene at the LQT1 locus (11p15.5) is the most common cause of LQTS, which predisposes patients to lethal cardiac arrhythmias (Goldenberg, Zareba, & Moss, 2008; Wang et al., 1996). The SENP2-deficient mice also developed profound bradycardia and high-degree AV conduction blocks but not ventricular arrhythmia (Qi, Wang, et al., 2014). The cardiac AV conduction blocks are usually followed by seizure activity, and can be prevented by atropine but not propranolol (Qi, Wang, et al., 2014), indicating that the AV conduction blocks observed in SENP2-deficient mice are likely due to an increase in parasympathetic tone following seizures. Interestingly, the Kv7.1 channel is also expressed in forebrain and brainstem nuclei, and mice-carrying human LQTS mutations in the *Kcnq1* gene also develop seizures and sudden death due to an increase in parasympathetic tone (Goldman et al., 2009). It was identified that Kv7.1 also can be modified by SUMO2 (Qi, Wang, et al., 2014), which may affect its function. Taken together, the findings indicate that defective Kv7 channel function may explain both seizures and sudden death in SENP2-deficient mice.

Retigabine is a third-generation antiepileptic drug that is clinically approved for treatment of partial-onset seizures. This drug activates channels Kv7.2 through Kv7.5 but not Kv7.1 via a unique mechanism: a hyperpolarization shift of the voltage dependence of channel activation and an increase of the channel's maximal open probability (Tatulian, Delmas, Abogadie, & Brown, 2001). In SENP2-deficient mice, retigabine restored M-current, prevented spontaneous firings, and abolished current injection-evoked firings in hippocampal neurons (Qi, Wang, et al., 2014). Furthermore, retigabine effectively prevented auditory stimulation-induced seizures and post seizure AV conduction blocks in SENP2-deficient mice (Qi, Wang, et al., 2014), suggesting that SENP2 deficiency, leading to hyper-SUMOylation of Kv7.2/7.3 and M-current deduction, may be a primary cause of epileptogenesis. Notably, we did not observe repetitive bursting behavior in CA3 neurons of SENP2-deficient mice in the presence of retigabine. Taken together, these results strongly support the notion that seizures and sudden death in SENP2-deficient mice are due to hyper-SUMOylation of Kv7 channels.

In summary, our study demonstrated that SENP2 deficiency, resulting in hyper-SUMOylation of Kv7 channels, causes spontaneous seizures, cardiac abnormalities, and SUDEP in mice (Fig. 1). With our novel animal model of SUDEP, we obtained the first *in vivo* evidence for a functional effect of

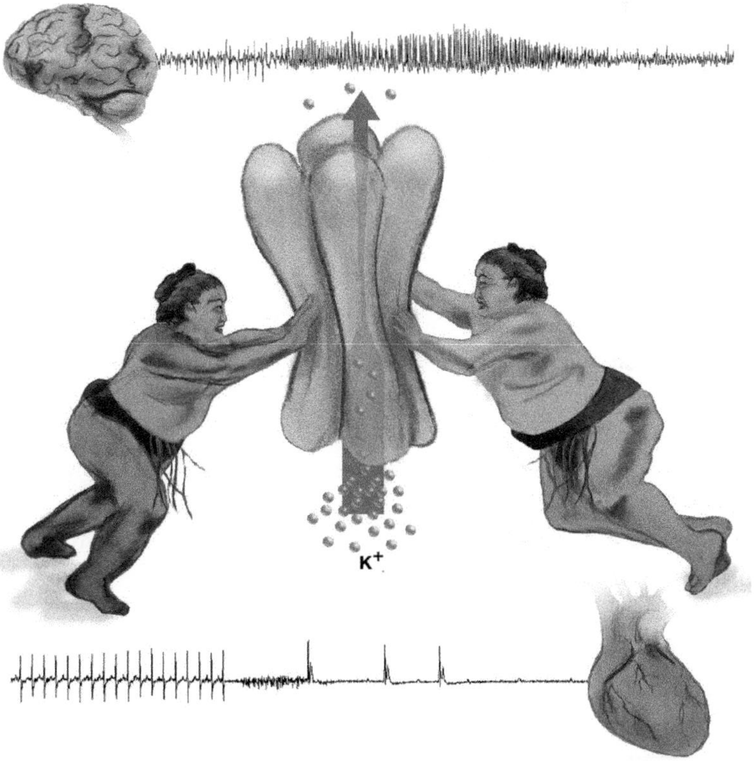

Figure 1 Hyper-SUMOylation of the Kv7.2/Kv7.3 causes seizure and sudden death. The cartoon depicts two SUMO wrestlers squeezing the potassium channel to turn off the flow of potassium ions. As a consequence, neurons become hyperexcitable, causing seizures, as indicated by spikes in EEG at the top. The seizure triggers high-degree cardiac conduction blocks, as indicated by ECG at the bottom.

SUMOylation on ion channel activity and demonstrated that hyper-SUMOylation of Kv7 channel is an important molecular mechanism and a potential target for novel epilepsy therapeutics (Qi, Wang, et al., 2014).

5. CONCLUSIONS AND OUTLOOK

SUMO conjugation has been reported to modulate the function of several neuronal potassium channels, including K2P1 (Rajan et al., 2005), Kv1.5 (Benson et al., 2007), and Kv2.1 (Dai et al., 2009; Plant et al., 2011), suggesting that SUMOylation is a general mechanism for regulating ion channel physiology (Martin, Nishimune, Mellor, & Henley, 2007). However, these studies were all based on *in vitro* findings, and the functional effect of SUMOylation on ion channel activity in live animals has only recently been explored and described. The epileptic and sudden death phenotype of SENP2-deficient mice closely resembles that of Kv1.1-null mice (Glasscock et al., 2010). However, despite the colocalization of SENP2 and Kv1.1 in hippocampal neurons, hyper-SUMOylation of Kv1.1 due to SENP2 deficiency does not significantly affect the biophysical function of Kv1.1 channel. Remarkably, the M-current, which plays a critical role in controlling membrane potentials and firing activity (Brown & Passmore, 2009; Hansen et al., 2008), was significantly diminished in hippocampal neurons of SENP2-deficient mice (Qi, Wang, et al., 2014). Characterization of the M-current as the main causal factor in SENP2 deficiency is supported by the observations that both seizures and AV conduction blocks can be prevented by the Kv7 channel opener retigabine. These findings reveal an important physiological role of SENP2 in regulation of the M-current, and reversal of SUMOylation may be a promising approach for treating subsets of individuals at risk for SUDEP. However, the reason why SUMO modification inhibits the M-current but has no effect on the Kv1 current is unknown, and the mechanism by which SUMOylation regulates the M-current needs further investigation.

Prevention of SUDEP and development of treatments for epilepsy without side effects is critical unmet needs, and SUMOylation can serve as a potential target for development of novel epilepsy therapeutics. We have developed an animal model of SUDEP caused by SUMOylation of Kv7.2/Kv7.3. Our findings expand the number of biological pathways leading to critical ion channel disorders in the central nervous system and validate SENP2 as a factor in SUDEP, adding to the growing list of potential SUDEP mechanisms. We propose that genes in the ion channel protein interactome play an important role in epileptic disorders and that it may be informative to screen patients with epilepsy for genetic variations in *SENP2* that reduce either expression or activity of SENP2. Pharmacological modulation of M-channel activity may also be useful in preventing SUDEP. Our specific *SENP2*$^{fxN/fxN}$ mice model is suitable for screening new antiepilepsy drugs, and other Kv7 channel openers should also be explored for their utility in prevention of seizures and sudden death.

ACKNOWLEDGMENT

We thank the National Natural Science Foundation of China (81502531 to H.W.) for financial support.

REFERENCES

Anderson, D. B., Wilkinson, K. A., & Henley, J. M. (2009). Protein SUMOylation in neuropathological conditions. *Drug News & Perspectives*, *22*(5), 255–265. http://dx.doi.org/10.1358/dnp.2009.22.5.1378636.

Bawa-Khalfe, T., Lu, L. S., Zuo, Y., Huang, C., Dere, R., Lin, F. M., et al. (2012). Differential expression of SUMO-specific protease 7 variants regulates epithelial-mesenchymal transition. *Proceedings of the National Academy of Sciences of the United States of America*, *109*(43), 17466–17471. http://dx.doi.org/10.1073/pnas.1209378109.

Bencsath, K. P., Podgorski, M. S., Pagala, V. R., Slaughter, C. A., & Schulman, B. A. (2002). Identification of a multifunctional binding site on Ubc9p required for Smt3p conjugation. *The Journal of Biological Chemistry*, *277*(49), 47938–47945. http://dx.doi.org/10.1074/jbc.M207442200.

Benson, M. D., Li, Q. J., Kieckhafer, K., Dudek, D., Whorton, M. R., Sunahara, R. K., et al. (2007). SUMO modification regulates inactivation of the voltage-gated potassium channel Kv1.5. *Proceedings of the National Academy of Sciences of the United States of America*, *104*(6), 1805–1810. http://dx.doi.org/10.1073/pnas.0606702104.

Berndt, A., Wilkinson, K. A., & Henley, J. M. (2012). Regulation of neuronal protein trafficking and translocation by SUMOylation. *Biomolecules*, *2*(2), 256–268. http://dx.doi.org/10.3390/biom2020256.

Best, J. L., Ganiatsas, S., Agarwal, S., Changou, A., Salomoni, P., Shirihai, O., et al. (2002). SUMO-1 protease-1 regulates gene transcription through PML. *Molecular Cell*, *10*(4), 843–855.

Boddy, M. N., Howe, K., Etkin, L. D., Solomon, E., & Freemont, P. S. (1996). PIC 1, a novel ubiquitin-like protein which interacts with the PML component of a multiprotein complex that is disrupted in acute promyelocytic leukaemia. *Oncogene*, *13*(5), 971–982.

Bohren, K. M., Nadkarni, V., Song, J. H., Gabbay, K. H., & Owerbach, D. (2004). A M55V polymorphism in a novel SUMO gene (SUMO-4) differentially activates heat shock transcription factors and is associated with susceptibility to type I diabetes mellitus. *The Journal of Biological Chemistry*, *279*(26), 27233–27238. http://dx.doi.org/10.1074/jbc.M402273200.

Bou-Abboud, E., Li, H., & Nerbonne, J. M. (2000). Molecular diversity of the repolarizing voltage-gated K+ currents in mouse atrial cells. *The Journal of Physiology*, *529*(Pt. 2), 345–358.

Brenner, R., & Wilcox, K. S. (2012). Potassium channelopathies of epilepsy. In J. L. Noebels, M. Avoli, M. A. Rogawski, R. W. Olsen, & A. V. Delgado-Escueta

(Eds.), *Jasper's basic mechanisms of the epilepsies [Internet]*. (4th ed.). Bethesda, MD: National Center for Biotechnology Information (US).

Brew, H. M., Gittelman, J. X., Silverstein, R. S., Hanks, T. D., Demas, V. P., Robinson, L. C., et al. (2007). Seizures and reduced life span in mice lacking the potassium channel subunit Kv1.2, but hypoexcitability and enlarged Kv1 currents in auditory neurons. *Journal of Neurophysiology, 98*(3), 1501–1525. http://dx.doi.org/10.1152/jn.00640.2006.

Brown, D. A., & Passmore, G. M. (2009). Neural KCNQ (Kv7) channels. *British Journal of Pharmacology, 156*(8), 1185–1195. http://dx.doi.org/10.1111/j.1476-5381.2009.00111.x. BPH111 [pii].

Brunet, S., Aimond, F., Li, H., Guo, W., Eldstrom, J., Fedida, D., et al. (2004). Heterogeneous expression of repolarizing, voltage-gated K+ currents in adult mouse ventricles. *The Journal of Physiology, 559*(Pt. 1), 103–120. http://dx.doi.org/10.1113/jphysiol.2004.063347.

Buschmann, T., Fuchs, S. Y., Lee, C. G., Pan, Z. Q., & Ronai, Z. (2000). SUMO-1 modification of Mdm2 prevents its self-ubiquitination and increases Mdm2 ability to ubiquitinate p53. *Cell, 101*(7), 753–762.

Bylebyl, G. R., Belichenko, I., & Johnson, E. S. (2003). The SUMO isopeptidase Ulp2 prevents accumulation of SUMO chains in yeast. *The Journal of Biological Chemistry, 278*(45), 44113–44120. http://dx.doi.org/10.1074/jbc.M308357200.

Cai, R., Yu, T., Huang, C., Xia, X., Liu, X., Gu, J., et al. (2012). SUMO-specific protease 1 regulates mitochondrial biogenesis through PGC-1alpha. *The Journal of Biological Chemistry, 287*(53), 44464–44470. http://dx.doi.org/10.1074/jbc.M112.422626.

Casimiro, M. C., Knollmann, B. C., Yamoah, E. N., Nie, L. P., Vary, J. C., Sirenko, S. G., et al. (2004). Targeted point mutagenesis of mouse Kcnq1: Phenotypic analysis of mice with point mutations that cause Romano-Ward syndrome in humans. *Genomics, 84*(3), 555–564. http://dx.doi.org/10.1016/j.ygeno.2004.06.007.

Charlier, C., Singh, N. A., Ryan, S. G., Lewis, T. B., Reus, B. E., Leach, R. J., et al. (1998). A pore mutation in a novel KQT-like potassium channel gene in an idiopathic epilepsy family. *Nature Genetics, 18*(1), 53–55. http://dx.doi.org/10.1038/ng0198-53.

Chauchereau, A., Amazit, L., Quesne, M., Guiochon-Mantel, A., & Milgrom, E. (2003). Sumoylation of the progesterone receptor and of the steroid receptor coactivator SRC-1. *The Journal of Biological Chemistry, 278*(14), 12335–12343. http://dx.doi.org/10.1074/jbc.M207148200.

Cheng, J., Kang, X., Zhang, S., & Yeh, E. T. (2007). SUMO-specific protease 1 is essential for stabilization of HIF1alpha during hypoxia. *Cell, 131*(3), 584–595. http://dx.doi.org/10.1016/j.cell.2007.08.045.

Cooper, E. C. (2012). Potassium channels (including KCNQ) and epilepsy. In J. L. Noebels, M. Avoli, M. A. Rogawski, R. W. Olsen, & A. V. Delgado-Escueta (Eds.), *Jasper's basic mechanisms of the epilepsies [Internet]* (4th ed.). Bethesda, MD: National Center for Biotechnology Information (US).

Cooper, E. C., Harrington, E., Jan, Y. N., & Jan, L. Y. (2001). M channel KCNQ2 subunits are localized to key sites for control of neuronal network oscillations and synchronization in mouse brain. *Journal of Neuroscience, 21*(24), 9529–9540.

Dai, X. Q., Kolic, J., Marchi, P., Sipione, S., & Macdonald, P. E. (2009). SUMOylation regulates Kv2.1 and modulates pancreatic beta-cell excitability. *Journal of Cell Science, 122*(Pt. 6), 775–779. http://dx.doi.org/10.1242/jcs.036632.

David, G., Neptune, M. A., & DePinho, R. A. (2002). SUMO-1 modification of histonedeacetylase 1 (HDAC1) modulates its biological activities. *The Journal of Biological Chemistry, 277*(26), 23658–23663. http://dx.doi.org/10.1074/jbc.M203690200.

Delmas, P., & Brown, D. A. (2005). Pathways modulating neural KCNQ/M (Kv7) potassium channels. *Nature Reviews. Neuroscience*, *6*(11), 850–862. http://dx.doi.org/10.1038/nrn1785.

Desterro, J. M., Rodriguez, M. S., & Hay, R. T. (1998). SUMO-1 modification of IkappaBalpha inhibits NF-kappaB activation. *Molecular Cell*, *2*(2), 233–239.

Di Bacco, A., Ouyang, J., Lee, H. Y., Catic, A., Ploegh, H., & Gill, G. (2006). The SUMO-specific protease SENP5 is required for cell division. *Molecular and Cellular Biology*, *26*(12), 4489–4498. http://dx.doi.org/10.1128/MCB.02301-05.

Dodson, P. D., & Forsythe, I. D. (2004). Presynaptic K+ channels: Electrifying regulators of synaptic terminal excitability. *Trends in Neurosciences*, *27*(4), 210–217. http://dx.doi.org/10.1016/j.tins.2004.02.012.

Dou, H., Huang, C., Singh, M., Carpenter, P. B., & Yeh, E. T. H. (2010). Regulation of DNA repair through deSUMOylation and SUMOylation of replication protein A complex. *Molecular Cell*, *39*(3), 333–345. http://dx.doi.org/10.1016/j.molcel.2010.07.021.

Feliciangeli, S., Bendahhou, S., Sandoz, G., Gounon, P., Reichold, M., Warth, R., et al. (2007). Does SUMOylation control K2P1/TWIK1 background K+ channels? *Cell*, *130*(3), 563–569. http://dx.doi.org/10.1016/j.cell.2007.06.012.

Gareau, J. R., & Lima, C. D. (2010). The SUMO pathway: Emerging mechanisms that shape specificity, conjugation and recognition. *Nature Reviews. Molecular Cell Biology*, *11*(12), 861–871. http://dx.doi.org/10.1038/nrm3011.

Geiss-Friedlander, R., & Melchior, F. (2007). Concepts in SUMOylation: A decade on. *Nature Reviews. Molecular Cell Biology*, *8*(12), 947–956. http://dx.doi.org/10.1038/nrm2293.

Gill, G. (2003). Post-translational modification by the small ubiquitin-related modifier SUMO has big effects on transcription factor activity. *Current Opinion in Genetics & Development*, *13*(2), 108–113.

Girdwood, D., Bumpass, D., Vaughan, O. A., Thain, A., Anderson, L. A., Snowden, A. W., et al. (2003). P300 transcriptional repression is mediated by SUMO modification. *Molecular Cell*, *11*(4), 1043–1054.

Glasscock, E., Qian, J., Yoo, J. W., & Noebels, J. L. (2007). Masking epilepsy by combining two epilepsy genes. *Nature Neuroscience*, *10*(12), 1554–1558. http://dx.doi.org/10.1038/nn1999.

Glasscock, E., Yoo, J. W., Chen, T. T., Klassen, T. L., & Noebels, J. L. (2010). Kv1.1 potassium channel deficiency reveals brain-driven cardiac dysfunction as a candidate mechanism for sudden unexplained death in epilepsy. *The Journal of Neuroscience*, *30*(15), 5167–5175. http://dx.doi.org/10.1523/JNEUROSCI.5591-09.2010.

Goldenberg, I., Zareba, W., & Moss, A. J. (2008). Long QT syndrome. *Current Problems in Cardiology*, *33*(11), 629–694. http://dx.doi.org/10.1016/j.cpcardiol.2008.07.002.

Goldman, A. M., Glasscock, E., Yoo, J., Chen, T. T., Klassen, T. L., & Noebels, J. L. (2009). Arrhythmia in heart and brain: KCNQ1 mutations link epilepsy and sudden unexplained death. *Science Translational Medicine*, *1*(2), 2ra6. http://dx.doi.org/10.1126/scitranslmed.3000289.

Gong, L., Kamitani, T., Fujise, K., Caskey, L. S., & Yeh, E. T. (1997). Preferential interaction of sentrin with a ubiquitin-conjugating enzyme, Ubc9. *The Journal of Biological Chemistry*, *272*(45), 28198–28201.

Gong, L., Li, B., Millas, S., & Yeh, E. T. (1999). Molecular cloning and characterization of human AOS1 and UBA2, components of the sentrin-activating enzyme complex. *FEBS Letters*, *448*(1), 185–189.

Gong, L., Millas, S., Maul, G. G., & Yeh, E. T. (2000). Differential regulation of sentrinized proteins by a novel sentrin-specific protease. *The Journal of Biological Chemistry*, *275*(5), 3355–3359.

Gong, L., & Yeh, E. T. (2006). Characterization of a family of nucleolar SUMO-specific proteases with preference for SUMO-2 or SUMO-3. *The Journal of Biological Chemistry, 281*(23), 15869–15877. http://dx.doi.org/10.1074/jbc.M511658200.

Gostissa, M., Hengstermann, A., Fogal, V., Sandy, P., Schwarz, S. E., Scheffner, M., et al. (1999). Activation of p53 by conjugation to the ubiquitin-like protein SUMO-1. *The EMBO Journal, 18*(22), 6462–6471. http://dx.doi.org/10.1093/emboj/18.22.6462.

Gu, N., Hu, H., Vervaeke, K., & Storm, J. F. (2008). SK (KCa2) channels do not control somatic excitability in CA1 pyramidal neurons but can be activated by dendritic excitatory synapses and regulate their impact. *Journal of Neurophysiology, 100*(5), 2589–2604. http://dx.doi.org/10.1152/jn.90433.2008.

Gu, N., Vervaeke, K., Hu, H., & Storm, J. F. (2005). Kv7/KCNQ/M and HCN/h, but not KCa2/SK channels, contribute to the somatic medium after-hyperpolarization and excitability control in CA1 hippocampal pyramidal cells. *The Journal of Physiology, 566*(Pt. 3), 689–715. http://dx.doi.org/10.1113/jphysiol.2005.086835.

Han, S., Tai, C., Westenbroek, R. E., Yu, F. H., Cheah, C. S., Potter, G. B., et al. (2012). Autistic-like behaviour in Scn1a+/− mice and rescue by enhanced GABA-mediated neurotransmission. *Nature, 489*(7416), 385–390. http://dx.doi.org/10.1038/nature11356.

Hang, J., & Dasso, M. (2002). Association of the human SUMO-1 protease SENP2 with the nuclear pore. *The Journal of Biological Chemistry, 277*(22), 19961–19966. http://dx.doi.org/10.1074/jbc.M201799200.

Hansen, H. H., Waroux, O., Seutin, V., Jentsch, T. J., Aznar, S., & Mikkelsen, J. D. (2008). Kv7 channels: Interaction with dopaminergic and serotonergic neurotransmission in the CNS. *The Journal of Physiology, 586*(7), 1823–1832. http://dx.doi.org/10.1113/jphysiol.2007.149450.

Hay, R. T. (2005). SUMO: A history of modification. *Molecular Cell, 18*(1), 1–12. http://dx.doi.org/10.1016/j.molcel.2005.03.012.

Hay, R. T., Vuillard, L., Desterro, J. M., & Rodriguez, M. S. (1999). Control of NF-kappa B transcriptional activation by signal induced proteolysis of I kappa B alpha. *Philosophical Transactions of the Royal Society of London. Series B, Biological Sciences, 354*(1389), 1601–1609. http://dx.doi.org/10.1098/rstb.1999.0504.

Henley, J. M., Craig, T. J., & Wilkinson, K. A. (2014). Neuronal SUMOylation: Mechanisms, physiology, and roles in neuronal dysfunction. *Physiological Reviews, 94*(4), 1249–1285. http://dx.doi.org/10.1152/physrev.00008.2014.

Herren, A. W., Bers, D. M., & Grandi, E. (2013). Post-translational modifications of the cardiac Na channel: Contribution of CaMKII-dependent phosphorylation to acquired arrhythmias. *American Journal of Physiology. Heart and Circulatory Physiology, 305*(4), H431–H445. http://dx.doi.org/10.1152/ajpheart.00306.2013.

Hershko, A., Ciechanover, A., Heller, H., Haas, A. L., & Rose, I. A. (1980). Proposed role of ATP in protein breakdown: Conjugation of protein with multiple chains of the polypeptide of ATP-dependent proteolysis. *Proceedings of the National Academy of Sciences of the United States of America,* 77(4), 1783–1786.

Hershko, A., Ciechanover, A., & Rose, I. A. (1979). Resolution of the ATP-dependent proteolytic system from reticulocytes: A component that interacts with ATP. *Proceedings of the National Academy of Sciences of the United States of America, 76*(7), 3107–3110.

Hochstrasser, M. (2000). Evolution and function of ubiquitin-like protein-conjugation systems. *Nature Cell Biology, 2*(8), E153–E157. http://dx.doi.org/10.1038/35019643.

Hochstrasser, M. (2009). Origin and function of ubiquitin-like proteins. *Nature, 458*(7237), 422–429. http://dx.doi.org/10.1038/nature07958.

Huang, C., Han, Y., Wang, Y., Sun, X., Yan, S., Yeh, E. T., et al. (2009). SENP3 is responsible for HIF-1 transactivation under mild oxidative stress via p300 de-SUMOylation. *The EMBO Journal, 28*(18), 2748–2762. http://dx.doi.org/10.1038/emboj.2009.210.

Itahana, Y., Yeh, E. T., & Zhang, Y. (2006). Nucleocytoplasmic shuttling modulates activity and ubiquitination-dependent turnover of SUMO-specific protease 2. *Molecular and Cellular Biology*, *26*(12), 4675–4689. http://dx.doi.org/10.1128/MCB.01830-05.

Jacobson, D. A., Kuznetsov, A., Lopez, J. P., Kash, S., Ammala, C. E., & Philipson, L. H. (2007). Kv2.1 ablation alters glucose-induced islet electrical activity, enhancing insulin secretion. *Cell Metabolism*, *6*(3), 229–235. http://dx.doi.org/10.1016/j.cmet.2007.07.010.

Jentsch, T. J. (2000). Neuronal KCNQ potassium channels: Physiology and role in disease. *Nature Reviews. Neuroscience*, *1*(1), 21–30. http://dx.doi.org/10.1038/35036198.

Jiang, Z., Fan, Q., Zhang, Z., Zou, Y., Cai, R., Wang, Q., et al. (2012). SENP1 deficiency promotes ER stress-induced apoptosis by increasing XBP1 SUMOylation. *Cell Cycle*, *11*(6), 1118–1122. http://dx.doi.org/10.4161/cc.11.6.19529.

Johnson, E. S. (2004). Protein modification by SUMO. *Annual Review of Biochemistry*, *73*, 355–382. http://dx.doi.org/10.1146/annurev.biochem.73.011303.074118.

Johnson, E. S., & Blobel, G. (1997). Ubc9p is the conjugating enzyme for the ubiquitin-like protein Smt3p. *The Journal of Biological Chemistry*, *272*(43), 26799–26802.

Johnson, E. S., & Gupta, A. A. (2001). An E3-like factor that promotes SUMO conjugation to the yeast septins. *Cell*, *106*(6), 735–744.

Kagey, M. H., Melhuish, T. A., & Wotton, D. (2003). The polycomb protein Pc2 is a SUMO E3. *Cell*, *113*(1), 127–137.

Kahyo, T., Nishida, T., & Yasuda, H. (2001). Involvement of PIAS1 in the SUMOylation of tumor suppressor p53. *Molecular Cell*, *8*(3), 713–718.

Kalume, F., Westenbroek, R. E., Cheah, C. S., Yu, F. H., Oakley, J. C., Scheuer, T., et al. (2013). Sudden unexpected death in a mouse model of Dravet syndrome. *The Journal of Clinical Investigation*, *123*(4), 1798–1808. http://dx.doi.org/10.1172/JCI66220.

Kamitani, T., Kito, K., Nguyen, H. P., Fukuda-Kamitani, T., & Yeh, E. T. (1998a). Characterization of a second member of the sentrin family of ubiquitin-like proteins. *The Journal of Biological Chemistry*, *273*(18), 11349–11353.

Kamitani, T., Kito, K., Nguyen, H. P., Wada, H., Fukuda-Kamitani, T., & Yeh, E. T. (1998b). Identification of three major sentrinization sites in PML. *The Journal of Biological Chemistry*, *273*(41), 26675–26682.

Kamitani, T., Nguyen, H. P., Kito, K., Fukuda-Kamitani, T., & Yeh, E. T. (1998). Covalent modification of PML by the sentrin family of ubiquitin-like proteins. *The Journal of Biological Chemistry*, *273*(6), 3117–3120.

Kang, X., Qi, Y., Zuo, Y., Wang, Q., Zou, Y., Schwartz, R. J., et al. (2010). SUMO-specific protease 2 is essential for suppression of polycomb group protein-mediated gene silencing during embryonic development. *Molecular Cell*, *38*(2), 191–201. http://dx.doi.org/10.1016/j.molcel.2010.03.005.

Kim, K. I., Baek, S. H., Jeon, Y. J., Nishimori, S., Suzuki, T., Uchida, S., et al. (2000). A new SUMO-1-specific protease, SUSP1, that is highly expressed in reproductive organs. *The Journal of Biological Chemistry*, *275*(19), 14102–14106.

Kodirov, S. A., Brunner, M., Nerbonne, J. M., Buckett, P., Mitchell, G. F., & Koren, G. (2004). Attenuation of I(K, slow1) and I(K, slow2) in Kv1/Kv2DN mice prolongs APD and QT intervals but does not suppress spontaneous or inducible arrhythmias. *American Journal of Physiology. Heart and Circulatory Physiology*, *286*(1), H368–H374. http://dx.doi.org/10.1152/ajpheart.00303.2003.

Kotaja, N., Karvonen, U., Janne, O. A., & Palvimo, J. J. (2002). PIAS proteins modulate transcription factors by functioning as SUMO-1 ligases. *Molecular and Cellular Biology*, *22*(14), 5222–5234.

Li, H., Guo, W., Yamada, K. A., & Nerbonne, J. M. (2004). Selective elimination of I(K, slow1) in mouse ventricular myocytes expressing a dominant negative Kv1.5alpha subunit. *American Journal of Physiology. Heart and Circulatory Physiology, 286*(1), H319–H328. http://dx.doi.org/10.1152/ajpheart.00665.2003.

Lima, C. D., & Reverter, D. (2008). Structure of the human SENP7 catalytic domain and poly-SUMO deconjugation activities for SENP6 and SENP7. *The Journal of Biological Chemistry, 283*(46), 32045–32055. http://dx.doi.org/10.1074/jbc.M805655200.

Mahajan, R., Delphin, C., Guan, T., Gerace, L., & Melchior, F. (1997). A small ubiquitin-related polypeptide involved in targeting RanGAP1 to nuclear pore complex protein RanBP2. *Cell, 88*(1), 97–107.

Mao, Y., Desai, S. D., & Liu, L. F. (2000). SUMO-1 conjugation to human DNA topoisomerase II isozymes. *The Journal of Biological Chemistry, 275*(34), 26066–26073. http://dx.doi.org/10.1074/jbc.M001831200.

Mao, Y., Sun, M., Desai, S. D., & Liu, L. F. (2000). SUMO-1 conjugation to topoisomerase I: A possible repair response to topoisomerase-mediated DNA damage. *Proceedings of the National Academy of Sciences of the United States of America, 97*(8), 4046–4051. http://dx.doi.org/10.1073/pnas.080536597.

Martin, S., Nishimune, A., Mellor, J. R., & Henley, J. M. (2007). SUMOylation regulates kainate-receptor-mediated synaptic transmission. *Nature, 447*(7142), 321–325. http://dx.doi.org/10.1038/nature05736.

Massey, C. A., Sowers, L. P., Dlouhy, B. J., & Richerson, G. B. (2014). Mechanisms of sudden unexpected death in epilepsy: The pathway to prevention. *Nature Reviews. Neurology, 10*(5), 271–282. http://dx.doi.org/10.1038/nrneurol.2014.64.

Matunis, M. J., Coutavas, E., & Blobel, G. (1996). A novel ubiquitin-like modification modulates the partitioning of the Ran-GTPase-activating protein RanGAP1 between the cytosol and the nuclear pore complex. *The Journal of Cell Biology, 135*(6 Pt. 1), 1457–1470.

Meinecke, I., Cinski, A., Baier, A., Peters, M. A., Dankbar, B., Wille, A., et al. (2007). Modification of nuclear PML protein by SUMO-1 regulates Fas-induced apoptosis in rheumatoid arthritis synovial fibroblasts. *Proceedings of the National Academy of Sciences of the United States of America, 104*(12), 5073–5078. http://dx.doi.org/10.1073/pnas.0608773104.

Mukhopadhyay, D., Ayaydin, F., Kolli, N., Tan, S. H., Anan, T., Kametaka, A., et al. (2006). SUSP1 antagonizes formation of highly SUMO2/3-conjugated species. *The Journal of Cell Biology, 174*(7), 939–949. http://dx.doi.org/10.1083/jcb.200510103.

Mukhopadhyay, D., & Dasso, M. (2007). Modification in reverse: The SUMO proteases. *Trends in Biochemical Sciences, 32*(6), 286–295. http://dx.doi.org/10.1016/j.tibs.2007.05.002.

Nishida, T., Kaneko, F., Kitagawa, M., & Yasuda, H. (2001). Characterization of a novel mammalian SUMO-1/Smt3-specific isopeptidase, a homologue of rat axam, which is an axin-binding protein promoting beta-catenin degradation. *The Journal of Biological Chemistry, 276*(42), 39060–39066. http://dx.doi.org/10.1074/jbc.M103955200.

Nishida, T., Tanaka, H., & Yasuda, H. (2000). A novel mammalian Smt3-specific isopeptidase 1 (SMT3IP1) localized in the nucleolus at interphase. *European Journal of Biochemistry, 267*(21), 6423–6427.

Noebels, J. L. (2003). The biology of epilepsy genes. *Annual Review of Neuroscience, 26*, 599–625. http://dx.doi.org/10.1146/annurev.neuro.26.010302.081210.

Okura, T., Gong, L., Kamitani, T., Wada, T., Okura, I., Wei, C. F., et al. (1996). Protection against Fas/APO-1- and tumor necrosis factor-mediated cell death by a novel protein, sentrin. *Journal of Immunology, 157*(10), 4277–4281.

Olson, T. M., Alekseev, A. E., Liu, X. K., Park, S., Zingman, L. V., Bienengraeber, M., et al. (2006). Kv1.5 channelopathy due to KCNA5 loss-of-function mutation causes human atrial fibrillation. *Human Molecular Genetics, 15*(14), 2185–2191. http://dx.doi.org/10.1093/hmg/ddl143.

Otto, J. F., Yang, Y., Frankel, W. N., White, H. S., & Wilcox, K. S. (2006). A spontaneous mutation involving Kcnq2 (Kv7.2) reduces M-current density and spike frequency adaptation in mouse CA1 neurons. *The Journal of Neuroscience*, *26*(7), 2053–2059. http://dx.doi.org/10.1523/JNEUROSCI.1575-05.2006.

Ouyang, J., Valin, A., & Gill, G. (2009). Regulation of transcription factor activity by SUMO modification. *Methods in Molecular Biology*, *497*, 141–152. http://dx.doi.org/10.1007/978-1-59745-566-4_9.

Owerbach, D., McKay, E. M., Yeh, E. T., Gabbay, K. H., & Bohren, K. M. (2005). A proline-90 residue unique to SUMO-4 prevents maturation and SUMOylation. *Biochemical and Biophysical Research Communications*, *337*(2), 517–520. http://dx.doi.org/10.1016/j.bbrc.2005.09.090.

Pannasch, U., Farber, K., Nolte, C., Blonski, M., Yan Chiu, S., Messing, A., et al. (2006). The potassium channels Kv1.5 and Kv1.3 modulate distinct functions of microglia. *Molecular and Cellular Neurosciences*, *33*(4), 401–411. http://dx.doi.org/10.1016/j.mcn.2006.08.009.

Peters, H. C., Hu, H., Pongs, O., Storm, J. F., & Isbrandt, D. (2005). Conditional transgenic suppression of M channels in mouse brain reveals functions in neuronal excitability, resonance and behavior. *Nature Neuroscience*, *8*(1), 51–60. http://dx.doi.org/10.1038/nn1375.

Pichler, A., Gast, A., Seeler, J. S., Dejean, A., & Melchior, F. (2002). The nucleoporin RanBP2 has SUMO1 E3 ligase activity. *Cell*, *108*(1), 109–120.

Pickart, C. M. (2001). Mechanisms underlying ubiquitination. *Annual Review of Biochemistry*, *70*, 503–533. http://dx.doi.org/10.1146/annurev.biochem.70.1.503.

Plant, L. D., Dementieva, I. S., Kollewe, A., Olikara, S., Marks, J. D., & Goldstein, S. A. (2010). One SUMO is sufficient to silence the dimeric potassium channel K2P1. *Proceedings of the National Academy of Sciences of the United States of America*, *107*(23), 10743–10748. http://dx.doi.org/10.1073/pnas.1004712107.

Plant, L. D., Dowdell, E. J., Dementieva, I. S., Marks, J. D., & Goldstein, S. A. (2011). SUMO modification of cell surface Kv2.1 potassium channels regulates the activity of rat hippocampal neurons. *The Journal of General Physiology*, *137*(5), 441–454. http://dx.doi.org/10.1085/jgp.201110604.

Plant, L. D., Rajan, S., & Goldstein, S. A. (2005). K2P channels and their protein partners. *Current Opinion in Neurobiology*, *15*(3), 326–333. http://dx.doi.org/10.1016/j.conb.2005.05.008.

Plant, L. D., Zuniga, L., Olikara, S., & Goldstein, S. A. N. (2010). One SUMO is sufficient to silence the dimeric background potassium channel K2P1. *Biophysical Journal*, *98*(3), 536a.

Poukka, H., Karvonen, U., Janne, O. A., & Palvimo, J. J. (2000). Covalent modification of the androgen receptor by small ubiquitin-like modifier 1 (SUMO-1). *Proceedings of the National Academy of Sciences of the United States of America*, *97*(26), 14145–14150. http://dx.doi.org/10.1073/pnas.97.26.14145.

Qi, Y., Wang, J., Bomben, V. C., Li, D. P., Chen, S. R., Sun, H., et al. (2014). Hyper-SUMOylation of the Kv7 potassium channel diminishes the M-current leading to seizures and sudden death. *Neuron*, *83*(5), 1159–1171. http://dx.doi.org/10.1016/j.neuron.2014.07.042.

Qi, Y., Zuo, Y., Yeh, E. T., & Cheng, J. (2014). An essential role of small ubiquitin-like modifier (SUMO)-specific Protease 2 in myostatin expression and myogenesis. *The Journal of Biological Chemistry*, *289*(6), 3288–3293. http://dx.doi.org/10.1074/jbc.M113.518282.

Rajan, S., Plant, L. D., Rabin, M. L., Butler, M. H., & Goldstein, S. A. (2005). Sumoylation silences the plasma membrane leak K+ channel K2P1. *Cell*, *121*(1), 37–47. http://dx.doi.org/10.1016/j.cell.2005.01.019.

Reverter, D., & Lima, C. D. (2006). Structural basis for SENP2 protease interactions with SUMO precursors and conjugated substrates. *Nature Structural & Molecular Biology, 13*(12), 1060–1068. http://dx.doi.org/10.1038/nsmb1168.

Rodriguez, M. S., Dargemont, C., & Hay, R. T. (2001). SUMO-1 conjugation in vivo requires both a consensus modification motif and nuclear targeting. *The Journal of Biological Chemistry, 276*(16), 12654–12659. http://dx.doi.org/10.1074/jbc.M009476200.

Ross, S., Best, J. L., Zon, L. I., & Gill, G. (2002). SUMO-1 modification represses Sp3 transcriptional activation and modulates its subnuclear localization. *Molecular Cell, 10*(4), 831–842.

Rytinki, M. M., & Palvimo, J. J. (2009). SUMOylation attenuates the function of PGC-1alpha. *The Journal of Biological Chemistry, 284*(38), 26184–26193. http://dx.doi.org/10.1074/jbc.M109.038943.

Schulman, B. A., & Harper, J. W. (2009). Ubiquitin-like protein activation by E1 enzymes: The apex for downstream signalling pathways. *Nature Reviews. Molecular Cell Biology, 10*(5), 319–331. http://dx.doi.org/10.1038/nrm2673.

Scott, R. C., & Holmes, G. L. (2012). Before epilepsy unfolds: Opening up the potassium door in neonatal seizures. *Nature Medicine, 18*(11), 1624–1625. http://dx.doi.org/10.1038/nm.2987.

Shen, L. N., Dong, C., Liu, H., Naismith, J. H., & Hay, R. T. (2006). The structure of SENP1-SUMO-2 complex suggests a structural basis for discrimination between SUMO paralogues during processing. *The Biochemical Journal, 397*(2), 279–288. http://dx.doi.org/10.1042/BJ20052030.

Shen, L. N., Geoffroy, M. C., Jaffray, E. G., & Hay, R. T. (2009). Characterization of SENP7, a SUMO-2/3-specific isopeptidase. *The Biochemical Journal, 421*(2), 223–230. http://dx.doi.org/10.1042/BJ20090246.

Shen, Z., Pardington-Purtymun, P. E., Comeaux, J. C., Moyzis, R. K., & Chen, D. J. (1996). UBL1, a human ubiquitin-like protein associating with human RAD51/RAD52 proteins. *Genomics, 36*(2), 271–279. http://dx.doi.org/10.1006/geno.1996.0462.

Shen, L., Tatham, M. H., Dong, C., Zagorska, A., Naismith, J. H., & Hay, R. T. (2006). SUMO protease SENP1 induces isomerization of the scissile peptide bond. *Nature Structural & Molecular Biology, 13*(12), 1069–1077. http://dx.doi.org/10.1038/nsmb1172.

Sillanpaa, M., & Shinnar, S. (2010). Long-term mortality in childhood-onset epilepsy. *The New England Journal of Medicine, 363*(26), 2522–2529. http://dx.doi.org/10.1056/NEJMoa0911610.

Singh, N. A., Otto, J. F., Dahle, E. J., Pappas, C., Leslie, J. D., Vilaythong, A., et al. (2008). Mouse models of human KCNQ2 and KCNQ3 mutations for benign familial neonatal convulsions show seizures and neuronal plasticity without synaptic reorganization. *The Journal of Physiology, 586*(14), 3405–3423. http://dx.doi.org/10.1113/jphysiol.2008.154971.

Surges, R., & Sander, J. W. (2012). Sudden unexpected death in epilepsy: Mechanisms, prevalence, and prevention. *Current Opinion in Neurology, 25*(2), 201–207. http://dx.doi.org/10.1097/WCO.0b013e3283506714.

Surti, T. S., & Jan, L. Y. (2005). A potassium channel, the M-channel, as a therapeutic target. *Current Opinion in Investigational Drugs, 6*(7), 704–711.

Tatham, M. H., Jaffray, E., Vaughan, O. A., Desterro, J. M., Botting, C. H., Naismith, J. H., et al. (2001). Polymeric chains of SUMO-2 and SUMO-3 are conjugated to protein substrates by SAE1/SAE2 and Ubc9. *The Journal of Biological Chemistry, 276*(38), 35368–35374. http://dx.doi.org/10.1074/jbc.M104214200.

Tatulian, L., Delmas, P., Abogadie, F. C., & Brown, D. A. (2001). Activation of expressed KCNQ potassium currents and native neuronal M-type potassium currents by the anticonvulsant drug retigabine. *The Journal of Neuroscience, 21*(15), 5535–5545.

Ulrich, H. D. (2008). The fast-growing business of SUMO chains. *Molecular Cell*, *32*(3), 301–305. http://dx.doi.org/10.1016/j.molcel.2008.10.010.
Ulrich, H. D. (2009). Regulating post-translational modifications of the eukaryotic replication clamp PCNA. *DNA Repair (Amst)*, *8*(4), 461–469. http://dx.doi.org/10.1016/j.dnarep.2009.01.006.
Van Nguyen, T., Angkasekwinai, P., Dou, H., Lin, F. M., Lu, L. S., Cheng, J., et al. (2012). SUMO-specific protease 1 is critical for early lymphoid development through regulation of STAT5 activation. *Molecular Cell*, *45*(2), 210–221. http://dx.doi.org/10.1016/j.molcel.2011.12.026.
Wang, Q., Curran, M. E., Splawski, I., Burn, T. C., Millholland, J. M., VanRaay, T. J., et al. (1996). Positional cloning of a novel potassium channel gene: KVLQT1 mutations cause cardiac arrhythmias. *Nature Genetics*, *12*(1), 17–23. http://dx.doi.org/10.1038/ng0196-17.
Wang, J., & Schwartz, R. J. (2010). Sumoylation and regulation of cardiac gene expression. *Circulation Research*, *107*(1), 19–29. http://dx.doi.org/10.1161/CIRCRESAHA.110.220491.
Weckhuysen, S., Mandelstam, S., Suls, A., Audenaert, D., Deconinck, T., Claes, L. R., et al. (2012). KCNQ2 encephalopathy: Emerging phenotype of a neonatal epileptic encephalopathy. *Annals of Neurology*, *71*(1), 15–25. http://dx.doi.org/10.1002/ana.22644.
Xu, Y., Zuo, Y., Zhang, H., Kang, X., Yue, F., Yi, Z., et al. (2010). Induction of SENP1 in endothelial cells contributes to hypoxia-driven VEGF expression and angiogenesis. *The Journal of Biological Chemistry*, *285*(47), 36682–36688. http://dx.doi.org/10.1074/jbc.M110.164236.
Yeh, E. T. (2009). SUMOylation and de-SUMOylation: Wrestling with life's processes. *The Journal of Biological Chemistry*, *284*(13), 8223–8227. http://dx.doi.org/10.1074/jbc.R800050200.
Yeh, E. T., Gong, L., & Kamitani, T. (2000). Ubiquitin-like proteins: New wines in new bottles. *Gene*, *248*(1–2), 1–14.
Zhang, H., Saitoh, H., & Matunis, M. J. (2002). Enzymes of the SUMO modification pathway localize to filaments of the nuclear pore complex. *Molecular and Cellular Biology*, *22*(18), 6498–6508.
Zhao, J. (2007). Sumoylation regulates diverse biological processes. *Cellular and Molecular Life Sciences*, *64*(23), 3017–3033. http://dx.doi.org/10.1007/s00018-007-7137-4.
Zhou, J., Kodirov, S., Murata, M., Buckett, P. D., Nerbonne, J. M., & Koren, G. (2003). Regional upregulation of Kv2.1-encoded current, IK, slow2, in Kv1DN mice is abolished by crossbreeding with Kv2DN mice. *American Journal of Physiology. Heart and Circulatory Physiology*, *284*(2), H491–H500. http://dx.doi.org/10.1152/ajpheart.00576.2002.
Zhou, L., Zhang, C. L., Messing, A., & Chiu, S. Y. (1998). Temperature-sensitive neuromuscular transmission in Kv1.1 null mice: Role of potassium channels under the myelin sheath in young nerves. *The Journal of Neuroscience*, *18*(18), 7200–7215.
Zuberi, S. M., Eunson, L. H., Spauschus, A., De Silva, R., Tolmie, J., Wood, N. W., et al. (1999). A novel mutation in the human voltage-gated potassium channel gene (Kv1.1) associates with episodic ataxia type 1 and sometimes with partial epilepsy. *Brain*, *122*(Pt 5), 817–825.

AUTHOR INDEX

Note: Page numbers followed by "*f*" indicate figures, and "*t*" indicate tables.

B

C

D

E

F

G

H

K

L

M

N

O

P

Q

R

S

T

U

X

Z

SUBJECT INDEX

Note: Page numbers followed by "*f*" indicate figures, and "*t*" indicate tables.

A

B

T

V

CPI Antony Rowe
Chippenham, UK
2016-11-11 21:14